Transition Metal Oxides

C.N.R. Rao
and
B. Raveau

C.N.R. Rao
Solid State & Structural Chemistry Unit
Indian Institute of Science
and Jawaharlal Nehru Center for
Advanced Scientific Research
Bangalore, 560 012
India

B. Raveau
Centre des Matériaux Superconducteurs
Laboratoire CRISMAT, ISMRA
6, blvd. du Maréchal Juin
14050 Caen, Cedex
France

Library of Congress Cataloging-in-Publication Data

Rao, C. N. R. (Chintamani Nagesa Ramachandra), 1934-
Transition metal oxides / C.N.R. Rao, B. Raveau.
p. cm.
Includes bibliographical references (p. -) and index.
ISBN 1-56081-647-3 (alk. paper)
1. Transition metal oxides. I. Raveau, B. (Bernard), 1940- .
II. Title.
QD172.T6R36 1995
546'.6--dc20 95-10717
CIP

Printed in the United States of America

ISBN 1-56081-647-3 VCH Publishers

Printing History:

10 9 8 7 6 5 4 3 2 1

Published jointly by

VCH Publishers, Inc.
220 East 23rd Street
New York, New York 10010

VCH Verlagsgesellschaft mbH
P.O. Box 10 11 61
69451 Weinheim, Germany

VCH Publishers (UK) Ltd.
8 Wellington Court
Cambridge CB1 1HZ
United Kingdom

Preface

Transition metal oxides constitute one of the most fascinating classes of inorganic solids, exhibiting a very wide variety of structures, properties, and phenomena. The unusual properties of transition metal oxides are due to the unique nature of the outer *d* electrons, the metal–oxygen bonding varying anywhere from nearly ionic to metallic. Transition metal oxides possessing several types of complex structure have been characterized in recent years. These include not only the well-known perovskite, spinel, pyrochlore, and hexagonal ferrite structures, but also the octahedral tunnel structures exhibited by bronzes and bronzoids, lamellar and low-dimensional structures, as well as three-dimensional mixed frameworks of octahedra and tetrahedra. Some of the oxides show ordered defect complexes or extended defects instead of isolated point defects; they also occur as shear and block structures or infinitely adaptive structures. Intergrowth of like structural units is commonly found in complex oxides. It has been possible to obtain the fine architectural details of complex transition metal oxides in terms of both the crystal structure and the ultramicrostructure because of the recent advances in diffraction and microscopic techniques.

The phenomenal range of electronic and magnetic properties exhibited by transition metal oxides is equally noteworthy. There are oxides with metallic properties (e.g., RuO_2, ReO_3, $LaNiO_3$) at one end of the range and oxides with insulating behavior (e.g., $BaTiO_3$) at the other. There are oxides that traverse both these regimes with change in temperature, pressure, or composition (e.g., V_2O_3, $La_{1-x}Sr_xVO_3$). Interesting electronic properties also arise from charge-density waves (e.g., $K_{0.3}MoO_3$), charge ordering (e.g., Fe_3O_4), and defect ordering (e.g., $Ca_2Mn_2O_5$, $Ca_2Fe_2O_5$). Examples are known of oxides with diverse magnetic properties: ferromagnetic (e.g., CrO_2, $La_{0.5}Sr_{0.5}MnO_3$), ferrimagnetic (Fe_3O_4, $MnFe_2O_4$), and antiferromagnetic (e.g., NiO, $LaCrO_3$). Many oxides possess switchable orientation states as in ferroelectric (e.g., $BaTiO_3$, $KNbO_3$) and ferroelastic [e.g., $Gd_2(MoO_4)_3$] materials. No discovery in the physical sciences has created as much sensation, however, as that of high temperature superconductivity in cuprates. Although superconductivity in transition metal oxides was known for some time, the highest T_c reached till 1987 was around 13 K; we now have oxides with T_c values in the region of 135 K. The discovery of high temperature superconducting oxides has focused worldwide interest on the chemistry and physics of transition metal oxides and has at the same time revealed the inadequacy of our understanding of these fascinating materials.

The unusual properties of transition metal oxides that distinguish them from metallic elements, covalent semiconductors, and ionic insulators arise from several factors. First, the oxides of *d*-block transition elements have narrow electronic bands, because of the small overlap between the metal *d* and oxygen *p* orbitals. The bandwidths are typically of the order of 1 or 2 eV (rather than 5–15 eV as in most metals). Electron correlation effects play an important role because of the narrow electronic bands. The local electronic structure can be described in terms of atomiclike states [e.g., Cu^{1+} (d^{10}), Cu^{2+} (d^9), and Cu^{3+} (d^8) for Cu in CuO] as in the Heitler–London limit. The polarizability of oxygen is also of importance. The oxide ion O^{2-} does not exactly describe the state of oxygen, and configurations such as O^- have to be included, especially in the solid state. Species such as O^-, which are oxygen holes with a p^5 configuration instead of the filled p^6 configuration of O^{2-}, can be mobile and correlated. Many transition metal oxides are not truly three-dimensional, but have low-dimensional features. For example, La_2CuO_4 and La_2NiO_4 with the K_2NiF_4 structure are quasi-two-dimensional compared to $LaCuO_3$ and $LaNiO_3$, which are three-dimensional perovskites. Because of their varied features, it has not been possible to establish fully satisfactory theoretical models for transition metal oxides. There have however been many convenient approaches to the understanding and description of their electronic structures and properties.

The extraordinary range of structures and properties of transition metal oxides makes them worthy of special attention. These oxides provide an excellent case study in solid state chemistry and an appropriate gateway to understanding the behavior of inorganic solids. This monograph provides a rather detailed presentation of structures of various classes of complex transition metal oxides, particularly in polyhedral representation, since an appreciation of the structures is essential to an understanding of the properties and to design synthesis. We have tried to put together succinctly the salient features of transition metal oxides with respect to their properties as well as the phenomena exhibited by them. We have then described some of the important strategies employed for the preparation of oxides. While it is difficult to do justice to every aspect of transition metal oxides because of limitation of space, the material presented here should provide a fairly comprehensive picture of the present state of the subject. We believe that the book serves to bring out the flavor of modern inorganic solid state chemistry, through the medium of oxides. By making use of standard texts in solid state physics and chemistry for explanations of some of the terms and concepts, where necessary, readers should be able to comprehend the essence of this book. We trust that the book will be found useful by students, teachers, and practitioners of inorganic chemistry and solid state chemistry as well as condensed matter science and materials science.

C.N.R. Rao
Bangalore, India

B. Raveau
Caen, France

March 1995

Contents

Part II Properties and Phenomena 211

Part III Preparation of Materials 289

PART I

Structure

PART

I

Structure

1 Introduction

The chemistry of transition metal oxides can be understood only when we have a sound knowledge of the crystal chemistry of these materials. Crystal chemistry represents not only the crystal structures of the oxides, but also the nature of bonding in them. Crystal chemistry is indeed a crucial constituent of solid state chemistry and provides the basis for designing and synthesizing new materials (for general review about inorganic crystal chemistry see for instance ref 1 to 9). Transition metal oxides are by far the most fascinating class of materials when it comes to crystal chemistry. Metal oxides crystallize in a variety of structures, and bonding in these materials can vary anywhere from ionic (e.g., MgO, $Fe_{1-x}O$) to metallic (TiO, ReO_3). Associated with such changes in bonding, these materials also show a gamut of fascinating properties. In recent years, it has been possible to determine the structures of complex transition metal oxides by employing some of the new techniques of crystallography. Today, we can obtain detailed structures not only of oxides in single crystal form, but also of powders employing methods of X-ray and neutron diffraction. These two techniques have become really powerful because of the availability of synchrotron X-rays and intense pulsed neutron sources.

To fully understand the structures of complex transition metal oxides, it becomes necessary to understand not only the crystal structure and bonding, but also the local or ultramicrostructures, often arising from defects or compositional changes. Local structure of oxides is best studied by high resolution electron microscopy. In the last two decades, this technique has been employed widely to establish the ultramicrostructures of a large number of complex transition metal oxides. With the availability of commercial electron microscopes of 1.7 Å resolution, it is possible to pinpoint the nature of defects or local ordering in atomistic detail.

In Part I we shall examine the structures of a variety of transition metal oxides and in doing so, we shall describe the structures by means of polyhedral representation. Since it is important to visualize structures to be able to design new oxides of novel structures or to understand the properties of known ones, we discuss in detail the structures of a large number of examples from the different families of transition metal oxides.

Before we go into the various structural families, we shall briefly discuss some elements of crystal chemistry and the notion of defects in crystals. The presence of defects and more precisely, the ordering of defects in transition metal oxides, gives rise to new structural principles such as the crystallographic shear. Structures of transition metal oxides are governed by such principles giving rise to novel features such as homology, superstructures, intergrowths, and tunnels. We shall also briefly discuss some aspects of the techniques of characterization and phase transitions. Clearly, each of these topics—defects, techniques, and phase transitions—is by itself so vast in scope that we can only present the highlights in a rudimentary fashion to provide the necessary background to look at the structures of transition metal oxides.

2 Basic Background Material

2.1 Description of Crystalline Oxides

Crystals are composed of infinite arrays of atoms, ions, or molecules in the three dimensions. Periodicity in crystals is generally represented by replacing the repeating unit by a point, the resulting array of such points in space being called a lattice. In a *space lattice,* the translation vectors, **a**, **b**, and **c** in the three crystallographic directions define a *primitive cell.* When a primitive cell or some other suitable combination is chosen as the repeating unit of the lattice, it is referred to as the *unit cell.* A crystallographic unit cell is defined by three translation vectors and three angles α, β, and γ. The seven *crystal systems,* based on the six parameters are cubic, tetragonal, orthorhombic, rhombohedral, hexagonal, monoclinic, and tetragonal. There are 14 independent ways of arranging points in three dimensions, giving rise to the 14 *Bravais lattices* listed in Table 1 and shown in Figure 1.

Based on extensive studies of the symmetry in crystals, it is found that crystals possess one or more of the 10 basic symmetry elements (five proper rotation axes, 1, 2, 3, 4, 6, and five inversion or improper axes). A set of symmetry elements intersecting at a common point within a crystal is called the *point group.* The 10 basic symmetry elements along with their 22 possible combinations constitute the 32 *crystal classes.* Additional symmetry elements in crystals are the *screw axis* and the *glide plane.* A *space group* is a combination of these

Table 1 Crystal Systems and Bravais Lattices

System	Unit Cell Specification	Essential Symmetry[a]	Bravais Lattice[b]
Cubic	$a = b = c$ $\alpha = \beta = \gamma = 90°$	Four 3*s*	P, I, F
Tetragonal	$a = b \neq c$ $\alpha = \beta = \gamma = 90°$	One 4 or $\bar{4}$	P, I
Orthorhombic	$a \neq b \neq c$ $\alpha = \beta = \gamma = 90°$	Three 2*s* mutually perpendicular or one 2 intersecting with two *m*'s	P, I, C, F
Rhombohedral	$a = b = c$ $\alpha = \beta = \gamma \neq 90°$	One 3	R (P)
Hexagonal	$a = b \neq c$ $\alpha = \beta = 90°$ $\gamma = 120°$	One 6	P
Monoclinic	$a \neq b \neq c$ $\alpha = \gamma = 90° \neq \beta$	One 2 or one *m*	P C
Triclinic	$a \neq b \neq c$ $\alpha \neq \beta \neq \gamma$	none	P

[a] Unmodified numbers (3, 4, etc.) are rotation axes; overbars indicate; inversion axes *m* is mirror plane.
[b] P, Primitive lattice containing lattice points at the corners of the unit cell; F, face-centered lattice; I, body-centered lattice.

elements involving translation with the point group symmetry. For example, while there are only two possible space groups, $P1$ and $P\bar{1}$ for a triclinic system, there are 13 possible space groups in a monoclinic system. In all, there are 230 possible space groups.

Based on bonding considerations, five types of crystals can be defined: ionic, covalent, metallic, molecular (van der Waals), and hydrogen-bonded. Ionic crystals are formed between highly electropositive and highly electronegative elements, which favor electron transfer to produce oppositely charged ions, generally with closed-shell electronic configurations. Following Born, the cohesive energy of ionic crystals, U, containing oppositely charged ions with charges Z_1 and Z_2 at a distance R, is written as the sum of two terms, one due to attraction and the other due to repulsion:

$$U = \frac{AZ_1Z_2e^2}{R} + B \exp\left(\frac{R}{\rho}\right)$$

Here, the *Madelung constant, A*, is characteristic of the geometric arrangement of ions in crystals, B is the repulsion constant, and ρ the repulsion exponent. The repulsion term accounts for the stability of ionic crystals without collapsing and arises from the ability of ions with closed electron shells to resist overlap of their electron clouds with neighboring ions. The constants B and ρ are respectively a measure of the strength and the range of the repulsive interaction. The Madelung constant is a function of crystal structure and is computed from the

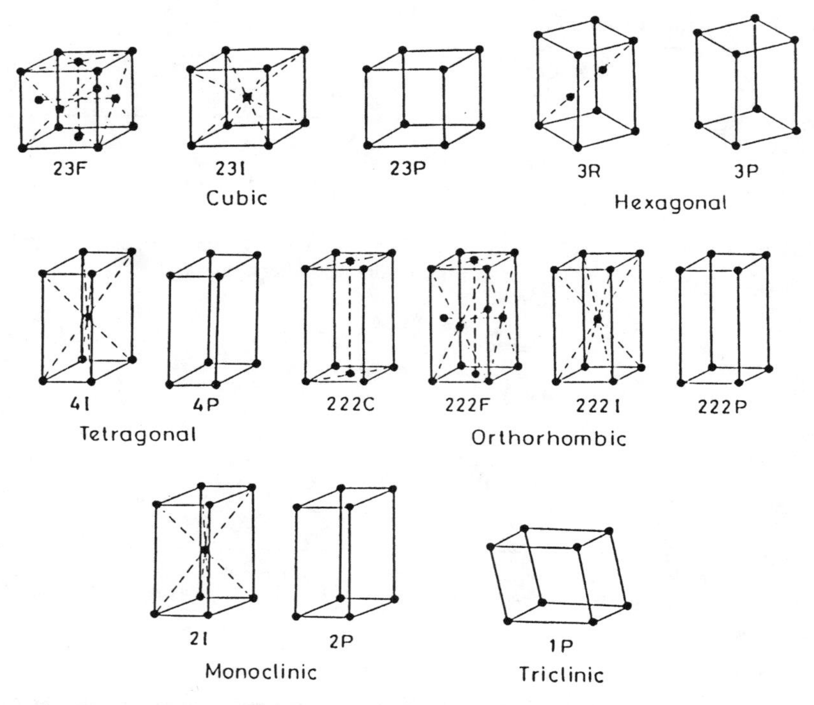

Figure 1. Fourteen Bravais lattices: letters P, F, I, and C represent type of lattice; numbers represent the symmetry axes.

geometrical arrangement of ions in the crystal. For example, the Madelung constant for the NaCl structure has a value of 1.74756. The preceding equation for U is only approximate, and it is necessary to include contributions from van der Waals forces and the correction for zero-point energy. These are, however, minor terms accounting for a small percentage of the total lattice energy. Cohesive energies of ionic solids have been extensively reviewed in the literature.[10,11] Experimental lattice energies of ionic solids are obtained from the *Born–Haber cycle.* The ionic model is a poor approximation for crystals containing large anions and small cations (e.g., oxides and sulfides), where the covalent contribution to bonding becomes significant. Furthermore, cohesive energy calculations cannot be used a priori to predict the structure of an ionic solid, since the method employs experimental interatomic distances in conjunction with formal ionic charges.

Typical covalent solids are formed by group IV elements such as carbon, silicon, and germanium. These elements crystallize in the diamond structure wherein each atom is bonded to four others through covalent bonds. Atoms in covalent solids tend to achieve closed-shell electronic configuration by electron sharing with neighbors. Unlike an ionic bond, a covalent bond has distinct directional character. Metal oxides can be nearly ionic as in MgO, but in most transition metal oxides the bonding is only partly ionic. In other words, there is a considerable overlap between the orbitals of the cations and anions. Many transition metal oxides also exhibit metallic properties. We shall discuss ways of describing electrons in transition metal oxides in Part II. In many transition metal oxides, especially those with layered structures, van der Waals interaction plays a crucial role. For example, interlayer interaction in many of the layered materials is governed by van der Waals forces. In many oxide hydrates or hydroxy oxides, hydrogen bonding also contributes to the cohesive energy.

Inorganic solids containing a metallic and a nonmetallic element form binary compounds (of the type AB, A_2B_3, AB_2, etc.). Bonding in such compounds is ionic if the electronegativity difference between A and B is sufficiently large. Crystal structures of these binary compounds are essentially determined by the nature of bonding and the relative sizes of A and B. When the bonding is mainly ionic as in fluorides and many of the oxides, the structures can be described in geometric terms on the basis of close packing of ions. (Note that structures of metals are readily described in terms of packing of spheres, the closest-packing arrangements being CCP (cubic: ABCABC type) or HCP (hexagonal: ABAB type). Since in metal oxides, anions are generally larger in size than the cations, the structures can be thought of consisting of a closest-packed array (HCP or CCP) of the oxide ions, the interstices (voids) being occupied by the cations. Where the cations are larger than anions, the closest-packed arrangement would be that of the cations, with the anions occupying the interstices. Crystal structures of such binary compounds are, hence, determined by the relative sizes of the ions and the voids in the closest-packing arrangement. Thus, if R is the radius of a large sphere forming the closest packing, a smaller sphere of radius $r = 0.414R$ can be snugly fitted into an octahedral void and a still smaller sphere of radius $r = 0.225R$ goes into a tetrahedral void. If the larger spheres are the anions and the smaller ones are the cations, the sizes of voids would define the limiting radius ratios, r_c/r_a, for octahedral and tetrahedral coordination of cations by anions. Table 2 gives the radius ratios for different coordination geometries. Because of limitations in the ionic model, however, the actual structures of ionic solids differ from the ones just predicted. The cation coordination number is determined by r_c/r_a ratio, which implies the larger the cation radius, the larger

Table 2 Radius Ratios for the Six Coordinations

Geometry and Coordination Number	r_c/r_a
Linear, 2	00.00–0.155
Triangular, 3	0.155–0.225
Tetrahedral, 4	0.225–0.414
Octahedral, 6	0.414–0.732
Cubic, 8	0.732–1.000
Cubic, 12	>1.000

is its coordination number. Many examples can be given to show that this assumption is wrong. In $MgAl_2O_4$ (spinel), it is the larger Mg^{2+} that has tetrahedral coordination and the smaller Al^{3+}, octahedral coordination.

The ionic model assumes that the cations are much smaller than the anions.[12] The radius of the oxide ion (1.44 Å) given by Shannon and Prewitt[13] is in fact larger than the radii of most cations. As it happens, however, ionic radii correspond to free ions and not to ions in crystals. The anions in crystals are subject to a positive Madelung potential giving rise to a contraction of the charge cloud, while cations are subject to a negative potential causing an opposite effect. When such effects are taken into consideration, the difference between the anion and the cation radii does not turn out to be significant. In the absence of a satisfactory procedure to divide bond lengths into cation and anion parts, tables of ionic radii can at best be regarded as tables of bond lengths for certain rational bond strengths. The ionic radii available in the literature are derived from bond length data of oxides and fluorides, but these values are however not of general applicability. Ionic radii values derived from oxides and fluorides do not appear to be useful even for oxyfluorides. It must be noted that Pauling's concept of electrostatic bond strength and bond strength–bond length relationships of the type developed by Brown are quite useful.[12]

In the ionic model, the anion array is taken to be in closest packing. In real crystals, however, anion packing densities are much smaller than the ideal value of 0.74. For example, in α-Al_2O_3, with an HCP array of anions, the anion packing density is only 0.595. Although the anions are topologically arrayed as if they are in closest packing, they are really not in contact with one another. O'Keeffe has suggested the term "eutaxy" to describe such a situation. O'Keeffe prefers to describe ionic structures as maximum volume (minimum electrostatic energy) structures for fixed cation–anion distances.[12] In spite of the many limitations, the ionic model continues to be useful and provides a basis for describing structures of several families of inorganic solids including metal oxides.

Inorganic compounds of the formula AB can have the rock salt (B1), CsCl (B2), zinc blende (B3), wurtzite (B4) or NiAs (B8) structure. Metal monoxides are found mainly in the rock salt and wurtzite structures. The rock salt structure with the 6:6 octahedral coordination (Fig. 2a) is exhibited by alkaline earth metal oxides such as MgO and monoxides of 3*d* transition metals as well as lanthanides and actinides such as TiO, NiO, EuO, and NpO (see Table 3). Several ternary compounds of the type $AA'B_2$ with two different cations statistically distributed (e.g., γ-$LiAlO_2$, α-$NaTlO_2$) or ordered in alternate (111) planes along the body diagonal (e.g., $LiNiO_2$) crystallize in rock salt related structures. Oxides of the type $Mg_6Mn^{4+}O_8$ crystallize in this structure with one-eighth of the cation sites vacant. The wurtzite structure (Fig. 2c), closely related to the zinc blende structure (Fig. 2b), has the same 4:4 tetrahedral coordination arising from an HCP arrangement of anions with half the tetrahedral sites occupied by the cations. Examples of oxides in wurtzite structure are BeO and ZnO.

Among the less common AB structures are those of PbO, SnS, TlI, and HgS. The first three structures illustrate the effect of the cation inert pair (5*s* or 6*s*)

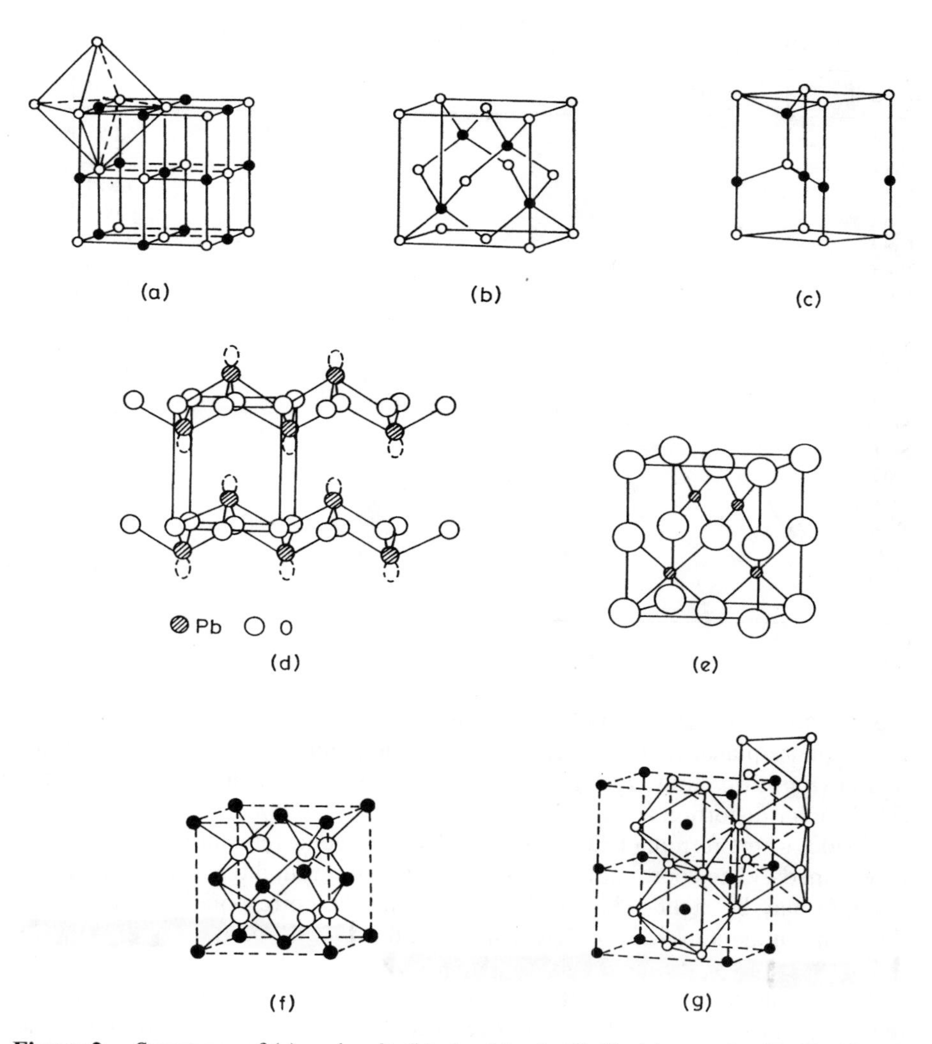

Figure 2. Structures of (a) rock salt, (b) zinc blende (ZnS), (c) wurtzite (ZnO), (d) red PbO, (e) PtO (PtS structure), (f) fluorite, and (g) rutile.

electrons on the structure. There is a pronounced directional character associated with these electrons which influences the structure adopted. PbO is dimorphous, the red variety (litharge: Fig. 2d) having a tetragonal layer structure in which Pb occupies the apex of a tetragonal pyramid and oxygen has a tetrahedral coordination of metal atoms. Each lead atom carries an inert pair that points in between two metal atoms in the adjacent layers. The blue-black modification of SnO is isomorphous with red PbO. Yellow PbO (massicot) has an orthorhombic chain structure consisting of zigzag chains of —Pb—O—Pb— parallel to the

Table 3 Oxides with the Rock Salt Structure

Crystal	a (Å)
AmO	5.05
BaO	5.523
CaO	4.810
CdO	4.695
CoO	4.267
EuO	5.144
$FeO_{1-\delta}$	4.27–4.31
MnO	4.445
NbO	4.209
NiO	4.168
PuO	4.959
SmO	4.988
SrO	5.160
TaO	4.422–4.439
TiO	4.176
WO	4.92
VO	4.062
YbO	4.86
ZrO	4.62

b-axis. PtS, PtO, PdO and PdS possess tetragonal structures in which the metal has a square-planar (dsp^2) coordination and the anion has a tetrahedral coordination (Fig. 2e). CuO crystallizes in a distorted PtO structure because of the Jahn–Teller distortion.

Oxides of the formula BO_2, which are distinctly ionic, crystallize in structures determined by the coordination geometry of anions around the cation. The coordination may be eight-, six-, or fourfold, fixing the corresponding anion coordination numbers to 4, 3, or 2. We thus get the fluorite (8:4), rutile (6:3), and silica (4:2) structures for the AB_2 compounds. The first two are shown in Figure 2. The fluorite structure (Fig. 2f) consists of a cubic close-packed array of cations in which all the tetrahedral sites are occupied by anions. The relation to the zinc blende structure (Fig. 2b) is obvious. Anions are tetrahedrally coordinated by cations; cations occur in cubic coordination. Oxides of fluorite structure are discussed later in Section 3. The antifluorite structure adopted by some oxides (e.g., Li_2O), is derived from the fluorite structure by the interchange of positions of the anions and the cations (4:8 coordination). The rutile (TiO_2) structure (Fig. 2g), discussed at length later in Section 3, is also tetragonal and is stable for ionic compounds with $r_c/r_a < 0.73$. It consists of an infinite array of TiO_6 octahedra linked through opposite edges along the *c*-axis. Many tetravalent metal oxides (Ti, V, Cr, Mn, Nb, Mo, W, Ru, Os, Ir, Sn, Pb) crystallize in this structure. Just as with the fluorite structure, it is possible to substitute in this structure stoichiometrically equivalent amounts of appropriate cations to obtain ternary compounds (e.g., $AlSbO_4$, $CrNbO_4$, $FeNbO_4$, $RhVO_4$). In Table 4, we list several oxides possessing the rutile structure.

The cuprate (Cu_2O) structure consists of a body-centered cubic array of oxygens, with the copper atoms occupying centers of four of the eight cubelets into which the BCC cell may be divided (see Section 3 for details). In this structure, copper has a linear coordination and oxygen has tetrahedral coordination (2:4). This structure is unique in the sense that it has two identical interpenetrating frameworks, not directly linked to each other.

Crystal structures of B_2O_3 type sesquioxides fall into two main categories depending on whether the B—O bond is predominantly ionic or covalent. Ionic B_2O_3 type oxides are typified by the sesquioxides of *d*-transition metals and lanthanides, which crystallize either in the Al_2O_3 (corundum) or C–rare earth oxide structure. The corundum structure ($D5_1$ type), adopted by many oxides (Table 5), consists of a hexagonal close-packed array of anions in which the cations occupy two-thirds of the octahedral interstices; the cations have octahedral coordination and the oxygens, nearly tetrahedral coordination (6:4) (see

Table 4 Oxides with the Tetragonal Rutile Structure

Crystal	Unit Cell Dimensions (Å)	
	a	*c*
CrO_2	4.41	2.91
$(Cr_{0.33},Mo_{0.67})O_2$	4.696	2.886
$CrO_{2.14}$	4.423	2.917
IrO_2	4.49	3.14
β-MnO_2	4.396	2.871
MoO_2	4.86	2.79
NbO_2	4.77	2.96
OsO_2	4.51	3.19
PbO_2	4.946	3.379
RuO_2	4.51	3.11
SnO_2	4.73727	3.186383 (20–23°C)
TaO_2	4.709	3.065
TeO_2	4.79	3.77
TiO_2 (rutile)	4.59373	2.95812 (25°C)
WO_2	4.86	2.77
$AlSbO_4$	4.510	2.961
$CrNbO_4$	4.635	3.005
$CrTaO_4$	4.626	3.009
$FeNbO_4$	4.68	3.05
$FeSbO_4$	4.623	3.011
$FeTaO_4$	4.672	3.042
$RhNbO_4$	4.686	3.014
$RhTaO_4$	4.684	3.020
$RhVO_4$	4.607	2.923

Table 5 Oxides with the Rhombohedral Corundum and Ilmenite Structures

Crystal	a_0 (Å)	α	a'_0 (A)	c'_0 (Å)
Corundum				
Al_2O_3 (corundum)	5.128	55°20′	4.76280	13.00320 (31°C)
$(Cr_{1.90}, V_{0.09}, Fe_{0.01})_2O_3$ (eskolaite)	5.361	55°5′	4.958	13.60
α-Fe_2O_3 (hematite)	5.4135	55°17′	5.035	13.72
Ga_2O_3	5.320	55°48′	4.9793	13.429 (24°C)
Rh_2O_3	5.47	55°40′	5.11	13.82
Ti_2O_3	5.431	56°36′	5.148	13.636
V_2O_3	5.647	53°45′	5.105	14.449
Ilmenites				
$FeTiO_3$ (ilmenite)	5.538	54°41′	5.08	14.026
$NiMnO_3$	5.343	54°39′	4.905	13.59
$CoMnO_3$	5.385	54°31′	4.933	13.91
$MnTiO_3$	5.610	54°30′	5.137	14.283
$NiTiO_3$	5.437	55°07′	5.044	13.819
$CdTiO_3$ (low)	5.82	53°36′	5.248	14.906
$CoTiO_3$	5.49	54°42′	5.044	13.961
$FeRhO_3$	5.46	55°26′	5.079	13.817
$FeVO_3$	5.42	55°14′	5.026	13.735
$LiNbO_3$	5.47	55°43′	5.112	13.816
$MgTiO_3$	5.54	54°39′	5.054	13.898

Section 3 for details). The ilmenite structure is derived from the corundum structure by replacing the cations in alternate layers with Fe and Ti (Table 5). Sesquioxides of iron and aluminum also exist in the metastable γ-form, which has a cation-deficient cubic spinel structure. The C–rare earth oxide structure, adopted by sesquioxides of lanthanides and actinides as well as Mn_2O_3, is derived from the fluorite structure by the removal of one-fourth of the anions, to confer a sixfold coordination on the metal atoms; the coordination is however not exactly octahedral. Rare earth oxides crystallize in two other forms, A (hexagonal) and B (monoclinic), where the lanthanide ion is heptacoordinated. Among the sesquioxides, Bi_2O_3 is unique in that it crystallizes in at least four polymorphic modifications: α (monoclinic), β (tetragonal), γ (BCC), and δ (face-centered cubic), of which the δ (FCC) phase, having a defect fluorite structure exhibits high ionic conductivity.

Structures of several metal oxides and fluorides are based on the structure of ReO_3 (DO_9-type). The structure of ReO_3 is cubic, comprising corner-shared ReO_6 with linear Re—O—Re links (see Section 3 for details). WO_3 possesses a distorted ReO_3 structure, and many of the niobium oxides contain ReO_3-type blocks. Perovskites are derived from the ReO_3 structure, and we discuss these and other structures of complex oxides such as spinels, pyrochlores, bronzes, and tunnel structures later in Part I.

Crystal structures of oxides are generally described on the basis of the linkages between cation-centered coordination polyhedra through corners, edges, and/or faces. For example, the ReO_3 structure is described as ReO_6 octahedra connected through all corners. The structure of NaCl is described as $NaCl_6$ octahedra sharing all edges. The cluster compound $NaMo_4O_6$ is built up of Mo_6O_{12} clusters by trans edge-linking of Mo_6 octahedra. The structure consists of Mo_4O_6 chains running parallel to the tetragonal **c**-axis, the alkali metal atoms being present in between the chains. The structure of NbO is considered to be made up of Nb_6O_{12} clusters linked three-dimensionally through the corners of Nb_6 octahedra, $Nb_{6/2}O_{12/4}$.

2.2 Defects in Oxides

A majority of oxides have definite compositions, but some exhibit a wide range of compositions or show no simple relation between the composition and the detailed structure. For example, it has been known for some time that stoichiometric $FeO_{1.00}$ does not fall in the stability range of ferrous oxides ($FeO_{1.05}$–$FeO_{1.15}$). Point defects in crystals such as vacancies and interstitials described by Schottky and Frenkel account for the transport properties of ionic solids, but it is not possible to apply the point defect model to oxides possessing a wide stoichiometric range or to solids exhibiting ordering of defects or extended defects (such as *crystallographic shear* planes). It appears that the point defect model is valid only when the defect concentration (or the deviation from stoichiometry) is extremely small.[6,8,9]

The various types of defect that occur in the ionic solids are called *point, linear, planar, and volumetric.* Point defects arise from the absence of the constitutent atoms (or ions) on the lattice sites or their presence in interstitial positions. The presence of alien atoms or ions constitutes another type of point defect. Point defects cause polarization of the surrounding region in the crystal and give rise to small displacements on neighboring atoms or ions. Thus, a cation vacancy in an ionic solid will have an effective negative charge, causing displacements of neighboring anions. The energy of formation of a point defect depends mainly on the atomic arrangement in the immediate neighborhood of the point defect. Linear defects in crystals are dislocations, corresponding to rows of atoms that do not possess the right coordination. Planar defects are boundaries between small crystallites *(grain boundaries), stacking faults,* crystallographic shear planes, *twin boundaries,* and *antiphase boundaries.* Segregation of point defects give rise to three-dimensional volumetric defects.

The term ''nonstoichiometric'' has been used to describe the oxides such as $FeO_{1.05}$, Pr_6O_{11}, or Ti_3O_5, although it would not have the same meaning in these different oxides. A *nonstoichiometric oxide* can be defined as a crystalline oxide in equilibrium with its environment, behaving as a thermodynamically bivariant system. Although properties such as the cell dimensions may change with composition, the symmetry remains the same throughout the composition range of stability. Note that the stoichiometry of a crystal is itself uniquely defined by the composition of the repeating unit cell. The formula of the unit cell may be complex, with seemingly irrational ratios of the constituent atoms as in $NbO_{2.4906}$, $NbO_{2.4681}$, $PrO_{1.833}$, and $PrO_{1.714}$, but these oxides are crystallographically well-defined compounds with the formulas $Nb_{53}O_{132}$, $Nb_{47}O_{116}$, $Pr_{12}O_{22}$, and Pr_7O_{12}, respectively.[6,9] An understanding of such oxides has thrown light on the nature of nonstoichiometry in oxides. In nonstoichiometric oxides, the average number of atoms per unit cell is not equal to the number of sites; in the anion or the cation sublattice, there is a deficiency or an excess of the species.

Nonstoichiometric oxides are generally mixed valent with nonintegral electron/atom ratios, and their electronic properties depend on the nature and magnitude of nonstoichiometry. Electronic conduction in some of the oxides such as Pr_6O_{11} occurs by hopping between the cations of different valencies (Pr^{3+} and Pr^{4+}), while in some others electrons are delocalized over the lattice (e.g., some of the oxides with metallic properties). Nonstoichiometric oxides are different from solid solutions such as $Ni_{1-x}Li_xO$ with Ni in +2 and +3 states depending on x or $LaNi_{1-x}Mn_xO_3$ where Mn^{2+}, Mn^{3+}, or Mn^{4+} is possible.

In ionic solids, the common point defects are *Schottky pairs* (pairs of cation and anion vacancies) and *Frenkel defects* (cation or anion interstitial plus a vacancy). When there is a large concentration of Schottky pairs, the measured pyknometric density of the solid is considerably lower than the density calculated from the X-ray unit cell dimensions (e.g., VO). Creation of defects is generally an endothermic process. Thus, the formation energies of vacancies in ionic solids are generally 2 eV or more. The intrinsic defect concentration in these solids is therefore extremely small even at high temperatures. The intrinsic defect concentration in binary solids around $0.8T_m$ is $\sim 10^{-5}$.

Energies of formation, interaction, and migration of point defects have been estimated theoretically. A vanancy in an ionic solid carries a virtual charge of opposite sign and the polarization energy therefore becomes significant. The first successful evaluation of vacancy formation energies in ionic solids was made by Mott and Littleton, who evaluated the polarization energy after allowing the lattice to relax. This approach has provided the basis for a variety of calculations of defect energies by Catlow, Jacobs, Stoneham, and others.[14–16]

It is useful to represent all the processes involved in the intrinsic *defect equilibria* in a solid (with a low concentration of defects) as well as its equilibrium with its external environment by a set of coupled quasi-chemical reactions. These equilibrium reactions are then handled by the law of mass action. The free energy and the equilibrium constants for each process can be obtained if we know the enthalpies and entropies of the reactions from theory or experiment. Such an approach has been used to investigate a variety of defect equilibria by Kröger.[17] Let us briefly examine the case of metal oxides. The partial pressure of oxygen controls the nature of defects and nonstoichiometry in oxides. The defects responsible for nonstoichiometry and the corresponding oxidation or reduction of cations can be described in terms of quasi-chemical defect reactions. Let us consider the example of transition metal monoxides, $M_{1-x}O$ (M = Mn, Fe, Co,

Ni), which exhibit metal-deficient nonstoichiometry. For the formation of metal vacancies in $M_{1-x}O$ (M = divalent ion), the following equations can be written:

$$\tfrac{1}{2}O_2 \rightleftharpoons V_M + O_0$$

where V_M is a vacancy on the metal site, O_0 is oxygen on the anion site, and

$$V_M + e^- \rightleftharpoons V_M^- + h^+$$

where H^+ is a hole in the valence band which can be taken as equivalent to the M^{3+} ion. Then we can write

$$V_M^- + e^- \rightleftharpoons V_M^{2-} + h^+$$

$$2M_M + 2h^+ \rightleftharpoons 2M_M^+$$

where M_M represents the metal on the metal site.

The total reaction is

$$\tfrac{1}{2}O_2 \rightleftharpoons V_M^{2-} + 2M_M^+ + O_0$$

Here, V_M^{2-} denotes a doubly ionized metal vacancy, M_M^+, the association of M^{2+} and h^+ (which may be regarded as the M^{3+} ion). Using the law of mass action, introducing the neutrality condition $[h^+] = 2\,[V_M^{2-}]$, and expressing the cation vacancies as x in $M_{1-x}O$, the equlibrium constant for the reaction can be written as follows:

$$K_{V_M^{2+}} = 4x^3 P_{O_2}^{-1/2}$$

The composition of the oxide therefore depends on the oxygen partial pressure as given by, $x \propto P_{O_2}^{1/6}$. If the defects are singly ionized vacancies, V_M^-, then, $x \propto P_{O_2}^{1/4}$.

It is convenient to discuss the thermodynamics of nonstoichiometry of oxides in terms of the relative partial molar free energy of oxygen, ΔG_{O_2}, which is defined

$$\Delta\overline{G}_{O_2} = \mu_{O_2} - \mu_{O2}^{\circ} = RT \ln P_{O_2}$$

Here, μ_{O_2} is the chemical potential of oxygen in the solid, μ_{O2}° the chemical potential of oxygen in its standard state, and P_{O_2} the equilibrium partial pressure of oxygen in the atmosphere around the nonstoichiometric oxide. For $M_{1-x}O$, the relation between $\Delta\overline{G}_{O_2}$ and composition x can be expressed as $\Delta\overline{G}_{O_2} = nRT \ln x$. A plot of $\Delta\overline{G}_{O_2}$ versus $\ln x$ would be linear, with the slope giving n; the value of n gives information about the types of defect involved. For doubly ionized metal vacancies, V_M^{2-}, $n = 6$, while for V_M^-, $n = 4$. Isothermal plots of $\Delta\overline{G}_{O_2}$ versus $\ln x$ can be constructed from equilibrium thermogravimetric data or electrochemical measurements. A typical plot of $\Delta\overline{G}_{O_2}$ versus $\log x$ for the $Fe_{1-x}O$ system[18] is shown in Figure 3, where n values corresponding to various regions are indicated. We see that $n = 6$ occurs when $x \geqslant 0.09$, while $n = 5$, corresponding to V_M—$V_M)^{4-}$ pairs occurs close to this region. Values of $n = 3$

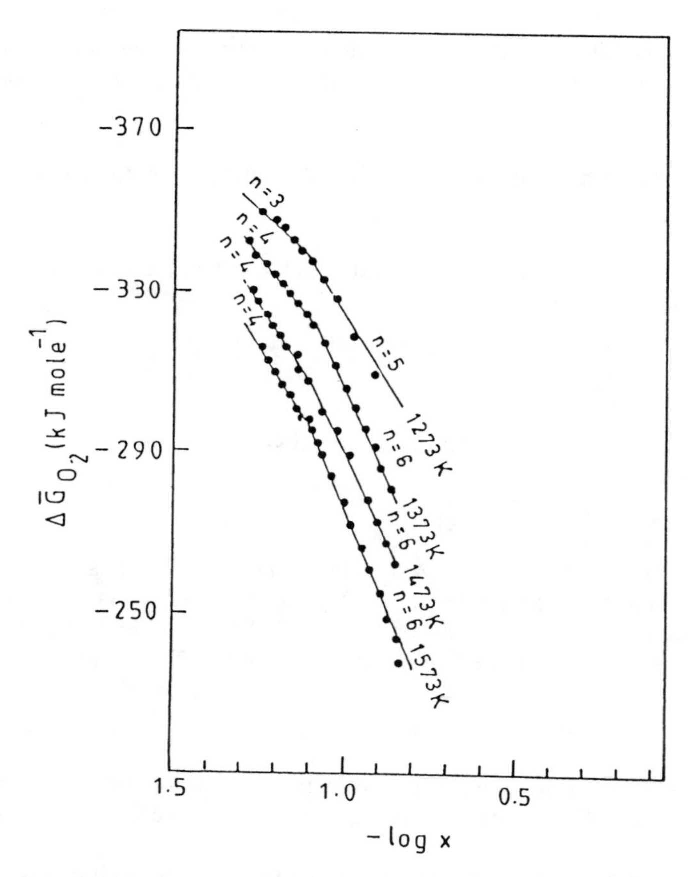

Figure 3. Plot of the relative partial molar free energy of oxygen, $\Delta\bar{G}_{O_2}$, of $Fe_{1-x}O$ against the oxygen stoichiometry (log x). (From the data of Sørensen.[18])

and 4 observed for $x < 0.09$ can be explained in terms of the $[4V_M^{2-}—Fe_i^{3+}]^{5-}$ and $[16V_M^{2-}—5Fe_i^{3+}]^{17-}$ defect complexes, respectively. There is structural evidence for the presence of both these defect complexes in $Fe_{1-x}O$. The 16:5 complex corresponds to an element of Fe_3O_4 structure. Thermodynamic data do not however indicate the existence of such a complex in $Fe_{1-x}O$. Thermodynamic data do show evidence for such a complex in $Mn_{1-x}O$.

Point defects such as vacancies order themselves to give rise to *superstructures* or *complexes.*[6,19–26] Such ordering has been observed, in many oxides, TiO and VO being good examples with ~20% vacancies. The nature of ordering is different in TiO and VO. Superstructures in the VO system are formed at the oxygen-rich end, where the oxygen sublattice is nearly filled and some vanadium atoms occupy tetrahedral sites (e.g., V_{52}, O_{64}, V_{244} O_{320}). Each tetrahedral cation has four vacant octahedral sites as nearest neighbors (Fig. 4a), and the cluster

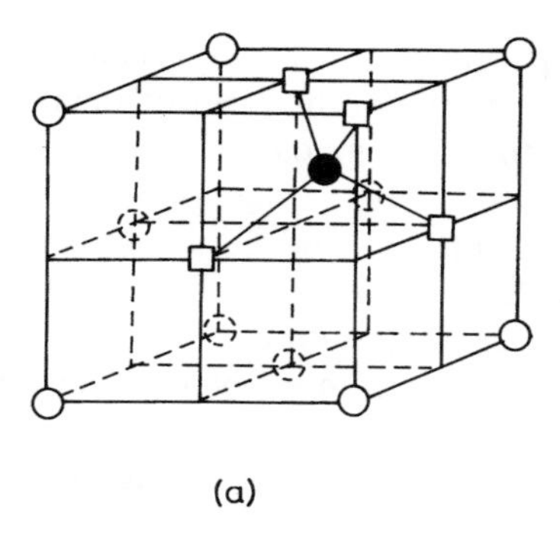

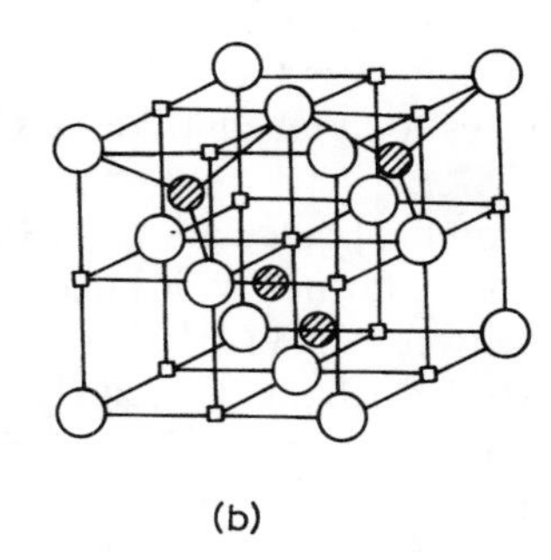

Figure 4. (a) Cluster of one tetrahedral cation and four vacant octahedral sites in one octant of the rock salt structure. (b) Koch–Cohen cluster of defects in $Fe_{1-x}O$: hatched circles, tetrahedral Fe atoms; open circles, oxide ions; square, Fe vacancies.

so formed is topologically similar to the rock salt structure, giving rise to a $2\sqrt{2} \times 2\sqrt{2} \times 2$ superstructure (seen readily in the electron diffraction pattern). In $V_{52}O_{62}$, all lattice sites (other than those of the defect cluster) are occupied as in the rock salt structure. In $VO_{1.3}$, however, some of the octahedral sites are also occupied.[19,20] The high temperature form of TiO_x has an averaged rock salt structure in the composition range of $x = 0.65$–1.25; the low temperature structure is monoclinic. We shall examine the defect structure of TiO in Section 3.

Wustite, $Fe_{1-x}O$, is an interesting cation-deficient system with the composition range $Fe_{0.85}O$–$Fe_{0.95}O$. The reaction for the formation of wustite can be written as follows:

$$2Fe^{2+} + \tfrac{1}{2}O_2 \rightarrow 2Fe^{3+}_{oct} + V_{Fe,oct} + O^{2-}$$

Since wustite has the NaCl structure, the oxidized Fe^{3+} and the newly created vacancies would be expected to remain in octahedral sites. Based on neutron diffraction measurements, Roth [21] suggested that a substantial proportion of the Fe^{3+} ions occupy tetrahedral sites according to the reaction

$$Fe^{3+}_{oct} + V_{int} \rightarrow Fe^{3+}_{tet} + V_{Fe,oct}$$

Roth proposed a model of (V_{Fe}—Fe^{3+}_{tet}—V_{Fe}) complexes in which the Fe^{3+} ion is tetrahedrally surrounded by oxygens. X-Ray crystallographic studies on quenched $Fe_{0.9}O$ by Koch and Cohen[22] showed that the defect complex is as large as the unit cell of FeO, consisting of 4 tetrahedral Fe^{3+} ions in alternate tetrahedral sites (Fig. 4*b*) and 13 vacancies on Fe^{2+} sites. Such a cluster corresponds to the chemical formula FeO; the net charge of the complex (including the virtual charges on the vacancies) is negative, and it has an antisphalerite structure. Since the packing of oxide ions remains undisturbed, these defect clusters can exist with coherent interfaces with the matrix. The ratio of vacancies to tetrahedral Fe^{3+} ions in the *Roth cluster* is $\frac{2}{1} = 2$, whereas in the *Koch–Cohen* *cluster* it is $\frac{13}{4} = 3.25$. Neutron diffraction experiments of Cheetham et al.[23] have shown that over the 1073–1473 K temperature range, the ratio of octahedral vacancies to tetrahedral Fe^{3+} is around 3.25 for the composition range $Fe_{0.9}O$–$Fe_{0.95}O$.

Theoretical calculations of Catlow and co-workers[14] indeed show that such clusters are formed with a net gain in energy. Binding energies of defect clusters vary between 1.98 (4:1 cluster) and 2.52 (8:3 cluster), with 2.1 eV for the 13:4 cluster of Koch and Cohen. The 16:5 cluster, which is an element of the spinel structure of Fe_3O_4, is also quite stable (binding energy 2.38 eV) indicating that it may be a precursor in the disproportionation of wustite. As suggested by Anderson,[24] starting from the rock salt structure, oxidation through the nonstoichiometric range could involve the following changes: isolated vacancies → dipolar associates → 4:1 clusters → 6:2, 8:3, 13:4, and such complex defect clusters → corner-shared 16:5 clusters → Fe_3O_4 (inverse spinel). Such clusters could be present in the $Mn_{1-x}O$ system as well; the cluster-binding energies calculated by Catlow and co-workers are lower in $Mn_{1-x}O$ than in $Fe_{1-x}O$. Free point defects, if any, may be present in $Mn_{1-x}O$, but not in $Fe_{1-x}O$.

The fluorite structure generates a variety of derivative structures with anion excess or anion deficiency; the cation sublattice remains essentially perfect. A unit cell of an oxide MO_2 having the fluorite structure corresponds to a face-centered cubic close packing of M^{2+} ions with all the eight available tetrahedral sites being occupied by oxide ions. An interstitial position of high symmetry is present at the center. In these structures, disorder arises either from anion vacancies or from interstitial anions. Anion-deficient MO_{2-x} and anion-excess MO_{2+x} have been investigated in detail. The nature of defect complexes in these systems seems to be independent of the chemical system studied. In oxides of the MO_{2-x} type, the defect clusters (referred to as *Bevan clusters*) correspond to tightly bound vacancies along a body diagonal, $\langle 111 \rangle$; the central cation is hexacoordinated and is surrounded by six heptacoordinated cations.[25] We discuss these defects in the case of Pr_7O_{12} later (Section 3). Oxides of the type MO_{2+x} with excess oxygen accommodate the excess O^{2-} in high symmetry interstitial positions when the extent of nonstoichiometry is low. When the nonstoichiometry is high, considerable distortions occur and the interstitial oxygen ions no longer occupy 1/2 1/2 1/2 positions in the fluorite structure. Thus, in U_4O_9, the additional oxide ions in the unit cell of UO_2 are displaced considerably from the 1/2 1/2 1/2 position in the $\langle 110 \rangle$ direction; furthermore, two fluorite oxygens nearest to the interstitial oxygen are displaced in the $\langle 111 \rangle$ direction, creating two vacancies. The cluster of two vacancies, one interstitial of one kind and two interstitials of another kind, is known as the 2:1:2 *Willis cluster.*[26] The more common 2:2:2 cluster involves another interstitial in a normally unoccupied fluorite octahedral site.

In ZrO_2/CaO and ZrO_2/Y_2O_3 solid solutions (which are good oxide ion conductors), there is evidence for ordering of defects (vacancies). In anion-excess ZrO_2/Nb_2O_5, the excess interstitial anions seem to aggregate on specific crystallographic planes.

We have seen that since defects in nonstoichiometric oxides are assimilated into the host lattice as structural elements, the defects cannot be considered to be point defects in the ordinary sense. One way by which anion-deficient nonstoichiometry is accommodated in some transition metal oxides is by eliminating point defects by the so-called *crystallographic shear*[6,9,27,28] (see Section 11 for details). A consequence of crystallographic shear is a decrease in the number of anion sites, compared to the parent structure. The plane in which octahedra share edges instead of corners (as in the ReO_3 structure) because oxygen sites have been eliminated is called the *shear plane.* An isolated shear plane or a random array of shear planes, referred to as a *Wadsley defect,* still gives rise to nonstoichiometry. It is interesting that even slightly reduced TiO_2 (e.g., $TiO_{1.997}$) shows the presence of shear planes. Wadsley employed crystallographic shear to explain structures of the homologous series oxides of tungsten, molybdenum, titanium, and vanadium discovered by Magnéli and co-workers. Since point defects are eliminated by this means, the process is nonconservative with respect to the lattice sites.

Dislocations are linear defects that play a major role in solid state phenomena and in determining the strength of materials. The simplest type of dislocation is the *edge dislocation,* which arises from the introduction of an extra layer of atoms in a crystal. Atoms in such a dislocation do not have their coordinations fully satisfied. Figure 5 shows an electron microscope image[29] of an edge dislocation in the $Ba_2Bi_4Ti_5O_{18}$. In a *screw dislocation,* two portions of a crystal are slipped past each other by an atomic spacing and the dislocation marks the boundary between the slipped and unslipped parts of the crystal. A screw dislocation transforms successive planes of atoms into the surface of a helix.

The surface of a crystal constitutes a planar, two-dimensional defect. The environment of atoms or ions on the surface of a crystal is considerably different from that in the bulk. In polycrystalline materials, there are *grain boundaries* between the particles. The interface between two solid phases is an important factor in determining the course of reactions, crystal growth, and so on. An interface may be coherent, incoherent, or semicoherent. The interface is coherent when there is perfect matching between the contact planes of two solid phases. *Epitaxial growth* occurs even when there is considerable mismatch (semicoherent interface). $MgFe_2O_4$ precipitates in MgO from coherent interfaces. The oxide ions in the two materials are cubic close-packed, and the precipitate and the matrix are therefore continuous.

In close-packed solids, one often encounters *stacking faults.* For example, in a solid with cubic close packing, ABC ABC ABC, there can be a fault such as ABC AB ABC. Other types of planar defects include *tilt boundary* (array of periodically spaced edge dislocations), *twist boundary* (array of screw dislocations), *twin boundary* (a layer with mirror plane symmetry with respect to the layers on either side), *coincidence boundary* (involving the rotation of one part of the crystal, on a specific plane, with respect to another), and *antiphase boundary* (across which the sublattice occupation gets interchanged).

Epitaxy is an example of *intergrowth* across a solid–solid interface. *Epitaxy*

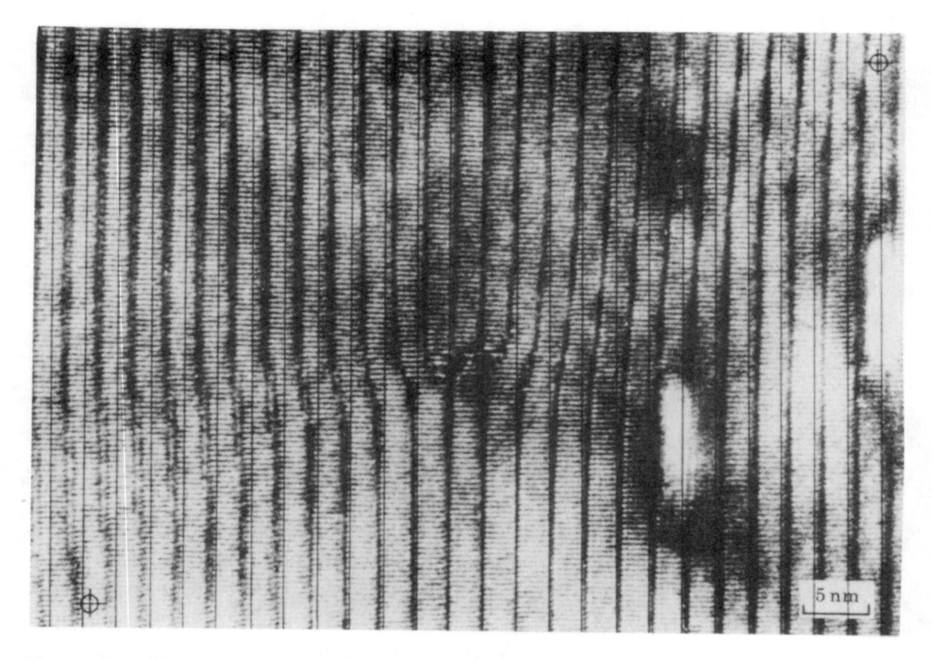

Figure 5. Electron microscopic lattice image of an edge dislocation in $Ba_2Ba_4Ti_5O_{18}$. (From Hutchison et al.[29])

is of considerable technological importance and involves laying down thin (single crystal) films of materials on an appropriate substrate (e.g., $YBa_2Cu_3O_7$, $LaNiO_3$, or $PbTiO_3$ on $SrTiO_3$). Intergrowth of one oxide composition with another is of common occurrence. Intergrowth can be recurrent (ordered) or nonrecurrent (disordered).[30] For example, β and β″ alumina interfrow nonrecurrently. If one examines the electron microscopic image of $Sr_3Ti_2O_7$ ($n = 2$, member of the $Sr_{n+1}Ti_nO_{3n+1}$ family), one sees the random occurrence of the $n = 4$, 5, 7, and other members. What is more interesting is the recurrent intergrowth found in many oxides, a good example being the alternate growth of the $n = 3$ and 4 members of oxides of the formula $Bi_2A_{n-1}B_nO_{3n+3}$ (Aurivillius family of oxides) shown in Figure 6. Siliconiobates, oxide bronzes, hexagonal barium ferrites, and many other oxides show intergrowth structures, and we shall discuss some of these in later sections. The long-range periodicity exhibited by some of the oxide intergrowth structures is truly amazing (see Fig. 6).[31]

2.3 Characterization Techniques

Characterization is an inescapable part of all investigations of solids. The relevant aspects of characterization are chemical composition and compositional homogeneity, structure (including crystal system and, where possible, atomic

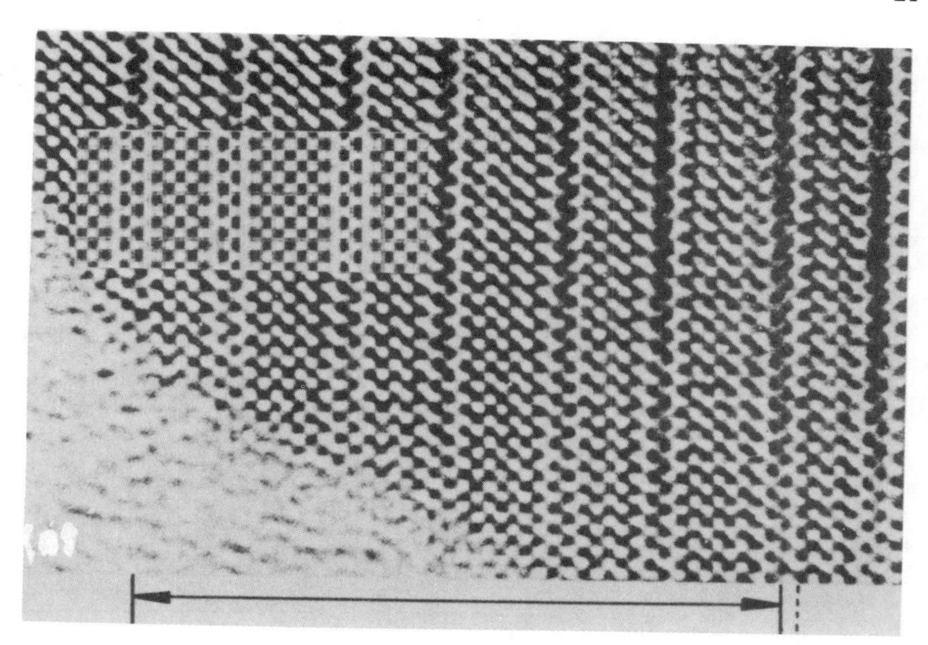

Figure 6. High resolution electron microscopic image of $Bi_9Ti_6CrO_{27}$. In this computer-simulated image, marker indicates the length of three unit cells. (From Jefferson et al.[31])

coordinates, bonding, and ultrastructure), and finally identification and analysis of defects and impurities influencing the properties. Characterization, therefore, describes all the aspects of composition and structure of a particular preparation of a material that enable one to reproducibly prepare the material. Recent advances in characterization techniques, especially in structure elucidation, have been stupendous and have opened new vistas in solid state chemistry.[5,6]

X-Ray diffraction has played a central role in identifying and characterizing solids since the early part of this century. Our ideas on the nature of bonding and the working criteria for distinguishing short-range and long-range order of crystalline substances from amorphous substances are largely derived from X-ray diffraction. While X-ray diffraction still remains a most useful tool to obtain structural information averaged over a large number of unit cells, several newer methods have become part of the arsenal of chemists. Efforts will undoubtedly continue to develop newer and better techniques for the characterization of solids.

A complete description of the structure of a solid requires the determination, by means of diffraction involving X-rays, electrons, or neutrons, of the crystal system, space group, unit cell dimensions, atomic coordinates, and finally the actual electron density distribution around the atoms. X-Ray diffraction has been commonly used for routine characterization as well as for detailed structural elucidation, but increasing use is being made of electron and neutron techniques to obtain information not readily provided by X-ray diffraction.

To obtain the detailed structure of a solid, a knowledge of X-ray diffraction intensities is essential, the intensities being related to the structure factor. Computer-controlled single crystal X-ray diffractometers with structure packages have rendered structure elucidation a routine matter. The availability of synchrotron X-radiation of continuously variable wavelength has made X-ray diffraction a still more powerful structural tool for the study of solids. A technique of great utility in determining the structures of oxides is the Rietveld treatment of powder X-ray diffraction profiles (see Fig. 7 for a typical profile). Automated structure packages for the determination of unknown structures by this method are available. Powder X-ray diffraction profiles obtained with synchrotron radiation are especially powerful in solving structures of oxides. In the profile analysis method, the structural parameters are fitted to the overall profile of the powder pattern instead of the traditional approach of fitting the intensities of individual peaks. The Rietveld method of X-ray profile analysis has been invaluable in solving structures of a variety of oxides that cannot be obtained in single crystal form.

Besides structure, X-ray diffraction gives a host of other valuable information. For example, powder diffraction has been widely used for phase identification, quantitative analysis of mixtures of phases, particle size analysis, characterization of physical imperfections (the last two being obtained from line broadening), and in situ studies of reactions. In the case of amorphous solids, X-ray diffraction data provide radial distribution functions, which give the number of atoms per unit volume at any distance from the reference atom. Variable temperature X-ray diffraction of crystalline materials is used to study phase transitions, thermal expansion, and thermal vibrational amplitude of atoms in solids. Similarly, X-ray diffraction at high pressures is employed to examine pressure-induced phase transitions. A recent innovation is the recording of X-ray diffraction patterns as a function of time (say every few seconds).

Electron beams differ from X-rays in two respects, rendering electron diffraction a valuable technique for studies in solids. The differences entail the small wavelength of electrons and the charge carried by them. The smaller wavelength leads to smaller Bragg angles in electron diffraction and makes it possible to record extensive sections of the reciprocal lattice directly with a small stationary crystal. Because of the charge, interaction of electrons with atoms is considerably stronger than that of X-rays, thus making it possible to record electron diffraction patterns almost instantaneously; the strong interaction, however, places stringent conditions on sample thickness. Employing electron diffraction, one can profitably investigate defect ordering, superstructures, fine particle samples, and so on.

Thermal neutrons with a velocity of about 4000 m/s associated with a wavelength of ~1.0 Å are used in neutron diffraction experiments. Since the neutron scattering amplitude does not show a smooth dependence on the atomic number of the atoms, neutron diffraction is particularly useful in locating light atoms

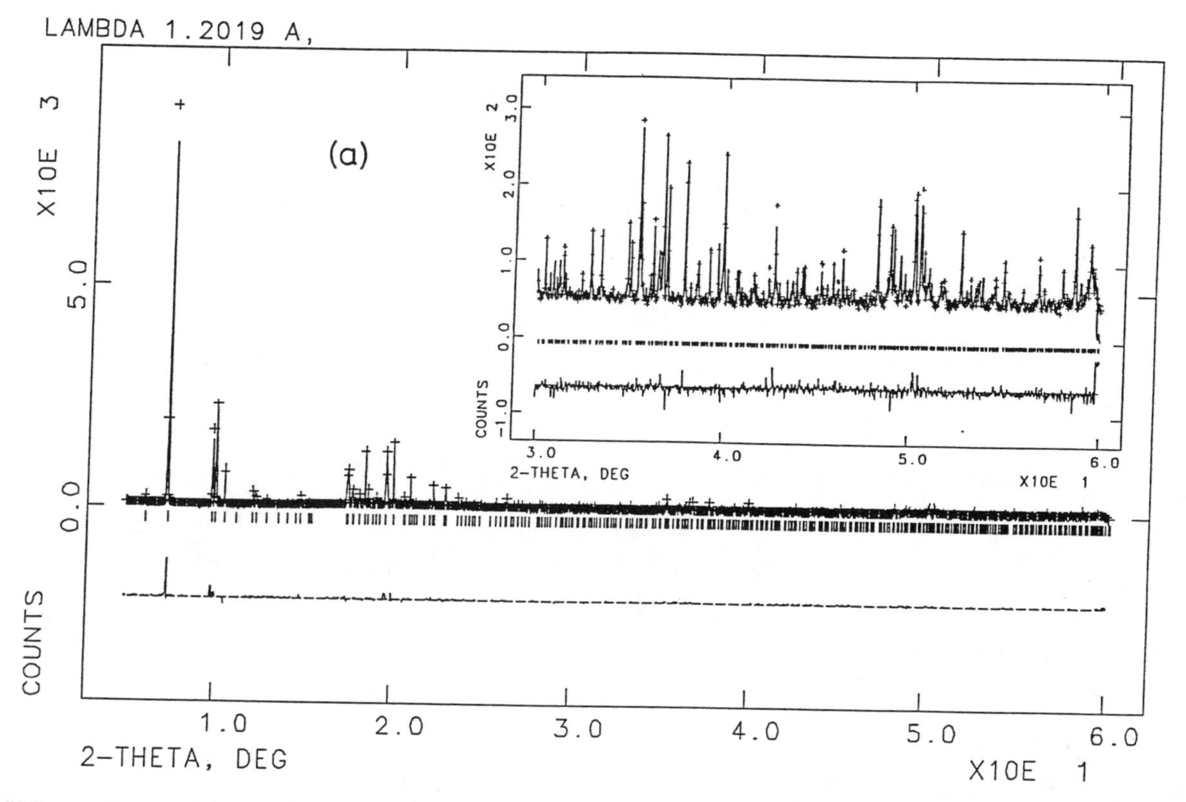

Figure 7. X-Ray and neutron diffraction profiles of silicon ferrierite. (a) Fit to the X-ray diffraction profile from data obtained at the National Synchrotron Light Source, Brookhaven National Laboratory. Inset: expanded 30–60° degree data. (b) Fit to the neutron diffraction profile data obtained at the National Institute for Standards and Technology. Inset: structure of the compound. (Profiles courtesy of A. K. Cheetham.)

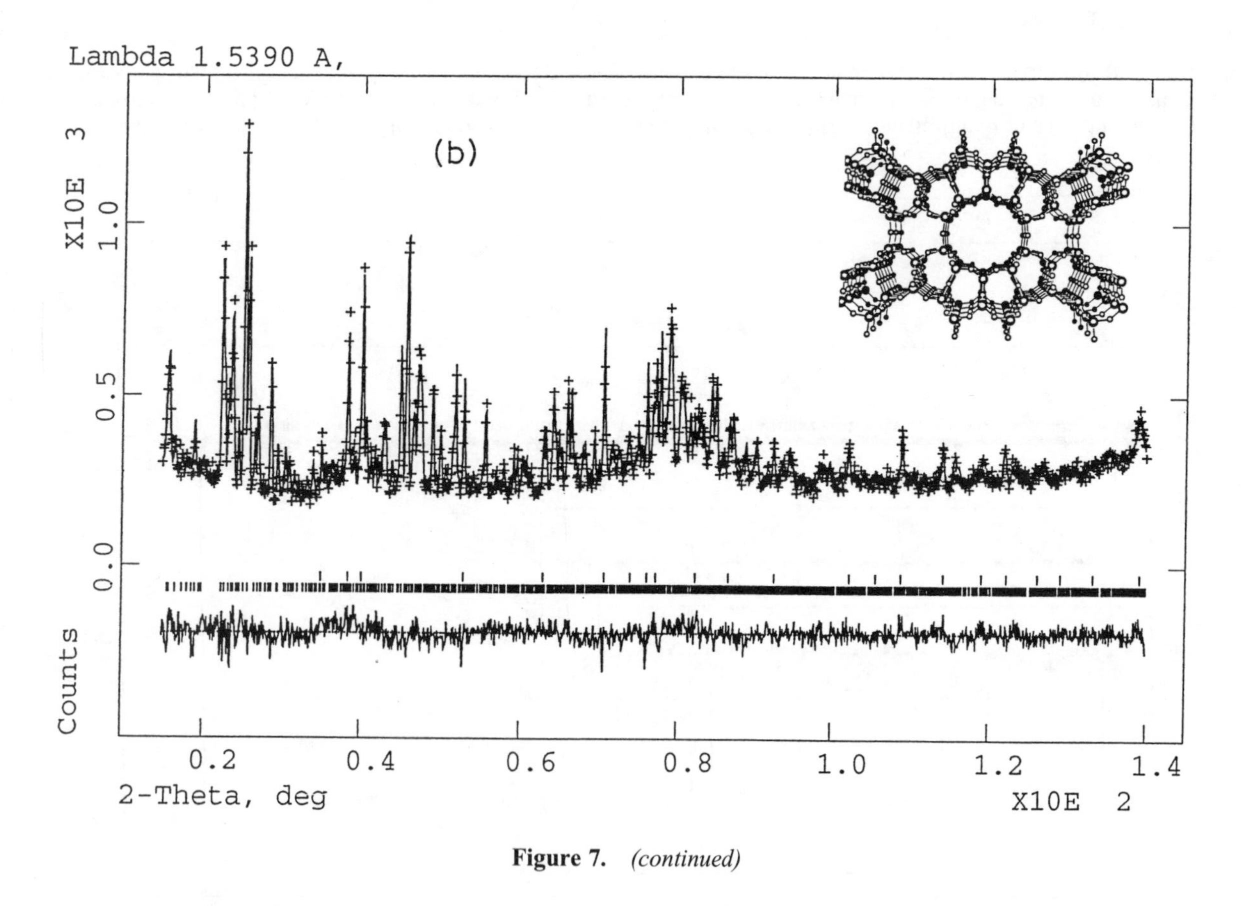

Figure 7. *(continued)*

such as hydrogen in crystals. Additional scattering of neutrons can arise from the magnetic moment of neutrons. Neutron diffraction provides an experimental means whereby the different magnetic structures can be determined. In addition to the two scattering effects, which are elastic, neutrons can also undergo inelastic scattering by crystals. Inelastic neutron scattering by crystals is used in the study of quantized vibrational modes (phonons) and dynamics in solids.

Because neutron beams are much weaker in intensity than X-rays, neutron diffraction requires large single crystals (10–100 mm^3 in volume vs. the ~0.1 mm^3 crystal volume used in X-ray diffraction work). It is however possible to obtain useful structural data by the analysis of neutron diffraction profiles from polycrystalline materials. Profile analysis is particularly suitable in neutron diffraction work because the peaks are accurately described by a Gaussian function and also because the positions of oxygen are readily determined by neutrons. Structures of a large number of oxides have been refined using powder neutron diffraction data, and excellent accuracy (see Fig. 7 for typical data) has been obtained in many instances by a combined use of X-ray and neutron diffraction profiles of powders. Several modifications of the Rietveld method are available, and these include allowance for anisotropic temperature factors and flexible molecular constants. The availability of intense pulsed neutron sources has made this technique still more effective.

The electron microscope is the most versatile instrument available to study the ultrastructure of materials, to identify new or known phases, and to simultaneously yield information on composition. Usually electron diffraction and imaging are employed together in transmission electron microscopy (TEM). Images of isolated atoms are well within the capability of modern electron microscopes. Recent advances in scanning instruments (which record signals due to back-scattered electrons or emitted secondary electrons) enable us to detect extremely small quantities of materials and also to map the topography of surfaces of a wide range of materials. These capabilities are further enhanced in a scanning transmission electron microscope (STEM).

The mechanism of image formation in TEM involves both elastic and inelastic scattering. In producing the image, the electron microscope carries out Fourier transformations. The crystal in real space is transformed to the diffraction pattern in reciprocal space, and the latter is then transformed to the image in real space. TEM generally produces a projection image wherein all depths of thin specimens are in focus at the same time, making it difficult to interpret surface features of specimens. Several important improvements have emerged recently to obviate this difficulty. With appropriate modifications, an electron microscope can be converted into a veritable laboratory to operate at different temperatures, in different atmospheres, and so on.

In scanning electron microscopy (SEM), a finely focused electron beam probe moves from one point on the specimen to the next to form a raster pattern, similar to television imaging. The intensity of the scattered (secondary) electrons is continuously measured and displayed on a TV monitor to form a raster image. With modern high intensity electron sources (LaB_6), a surface resolution of 20–

50 Å is readily attained. The resolution is limited by electron penetration underneath the surface and also by surface cleanliness. Magnetic domains can also be examined in the reflective mode in a modified scanning microscope.

Analytical electron microscopy today enables one to obtain both the high resolution structure and the elemental composition of a specimen. This is probably the best technique to obtain local elemental composition of small regions of heterogeneous solids. When a high energy electron beam is incident on a specimen, we get elastically and inelastically scattered electrons, characteristic emitted X-rays, back-scattered electrons, and secondary electrons. All these can be collected at the same time in the microscope from a small specimen area to carry out analysis of the structure and composition. Elastically scattered electrons give images and diffraction patterns. Electrons that have lost energy give information on composition, as does the mitted X-ray spectrum. Data from both spectra can be readily collected in a microcomputer, processed, and appropriately displayed. In some modern instruments, the operating vacuum has been improved greatly (10^{-9} torr) to maintain high intensity electron sources and to minimize contamination. Electron energy loss spectroscopy can be used for analysis and also for obtaining information on the oxidation states of metals in oxides and on the electronic structures of oxides and other materials. The L absorption edges of transition metals and K absorption edge of oxygen in transition metal oxides reveal considerable structural information.

The most important aspect of electron microscopy in the study of metal oxides is its ability to elucidate problems that are beyond the capability of X-ray or neutron crystallography. High resolution electron microscopic (HREM) images show local structures of crystals in great detail. Selected area diffraction, along with imaging, can often reveal atomic arrangements in regions containing only a few atoms. Since the scattering factors for electrons are a thousand times those for X-rays, structural information can be obtained with very small specimens (~5 Å diameter). HREM is also very useful in identifying new phases, some occurring hardly in one or two layers as imperfections or as intergrowths in other phases. Once the experimental image has been obtained using a thin specimen (≤100 Å), the interpretation is checked with the aid of computed images obtained by taking into account multiple scattering of electrons, lens characteristics, and specimen thickness. Applications of high resolution electron microscopy for investigating various phenomena, structures, defects, and reactions of importance to solid state chemistry have been reviewed adequately in the last few years. Examples of the use of HREM were cited earlier, and many more will be discussed in later sections.

Many other techniques are employed for the study of metal oxides. Some of the important ones are extended absorption fine structure (EXAFS) spectroscopy, solid state NMR spectroscopy, electron spin resonance (ESR) spectroscopy, and scanning tunneling microscopy (STM). The EXAFS technique is especially useful to obtain structural information on the first coordination of metals in oxides. STM provides images of oxide surfaces on an atomic scale and has been widely used recently to study superconducting cuprates. Various techniques of electron spectroscopy (e.g., X-ray and UV photoelectron spectroscopy, Auger spectroscopy) are employed to obtain information on the electronic structures of metal oxides.

2.4 Phase Transitions

A variety of metal oxides exhibit transformations from one crystal structure to another *(polymorphism)* as the temperature or pressure is varied. Besides phase transitions involving changes in atomic configuration, oxides undergo transformations in which the electronic or the spin configuration changes (see refs. 32 and 33 for details). During a phase transition, the free energy of the solid remains continuous, but thermodynamic quantities such as entropy, volume, and heat capacity exhibit discontinuous changes. Depending on which derivative of the Gibbs free energy, G, shows a discontinuous change at the transition, phase transitions are generally classified as first order or second order. In a *first-order transition,* where the $G(P,T)$ surfaces of the parent and product phases intersect sharply, the entropy and the volume show singular behavior. In *second-order transitions,* on the other hand, the heat capacity, compressibility, or thermal expansivity shows singular behavior.

We all know that when a liquid transforms to a crystal, there is a change in order: the crystal has greater order than the liquid. The symmetry also changes in such a transition: the liquid has more symmetry than a crystal, since the liquid remains invariant under all rotations and translations. Landau introduced the concept of an *order parameter,* ξ, which is a measure of the order resulting from a phase transition. In a first-order transition (e.g., liquid to crystal), the change in ξ is discontinuous, but in a second-order transition, where the change of state is continuous, the change in ξ is also continuous. Landau proposed that G in a second-order (or structural) phase transition is not only a function of P and T but also of ξ, and he expanded G as a series in powers of ξ around the transition point. In such transitions, the order parameter vanishes at the critical temperature T_c. Landau also considered the symmetry changes across phase transitions. Thus, a transition from a phase of high symmetry to one of low symmetry is accompanied by an order parameter. In a second-order transition, certain elements of symmetry appear or disappear across the transition. For example, when the tetragonal, ferroelectric $BaTiO_3$, in which the dipoles are all ordered, transforms to the cubic, paraelectric phase, where the dipoles are randomly oriented, there is an increase in symmetry (appearance of certain symmetry elements), but a decrease in order. In a ferroelectric–paraelectric transition, electric polarization is the order parameter. In ferromagnetic–paramagnetic transition, magnetization is the order parameter.

Many physical properties diverge near T_c; that is, they show anomalously large values as T_c is approached from either side. The divergences in different phase transitions are, however, strikingly similar. These divergences can be quantified in terms of *critical exponents,* λ:

$$\lambda = \lim_{\varepsilon \to 0} \left| \frac{\ln f(\varepsilon)}{\ln|\varepsilon|} \right|$$

where $\varepsilon = (T - T_c)/T_c$ and λ is the exponent, since $f(\varepsilon)$ is proportional to ε^λ. The most important exponents are those associated with the specific heat (α), the order parameter (β), the susceptibility (γ), and the range over which individual constituents like atoms and atomic moments are correlated (ν). It so happens that the individual exponents for many different transitions are roughly similar (e.g., β 0.33). More interesting, in most transitions, $\alpha + 2\beta + \gamma = 2$ independent of the detailed nature of the system. In other words, although individual values of exponents may vary from one transition to another, they all add up to 2. Such a universality in critical exponents is understood in the light of Kadanoff's concept of scale invariance associated with the fluctuations near T_c. The exponents themselves are calculated by employing the renormalization group method developed by Wilson. Because of these developments, we are able to characterize higher order phase transitions in terms of the physical dimensionality of the system, d, and the dimensionality of the order parameter, n. It is noteworthy that there can be no phase transitions in one dimension if short-range forces along operate.

An important aspect of phase transitions in solids is the presence of *soft modes.* Operationally, a soft mode is a collective excitation whose frequency decreases anomalously as the transition point is reached. In second-order transitions, the soft mode frequency goes to zero at T_c, but in first-order transitions, the phase change occurs before the mode frequency goes to zero. Soft modes have been found to accompany a variety of solid state transitions. Occurrence of soft modes in phase transitions can be inferred from Landau's treatment, wherein atomic displacements may themselves be considered to represent an order parameter. In mode softening, a vibration mode goes to zero frequency at T_c as given by

$$\omega = \omega_0(T - T_c)^{1/2}$$

The order parameter, ξ, varies with temperature similarly.

It has been found convenient to classify phase transitions in solids on the basis of the mechanism. Three important kinds of transition of common occurrence are as follows: *nucleation and growth* transitions, a typical example being the anatase–rutile transformatikon of TiO_2; positional and orientational *order–disorder* transitions; and *martensitic transitions.* A typical example of a positional order–disorder transition is that of AgI and TiO. Orientational order–disorder transitions are exhibited by many solids such as ammonium halides, plastic (orientationally disordered) crystals, and salts of di- or polyatomic anions. A martensitic transition is a structural change caused by atomic displacements (and not by diffusion) corresponding to a homogeneous deformation wherein the parent and product phases are related by a substitutional lattice correspondence, an irrational habit plane, and a precise orientational relationship. These transitions occur with high velocities of the order of sound velocity and were originally discovered in steel but are now known to occur in solids such as $KTa_{0.65}Nb_{0.35}O_3$ and ZrO_2.

Phase transitions and associated structural changes generally occur by a reconstructive mechanism in first-order transitions (e.g., anatase–rutile transition of TiO_2) or by means of a slight distortion of the lattice through small displacements or ordering of the constituent atoms or ions (e.g., α–β transition of SiO_2). Even in ordering transitions, there can be a displacive component, just as there can be an ordering component in a displacive transition. Structural phase transitions are also classified in the following manner:

Displacive: ferrodistortive, antiferrodistortive, and ferroelastic transitions, for example.

Order–disorder: vacancy ordering in oxides, tunneling of protons in hydrogen-bonded solids (e.g., KDP), and orientation of molecular ions (e.g., $NaNO_2$), for example.

Electronic: spin-state transitions, charge-density waves, metal–nonmetal transition.

Many solids, including oxides, exhibit *polytypism,* wherein the unit cells of the different polytypic forms differ from one another only in the c dimension. This arises because of differences in the sequence in which the atomic layers are stacked (i.e., different proportions of cubic and hexagonal layers).

3 Mother Structures of Some Binary Transition Metal Oxides

Some structure types play an important role in the identification of the complex structures of transition metal oxides. Two of them, the ReO_3 and spinel structures, will be examined in detail later (Sections 4 and 8). Five other structures may also be considered as basic structures in the crystal chemistry of these oxides. They are the rock salt, rutile, cuprite, fluorite, and corundum-type structures.

Many of the BO-type oxides (Table 3) crystallize in the cubic rock salt structure ($a_c \approx 4.2$–5.2 Å). This structure is generally described as a cubic close packing (ABC-type packing) of oxygen atoms along the $\langle 111 \rangle$ direction, forming octahedral cavities where the metallic B atoms are located (Fig. 8a). It can also be described in terms of layers of BO_6 octahedra sharing edges (Fig. 8b). This is true if the oxides are stoichiometric, as in alkaline earth oxides such as SrO; it is not however the case in many transition metal oxides. For instance, NiO can be Ni-deficient, leading to $Ni_{1-\delta}O$. FeO is characterized by complex nonstoichiometry with a wide homogeneity range, $Fe_{1-\delta}O$, involving the existence of microphases in the rock salt matrix. The most spectacular examples are those of TiO and NbO, which exhibit large deviations from stoichiometry. For instance, the high temperature cubic form of TiO contains 16% of vacancies on sites of both oxygen and the metal distributed in a statistical manner so that the cell formula is best represented as $(Ti_{3.3}\square_{0.7})(O_{3.3}\square_{0.7})$, where $\square$ represents a vacancy, instead of Ti_4O_4. The low temperature form of TiO obtained by anneal-

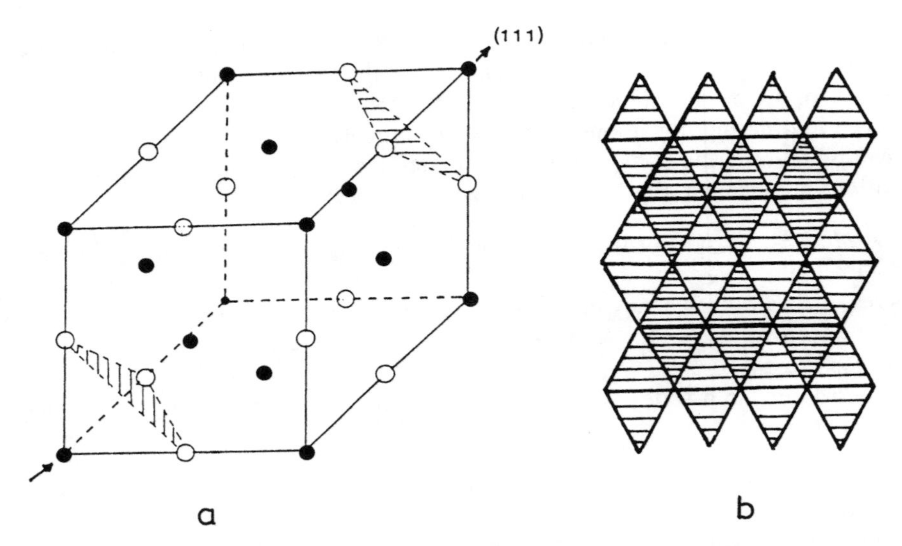

Figure 8. (a) The rock salt structure of NiO showing the oxygen planes perpendicular to the ⟨111⟩ direction: Ni atoms are represented by solid circles and O atoms by open circles. (b) Schematic structure of NiO viewed along ⟨110⟩, showing the edge-sharing NiO_6 octahedra (hatched lozenges).

ing shows the same density of vacancies, but distributed in an ordered manner leading to a monoclinic cell ($a \approx a_c\sqrt{2}$, $b \approx a_c\sqrt{5}$, $c \approx a_c$, $\gamma \approx 107°$) whose structure is shown schematically in Figure 9.[34] The superstructure with respect to the cubic rock salt cell is due to the ordering of oxygen and titanium vacancies. In one $[110]_c$ row out of three, half of the oxygens (or titaniums) are missing, with the result that one-sixth of the metallic as well as of the anionic sites are vacant. The structure of NbO is also cubic ($a \approx a_c$), but it is more lacunar (Fig. 10),[35] with the cubic cell containing only three NbO groups instead of four, leading to the formula $(Nb_3\square)(O_3\square)$; as a result, the ordered distribution of these vacancies makes the metal coordination square planar instead of octahedral. A homogeneity range of nonstoichiometry is found in some of the transition metal oxides with rock salt structures. Thus, titanium monoxide can be formulated as TiO_x with $0.70 \leq x \leq 1.25$. This system shows several microphases, all with oxygen- or metal-deficient ordered rock salt structures. Instead of interstitial oxygens, $TiO_{1.25}$ possesses a titanium-deficient structure. Its tetragonal structure ($a \approx a_c\sqrt{10}/2$ $c \approx a_c$) is a superstructure of the cubic rock salt type (Fig. 11).[34] All the oxygen sites are occupied, whereas two titanium sites out of 12 are vacant in an ordered way, leading to the structural formula $(Ti_{10}\square_2)O_{12}$. Many other intermediate microphases, ordered and semiordered, are observed in the Ti—O system and also in the VO_x oxides by electron microscopy. Owing to the presence of *d* electrons, these oxides exhibit unique semimetallic or semiconducting properties.

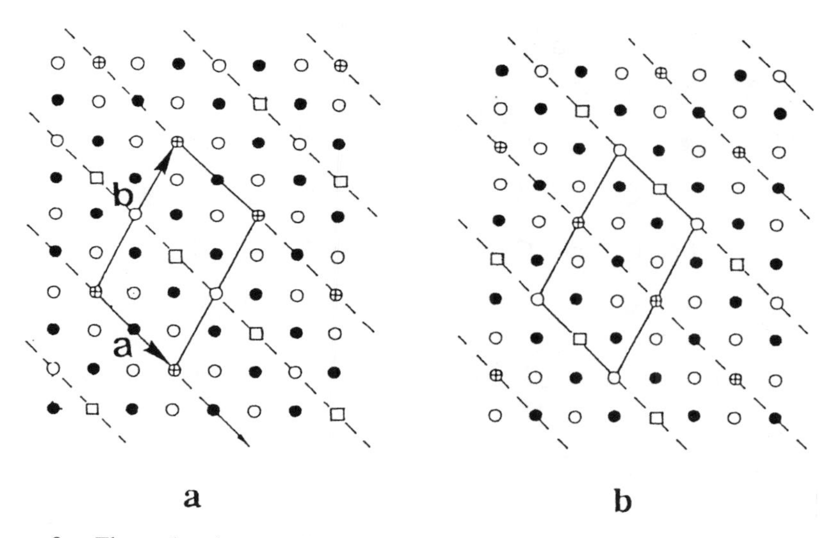

Figure 9. The ordered monoclinic structure of the nonstoichiometric rock salt type oxide TiO. View along $\vec{c}$ of two successive planes at $z = 0$ and $z = 1/2$: solid circles, Ti; open circles, O; crossed circles, Ti vacancy; open squares, O vacancy. (Following Watanabe.[34])

The rutile structure, represented by TiO_2, is exhibited by many transition element oxides (Table 4). The ideal structure (Fig. 12a)[36] is tetragonal ($a \approx 4.6$ Å, $c \approx 3$ Å). In the (001) plane, the TiO_6 octahedra share their corners (Fig. 12b),[37] forming very small square tunnels running along the $\vec{c}$ direction. In this direction, the TiO_6 octahedra share their edges (Fig. 12c),[37] forming $[TiO_2]_\infty$ ribbons, called rutile ribbons; the different rutile ribbons share the corners of their octahedra. As shown in Table 4, this structure is encountered in several BO_2 oxides as well as in ternary oxides $B'BO_4$. Many other oxides exhibit this structure, such as those of trirutile family $B'B_2O_6$ (B = Ta, Nb; B′ = Mg, Ni,

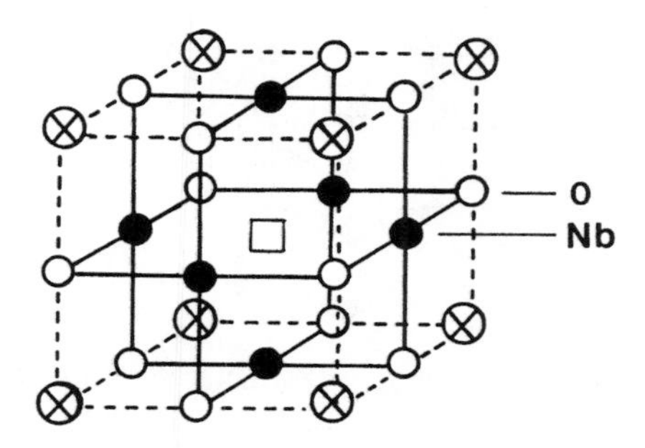

Figure 10. The cubic structure of NbO: solid circles, Nb; open circles, O; crossed circles, Nb vacancy; open squares, O vacancy. (After Collongues.[35])

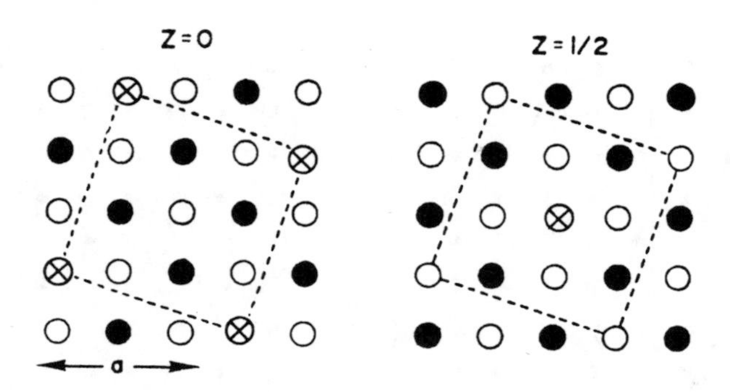

Figure 11. The tetragonal ordered structure of $TiO_{1.25}$: view along $\vec{c}$ of two successive planes at $z = 0$ and $z = 1/2$: solid circles, Ti; open circles, O; crossed circles, vacancies. (After Watanabe.[34])

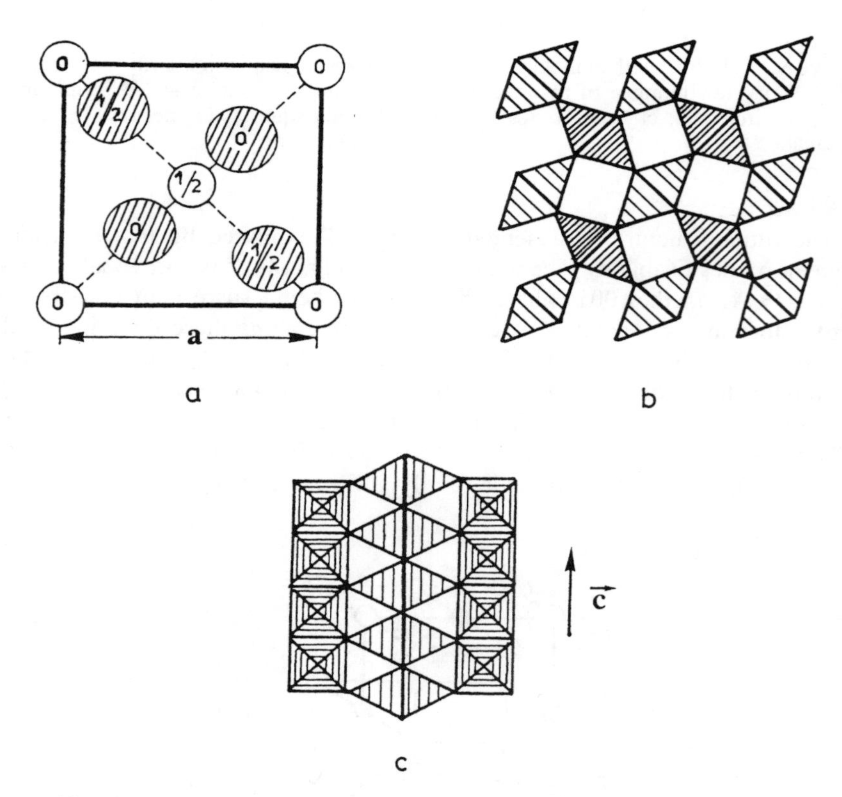

Figure 12. (a) Projection of the tetragonal structure of rutile along $\vec{c}$. The Ti atoms (small circles) are octahedrally surrounded by oxygen atoms (large hatched circles). [After Wycoff.[36]) (b) Assemblage of the TiO_6 octahedra in the (001) plane. [After Wadsley.[37]) (c) Edge-sharing TiO_6 octahedra forming rutile chains running along $\vec{c}$. (After Wadsley.[37])

Zn, Ca, Fe), where the B and B′ ions are distributed in an ordered manner along the $\vec{\mathbf{c}}$ direction, leading to a tripling of this parameter. Note that many oxides, such as the monoclinic VO_2, MoO_2, WO_2, TcO_2, and NbO_2, have a distorted rutile structure. Creation of oxygen deficiency in rutile-type oxides results in *shear structures* (see Section 11).

The fluorite structure, corresponding to the mineral CaF_2, is formed in several transition metal oxides (HfO_2, PrO_2, ThO_2, TbO_2, UO_2, and ZrO_2). The anti-fluorite structure is also obtained in oxides such as K_2O, Li_2O, and Na_2O. This cubic structure ($a \approx$ 5–5.5 Å) is characterized by a cubic coordination of the cations (Fig. 13a).[36] The three-dimensional structure of PrO_2 (Fig. 13b)[35] can be described as PrO_8 cubic groups sharing their edges in the three directions of the cubic cell. The praseodymium ions, owing to their ability to adopt several kinds of coordination, form a series of oxides with formula Pr_nO_{2n-2} derived from the fluorite structure. In these oxides, the coordination of Pr derives from the cubic coordination by the creation of anionic vacancies. Pr_2O_3 ($n = 4$) has the so-called bixbyte structure, also found in C-Mn_2O_3, Tl_2O_3, and other lanthanide oxides, Ln_2O_3. In this stricture, the metallic ions exhibit two sorts of distorted octahedral coordination (Fig. 14a,b)[38] derived from the cubic one. Thus the structure of the praseodymium oxide is derived from the fluorite structure by an ordered creation of anionic vacancies at the corners of the cubes (Fig. 14c); it results in a cubic structure (Fig. 14d) with $a \approx 2a_{\text{fluorite}}$. In the intermediate members, between $n = \infty$ (PrO_2, fluorite) and $n = 4$ (Pr_2O_3), PrO_7 groups derived from the cubic coordination are involved as shown in the case of Pr_7O_{12} ($n = 7$) in Figure 15a, where the rhombohedral cell exhibits ordered vacancies

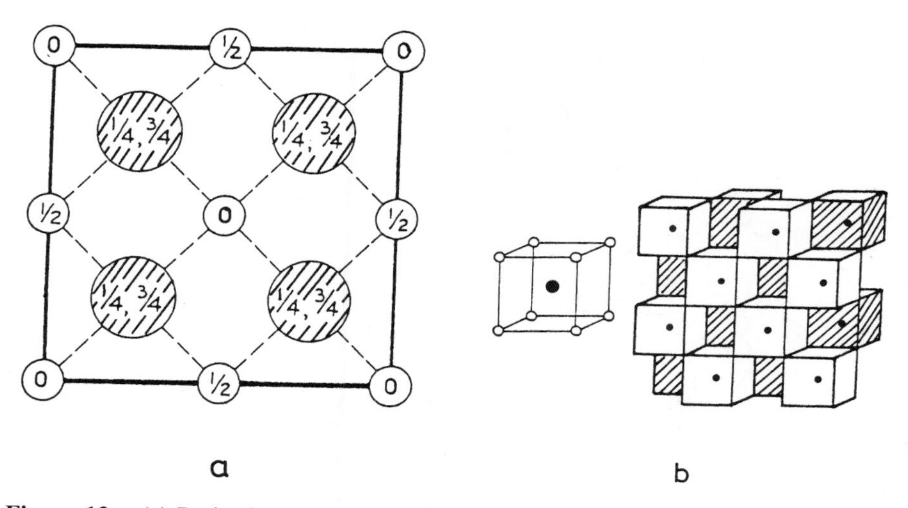

Figure 13. (a) Projection of the cubic structure of the fluorite type oxide PrO_2 along $\vec{\mathbf{a}}$: small circles, Pr; large hatched circles, O. (After Wycoff.[36]) (b) The three-dimensional fluorite structure of PrO_2 built up from edge-sharing PrO_8 cubic groups: dots, Pr. (After Collongues.[35])

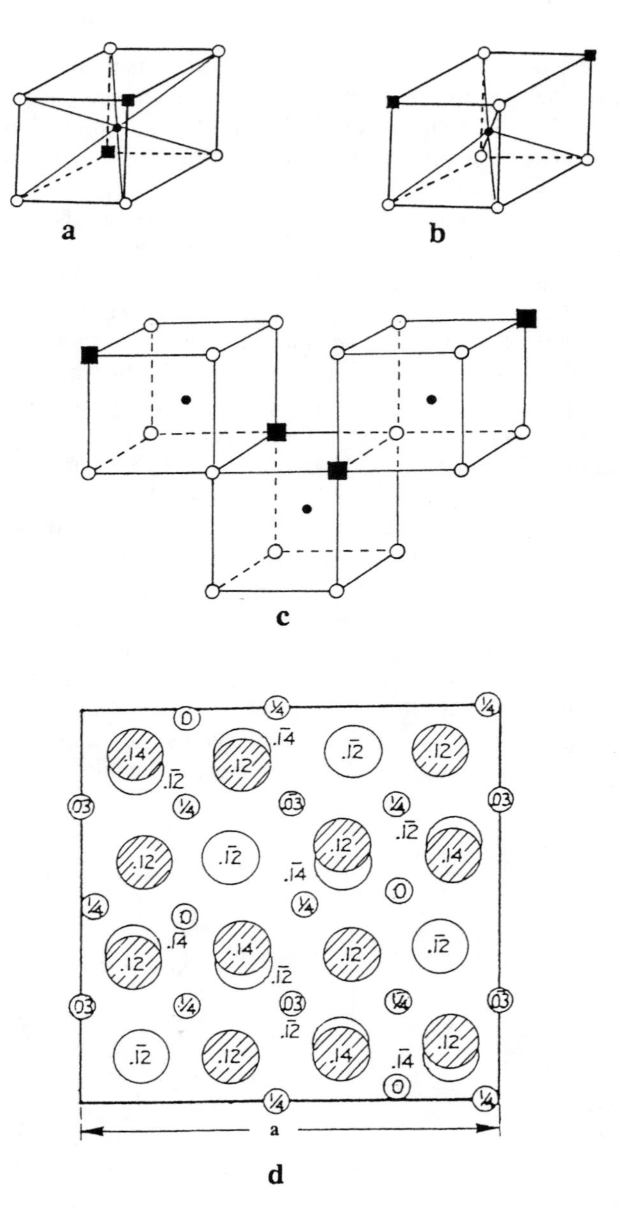

Figure 14. The bixbyite structure of Pr_2O_3: (a) and (b) distorted PrO_6 octahedra deduced from cubic PrO_8 groups by the creation of anionic vacancies (solid squares). (c) Assemblage of the PrO_6 octahedra. (After Eyring.[38]) (d) Projection of the structure along $\vec{\mathbf{a}}$. (After Wyckoff.[36])

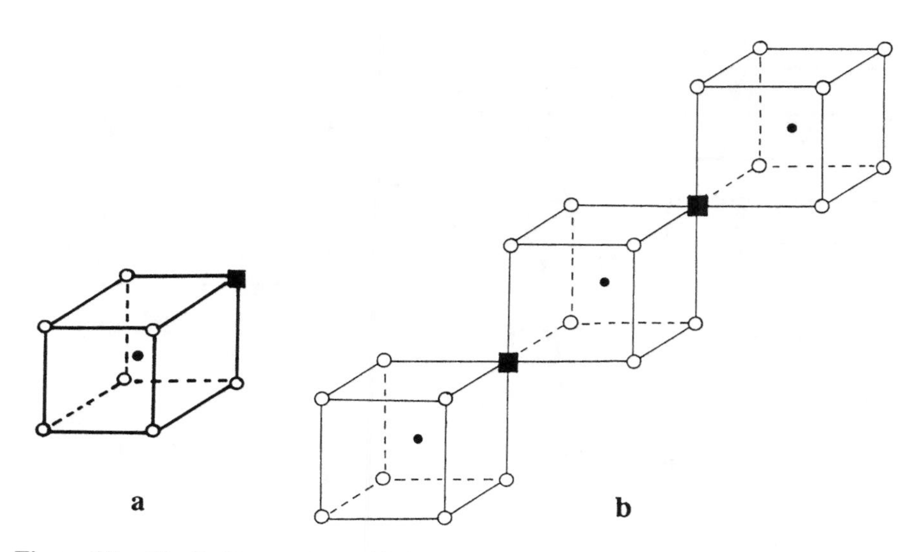

Figure 15. The Pr_7O_{12} structure: (a) the PrO_7 polyhedra and (b) the assemblage of PrO_6 and PrO_7 polyhedra. (After Eyring.[38])

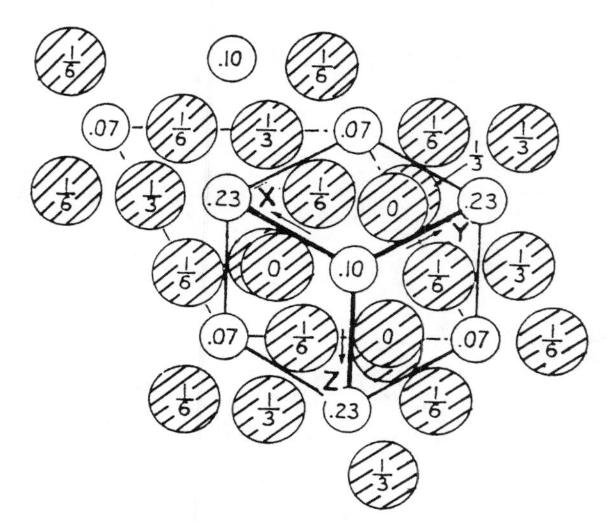

Figure 16. Projection of a portion of the α-Al_2O_3 structure of corundum along the threefold axis of the rhombohedral cell: small circles, Al; hatched circles, O. (After Wyckoff.[36])

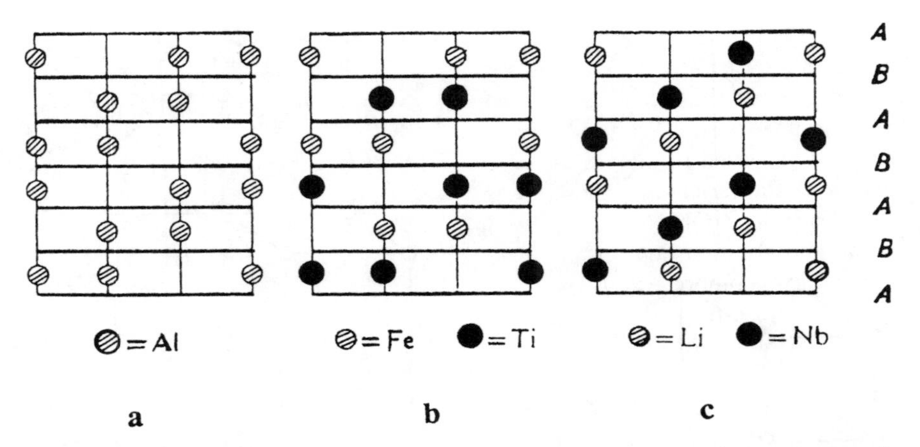

Figure 17. Hexagonal close-packed structures of (a) corundum (α-Al_2O_3), (b) ilmenite ($FeTiO_3$), and (c) $LiNbO_3$. Note the different distributions of the metallic atoms in the octahedral sites. (After Wells.[4])

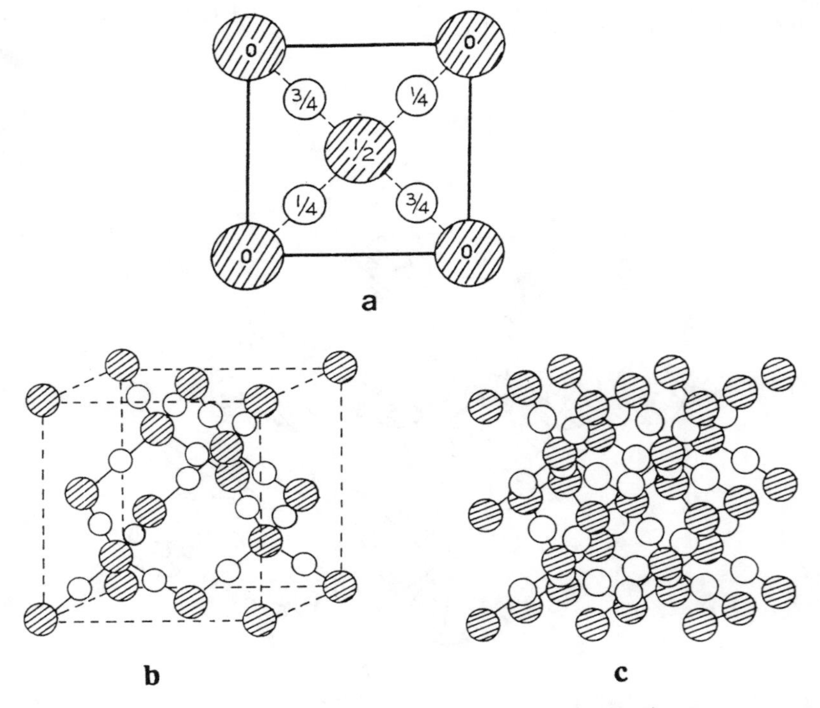

Figure 18. The cuprite structure of Cu_2O. (a) Projection along $\vec{a}$ (open and hatched circles correspond to copper and oxygen, respectively). (b) β-Cristobalite SiO_2 network (open and hatched circles represent oxygen and silicon, respectively). (c) The Cu_2O network involving two anti-β-cristobalite networks (open and hatched circles correspond to copper and oxygen, respectively). (After Wyckoff.[36])

along the $\langle 111 \rangle_c$ direction of the fluorite structure with one-seventh of cations forming PrO_6 distorted octahedra and six-sevenths forming PrO_7 polyhedra (Fig. 15b).[38] One can thus obtain different microphases built up from PrO_8, PrO_7, and PrO_6 groups in variable numbers according to the n value. Note that these structures are observed in other lanthanide oxides such as those of Ce and Tb and also in several ternary zirconium oxides and oxynitrides such as $Mg_2Zr_5O_{12}$ ($n = 6$), $CaZrO_9$, $ZrON_2$ ($n = 4$), $Zr_7O_8N_4$ ($n = 7$), and $Zr_7O_{11}N_2$ ($n = 13$).

The corundum structure is possessed by α-Fe_2O_3, V_2O_3, Ti_2O_3, Cr_2O_3, and Rh_2O_3. This rhombohedral structure ($a \approx 5.1$ Å, $\alpha \approx 55°$) has approximately a hexagonal close-packed array of oxygen atoms (Fig. 16), in which the B^{3+} ions occupy two-thirds of the octahedral holes. In this structure, the BO_6 octahedra are connected in a complex manner, sharing their vertices, edges, and faces. Ilmenite-type oxides, represented by $FeTiO_3$ and the $LiNbO_3$ family, exhibit a closely related structure. In both structures, one observes a similar hexagonal close packing of the oxygen atoms; since they differ from the corundum structure only in the distribution of the metallic atoms (Fig. 17) in the octahedral sites, the two structures can be described as superstructures of the corundum structure. Many oxides belong to this structural type (Table 5).

The cubic structure of cuprite, Cu_2O ($a \approx 4.3$ Å), is also found in Ag_2O and Pb_2O. A characteristic of this simple structure (Fig. 18a) is the linearity of the O—Cu—O bonds leading to a twofold coordination of Cu(I), whereas the oxygen is tetrahedrally coordinated. An interesting analogy with the β-cristobalite structure of SiO_2 must be pointed out. In the latter structure, (Fig. 18b), Si in tetrahedral coordination forms linear Si—O—Si bonds similar to the O—Cu—O bonds, and the Cu_2O structure can be described as two interpenetrating anti-β cristobalite networks (Fig. 18c).

4 Perovskites and Relatives

4.1 Stoichiometric Perovskites, ABO_3

Stoichiometric perovskites of the general formula ABO_3 form a large family of oxides whose structure is based on that of ReO_3. The latter oxide is characterized by a cubic cell parameter close to 3.8 Å, which is described in two different ways, by locating the rhenium atoms at the origin (Fig. 19a) or at the center (Fig. 19b). The first representation gives rise to a cage at the center of the cell, whereas the second has the merit of showing the regular octahedral coordination of Re by the oxygen atoms. This structure is generally described as an assemblage of corner-sharing octahedra (Fig. 20a). This structure can become more compact by the incorporation of a cation of a size similar to that of oxygen at the center of the cell (of Fig. 19a). One then obtains a perovskite, whose prototype $SrTiO_3$ has a cubic structure, with the titanium atoms replacing rhenium and strontium as the incorporated cation. We can consider the perovskites ABO_3, if they are close-packed structures, as having a host lattice of $[BO_3]_\infty$ with corner-sharing BO_6 octahedra, and forming cages bonded by 12 oxygen atoms where

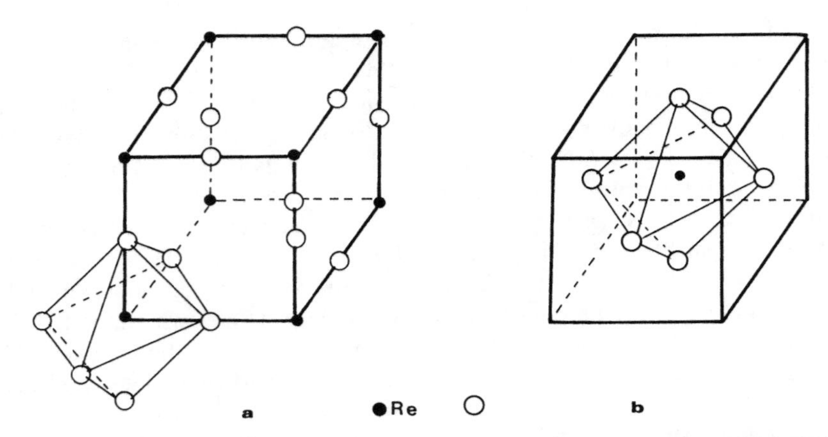

Figure 19. ReO_3-type structure: two representations of the cubic cell in which Re is at (a) at the corners of the cubic cell and (b) at the center.

the A cation is present (Fig. 20a). The superposition of the cages along the ⟨100⟩ direction makes it possible to describe these oxides by a tunnel structure with a square section (Fig. 20b). Stoichiometric perovskites are numerous. A systematic study of perovskites has been presented by Gallasso,[39] and tabulations of perovskites are available in the literature.[40,41]

Most of the ABO_3 perovskites exhibit a distortion of the cubic cell, as for instance in the $ATiO_3$ titanates ($SrTiO_3$ is an exception). Even the mineral $CaTiO_3$ is not cubic. Stoichiometric ABO_3 perovskites can be classified into three categories according to the valence of the A and B elements: $A^{I}B^{V}O_3$ (A = Na, Ag, K, and more rarely Rb, Tl, Cs; B = Nb, Ta, I), $A^{II}B^{IV}O_3$ (A = Ba,

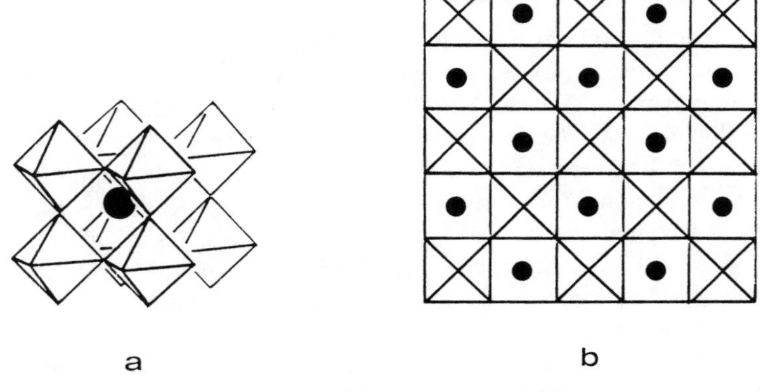

Figure 20. The ABO_3 perovskite structure. (a) Perspective of the assemblage of the corner-sharing BO_6 octahedra forming the O_{12} cages where the A cations (solid circles) are located. (b) Projection of the structures along ⟨100⟩ showing the existence of tunnels with a square section.

Sr, Ca, Pb, and more rarely Cd; B = Ti, Sn, Zr, Hf, Mn, Mo, Th, Fe, Ce, Pr, U), $A^{III}B^{III}O_3$ (A = Ln, Bi, Y; B = Fe, Cr, Co, Mn, Ti, V, Al, Sc, Ga, In, Rh). Many of the ferroelectric oxides belong to the first two series and are characterized by a small distortion of the cubic cell. A typical example is that of barium titanate, $BaTiO_3$, which has numerous technological applications. This oxide is characterized by a tetragonal cell at room temperature, which results from a shifting of the titanium from the center of the "O_6" octahedron leading to a noncoincidence of the gravity centers of positive charges (cations) and negative charges (anions) (Fig. 21)[42]; this rattling of titanium inside its octahedron gives rise to dipoles, which can be reversed in an electric field, hence to *ferroelectricity* (see Part II for details). Numerous perovskites that belong to these two first families exhibit distortions and superstructures that are due to small displacements of the atoms but are not ferroelectric. In the third family, most of the compounds exhibit the same kind of distortion and are orthorhombic; these perovskites are the $GdFeO_3$-type perovskites. Their orthorhombic cell (Fig. 22) has the following parameters:

$$a \approx a_p\sqrt{2},\ b \approx a_p\sqrt{2n},\ c \approx 2a_p$$

where a_p represents the parameter of the cubic cell of the perovskite.

Besides the ternary oxides with a perovskite structure, there exist numerous complex stoichiometric perovskites in which two or more metallic elements are located in the octahedral B sites. When the B sites are occupied by two elements simultaneously, one often observes ordering between the two ions, leading to multiple perovskite cells. This family can be classified in three categories: $A^{III}(B^{III}_{0.67}B^{VI}_{0.33})O_3$ (A = Ba, Sr, Pb; B^{III} = Ln, Sc, Fe, Cr, In; B^{VI} + W, U, Re), $A^{II}(B^{II}_{0.33}B^{V}_{0.67})O_3$ (A = Ba, Sr, Ca, Pb; B^{II} = Ca, Cd, Co, Ni, Cu, Fe, Mg, Mn, Zn; B^{V} = Nb, Ta), and $A(B'_{0.5}B''_{0.5})O_3$, in which A, B′, and B″ possess various oxidation states. The first category, which has twice the number of B(III) ions than B(VI) ions, is generally characterized by a cubic cell, with an *a* parameter

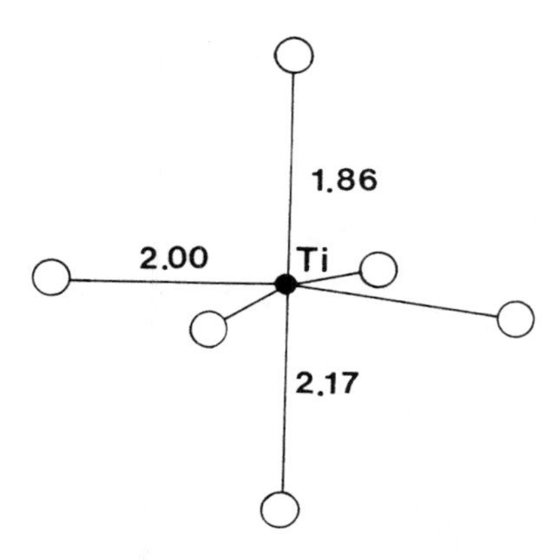

Figure 21. Tetragonal $BaTiO_3$ showing distortion of the TiO_6 octahedron. (After Jona and Shirane.[42])

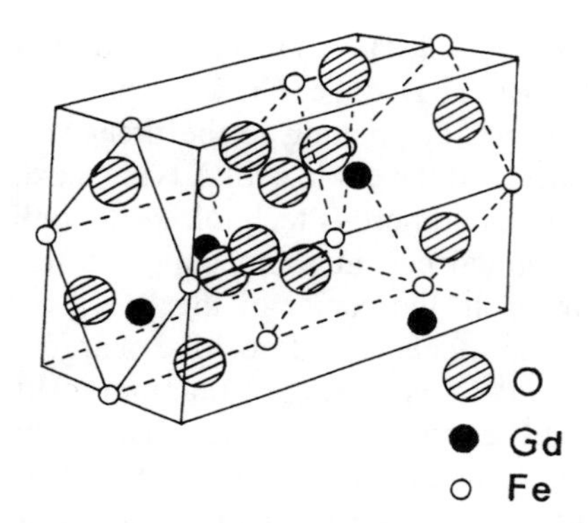

Figure 22. The orthorhombic structure of $GdFeO_3$. The Fe^{3+} ions are in octahedral coordination. (After Gallasso.[39])

(a') double that of the perovskite ($a' \approx 2a_p$). The structure of these oxides is derived from that of $(NH_4)_3FeF_6$, with the B site ions ordered in such a way that two different B ions are alternately at the corners of the perovskite subcell (Fig. 23). Several compounds of the $A^{II}(B^{II}_{0.33}B^{V}_{0.67})O_3$ series belong to the latter structure type, but most of them exhibit a different kind of ordering of the B site ions, leading to a hexagonal cell ($a \approx a_p\sqrt{2}$, $c \approx a_p\sqrt{3}$). Such an ordered perovskite belongs to the $Ba(Sr_{0.33}Ta_{0.67})O_3$ type. Its structure (Fig. 24a) can be described as a close-packed stacking of AO_3 layers parallel to the $(111)_c$ plane of the ideal cubic perovskite, the B ions being distributed in an ordered way in the octahedral cavities between these layers. The size difference between the

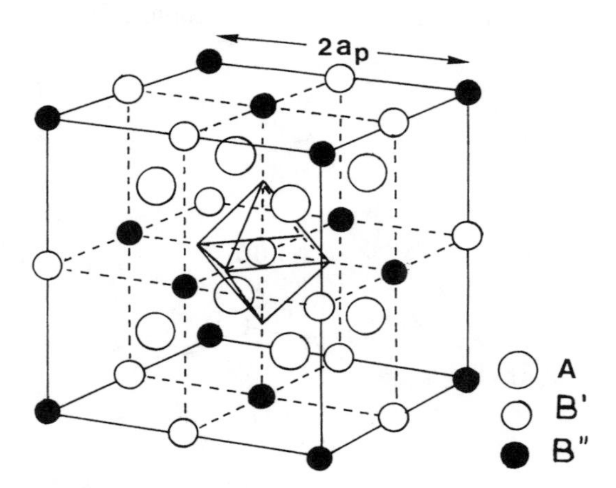

Figure 23. The $(NH_4)_3FeF_6$ structure of the oxides $A^{II}(B^{III}_{0.67}B^{IV}_{0.33})O_3$. The B′ sites are occupied by B(III), whereas the B″ sites are occupied by B(III) and B(IV). (After Gallasso.[39])

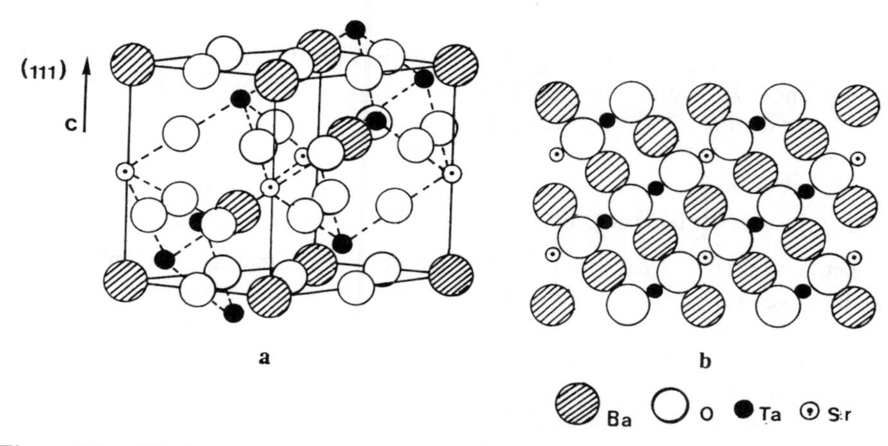

Figure 24. The hexagonal perovskite $Ba(Sr_{0.33}Ta_{0.67})O_3$ structure. (a) Three-dimensional view. (b) Three-layer repeat sequence: the (110) plane. (After Gallasso.[39])

B^{II}(Sr) and B^{V} ions (Ta) explains the ordering, where two B(V) planes alternate with one B(II) plane along $\mathbf{c}_H$ ($\langle 111 \rangle_c$), as shown in Figure 24b. The $A(B^{III}_{0.5}B^{V}_{0.5})O_3$ perovskites are large in number with (A = Ba, Sr, Ca, Pb; B^{III} = Ln, Y, Bi, Fe, Mn, Se, Rh, Ni, Al, Ga, In; B^{V} = Nb, Ta, Pa, Sb, Re, W. Os). Other such perovskites are $A^{II}(B^{II}_{0.5}B^{VI}_{0.5})O_3$ (A = Ba, Ln, Sr, Pb; B^{II} = Cu, Co, Fe, Mg, Mn, Ni, Zn, Cd, Ca; B^{VI} = W, U, Mo, Re, Os), $A^{II}(B^{I}_{0.5}B^{VII}_{0.5})O_3$ (A = Ba, Ca, Sr; B^{I} = Li, Na, Ag; B^{VII} = I, Os, Re), and $A^{III}(B^{II}_{0.5}B^{IV}_{0.5})O_3$ (A = La, Nd; B^{II} = Mg, Cu, Ni, Co, Zn; B^{IV} = Ir, Nb, Ti, Ru). The third category is characterized by various ordering of the cations on the B site. A great number of these oxides belong to the $(NH_4)_3FeF_6$ type.

The stability of the perovskite structure depends on the relative sizes of the A and B elements as well as on the electronic configuration of the B sites ions. The size factor is expressed by a *tolerance factor, t*, obtained by considering the structure as derived from a close-packed stacking of the A and oxide ions. It is generally defined by the relationship $R_A + R_0 = t\sqrt{2}(R_B + R_0)$, where R_A and R_B are the ionic radii (of the A and B ions taken from Shannon and Prewitt[13]). In this relation, the ideal close packing corresponds to $t = 1$. The perovskite structure is found for t values ranging from 0.8 to about 1.1. For t greater than 1, the space available for the B ion in its "O_6" octahedron is large so that it can rattle. This is the origin of the ferroelectric character of the tetragonal form $BaTiO_3$. Small t values ($t < 0.8$), correspond to A and B ions of similar size, leading to more close-packed structures such as ilmenite. Note that the tolerance factor is approximate and empirical, although useful to predict the occurrence of the perovskite structure. It must be carefully used, bearing in mind that the notion of ionic radius is itself approximate and depends on the coordination of the ions.

The influence of the electronic structure of the B ions on the perovskite structure is not obvious. Let us consider the stoichiometric perovskites ABO_3. Goodenough has shown, for instance, that since Sb(V) forms Sb–O–Sb bond angles of 90° and 130° rather than of 180°, $KSbO_3$ does not exhibit the perovskite structure (see Part II for a detailed discussion of the properties of perovskites).

4.2 Nonstoichiometric Perovskites, A_xBO_3

Nonstoichiometric perovskites, A_xBO_3, are not very numerous. Cationic defects in the cages (A site) can vary from zero (stoichiometric perovskites) to unity (ReO_3-type structure) depending on the nature of the A and B ions. The most important family is that of the *tungsten oxide bronzes,* A_xWO_3, which have been synthesized for A = Li, Na, Ca, Sr, U, Cd, Hf. The prototype of this family, the Na_xWO_3 bronzes, were described sometime ago by Hägg and Magnéli. The name *bronze* was given to such oxides because their physical properties were similar to those of alloys.[43] Indeed, these oxides exhibit a wide homogeneity range just like alloys (solid solutions) and varied metallic conductivity ($\sigma \approx 10^3$–$10^4\ \Omega^{-1}\ cm^{-1}$ at room temperature) or semiconducting behavior. The electrons introduced in the host lattice, $[WO_3]_\infty$, by the incorporation of the A cations give rise to the mixed valency of tungsten W(VI)–W(V). The electrons are delocalized in the octahedral framework (see Part II for details).

The extent of incorporation, x, in A_xWO_3 bronzes does not seem to be related directly to the size of the A ions (Table 6). The composition range is wide ($0 \leq x \leq 1$) for univalent cations such as lithium and sodium, but narrow for trivalent and tetravalent cations, such as lanthanides and hafnium ($x < 0.1$). The range is very narrow for bivalent cations ($x < 0.05$). The nature of the crystallographic cell depends on the nature of the A cation and also on the value of x. Lithium and lanthanide bronzes are cubic, whereas the sodium ones are cubic for high x values and tetragonal for low x. Different types of distortion of the perovskite cell are observed for divalent and tetravalent A cations.

The only $x = 0$ bronze, ReO_3 is characterized by metallic properties. Curi-

Table 6 Cubic and Pseudocubic Perovskite Bronzes, A_xBO_3

Bronze	Composition	Symmetry	Unit Cell Dimensions
Li_xWO_3	$0.31 < x < 0.57$	Cubic	a is variable (≈3.7 Å)
Na_xWO_3	$0.26 < x < 0.93$	Cubic	a is variable (≈3.8 Å)
Na_xWO_3	$x = 0.10$	Tetragonal	$a = 5.25$, $c = 3.90$ Å
Cu_xWO_3	$0.26 < x < 0.77$	Orthorhombic at $x = 0.26$	$a = 3.88$, $b = 3.73$, $c = 7.74$ Å
		Triclinic at $x = 0.77$	$a = 5.85$, $b = 6.65$, $c = 4.88$ Å $\alpha = 135.7°$, $\beta = 91.7°$, $\gamma = 93.6°$
La_xTiO_3	$0.67 < x < 1.0$	Cubic	a is variable (≈3.9 Å)
Sr_xNbO_3	$0.7 < x < 0.95$	Cubic	a is variable (≈4.0 Å)

ously, $Na_x ReO_3$ bronzes are difficult to synthesize; they can only be prepared under high pressures of about 65 kbar.

Besides the bronzes, nonstoichiometric perovskites A_xBO_3 with insulating properties are known. These dielectric materials, which are isotypic with the bronzes, have been called *bronzoids* by Magnéli and Kihlborg because of their wide homogeneity range and similarity of structure. Such bronzoids are known for lanthanides, with the general formula $Ln_x(B_{3x}W_{1-3x})O_3$ (B = Ta, Nb). These dielectric materials are white and exhibit a wider homogeneity range ($0 \leq x \leq 0.33$) than the dark blue bronzes Ln_xWO_3 ($x < 0.10$). The symmetry of the perovskite cell of these phases (Table 7) depends on the value of x. For instance, in $Gd_x(Ta_{3\infty}W_{1-3x})O_3$, one observes successively an orthorhombic cell (due to a monoclinic distortion of the cubic subcell), then a cubic ideal cell, and finally an orthorhombic cell with a doubling of one a parameter, as x increases from 0.0 to 0.33. Several other defect perovskites have been isolated (e.g., $La_{0.66}TiO_3$, $Th_{0.25}NbO_3$, $Hf_{0.25}NbO_3$). They are characterized by the presence of d^0 elements leading to dielectric properties.

WO_3 represents the $x = 0$ limit of bronzoids. It is different from ReO_3 and does not exhibit electronic delocalization owing to the d^0 electronic configuration of W(VI). The d^0 configuration is also at the origin of the ferroelectric properties of WO_3 at low temperatures. In fact, one observes several types of distortion of the structure, the cubic form of WO_3 being only stable at high temperatures. The different symmetries of the perovskite cell result from a distortion of the WO_6 octahedra, in which W tends to adopt a 4 + 2 coordination and also from the tilting of the octahedra in such a way that the perovskite cages are flattened (the latter do not form 90° angles as in the ideal cubic perovskite, but angles close to 60° and 120° between two neighboring octahedra). For instance, a view of structure (Fig. 25) of the monoclinic form of WO_3 ($a \approx 5.27$ Å, $b \approx 5.15$ Å, $c \approx 7.67$ Å, $\beta \approx 91.7°$), shows that in the (100), (010), and (001) planes, the O_6 octahedra are not strongly distorted; rather, the tungsten atoms are off-center and the WO_6 octahedra are tilted, forming diamond-shaped cages.[44]

Incorporation of A cations of small size, with Jahn–Teller distortion, can induce a distortion of the perovskite cage. In such systems, one observes a transition from square- to diamond-shaped windows. Numerous perovskites forming diamond-shaped windows, generated by Jahn–Teller cations such as Cu^{2+} or Mn^{3+}, have been synthesized. This is the case in stoichiometric perovskites of the types $(CaCu_3)B_4O_{12}$ (B = Ti, Mn) and $(CaMn_3)Mn_4O_{12}$ in which three A sites out of four are occupied by Jahn–Teller cations in an ordered manner. Derived from the latter, nonstoichiometric perovskites containing only Jahn–Teller cations and vacancies at the A sites distributed in an ordered manner have been synthesized. Examples of such perovskites are $(Cu\square)Ta_2O_6$ and $(Mn_3\square)Mn_4O_{12}$. These perovskites exhibit a cubic cell with $a \approx 2a_p$ and are deduced from the ideal perovskite by a compression and a tilting of the BO_6 octahedra (Fig. 26). The tilting is the consequence of the presence of Jahn–Teller

Table 7 Some Perovskite Bronzoids, A_xBO_3

Compounds	Symmetry	Parameters (Å)		
		a	*b*	*c*
$Ce_{0.33}NbO_3$	Orthorhombic	3.89	3.91	7.86
$Ce_{0.33}TaO_3$	Orthorhombic	3.90	3.91	7.86
	$\gamma = 90.8°$			
$Dy_{0.33}TaO_3$	Monoclinic	3.83	3.83	7.75
$Gd_{0.33}TaO_3$	Orthorhombic	3.87	3.89	7.73
$La_{0.33}NbO_3$	Tetragonal	3.91		7.90
$La_{0.33}TaO_3$	Tetragonal	3.92		7.88
$Nd_{0.33}NbO_3$	Orthorhombic	3.90	3.91	7.76
$Nd_{0.33}TaO_3$	Tetragonal	3.91		7.77
	$\gamma = 90.9°$			
$Y_{0.33}TaO_3$	Monoclinic	3.82	3.83	7.74
	$\gamma = 91.6°$			
$Yb_{0.33}TaO_3$	Monoclinic	3.79	3.80	7.70
$Ca_{0.5}TaO_3$	Orthorhombic	11.068	7.505	5.378
$Gd_xTa_{3x}W_{1-3x}O_3$				
$x = 0.29$	Orthorhombic	3.869	3.858	7.738
$x = 0.22$	Cubic	3.844		
$x = 0.145$	Orthorhombic	5.395	5.406	3.819
$x = 0.10$	Orthorhombic	5.354	5.394	3.80
$x = 0.07$	Orthorhombic	5.326	5.381	3.786
$Ln_{0.22}(Ta_{0.67}W_{0.33})O_3$				
Ln = La	Cubic	3.895		
Ln = Nd	Cubic	3.872		
Ln = Sm	Cubic	3.856		
Ln = Gd	Cubic	3.844		
Ln = Dy	Cubic	3.832		
$Sm_x(Ta_{3x}W_{1-3x})O_3$				
$x = 0.04$	Orthorhombic	5.304	5.370	3.774
$x = 0.06$	Orthorhombic	5.326	5.380	3.785
$x = 0.10$	Orthorhombic	5.375	5.394	3.8075
$x = 0.12$	Pseudocubic	3.818		
$x = 0.185$	Cubic	3.844		
$x = 0.255$	Cubic	3.866		
$x = 0.26$	Tetragonal	3.867		7.734
$x = 0.033$	Tetragonal	3.891		7.747
$Ln_{0.05}(Ta_{0.15}W_{0.85})O_3$				
Ln = Nb	Orthorhombic	5.311	5.391	3.784
Ln = Eu	Orthorhombic	5.313	5.378	3.780
Ln = Gd	Orthorhombic	5.309	5.376	3.778
Ln = Dy	Orthorhombic	5.297	5.374	3.774
Ln = Ho	Orthorhombic	5.291	5.370	3.770
Ln = Yb	Orthorhombic	5.280	5.366	3.764

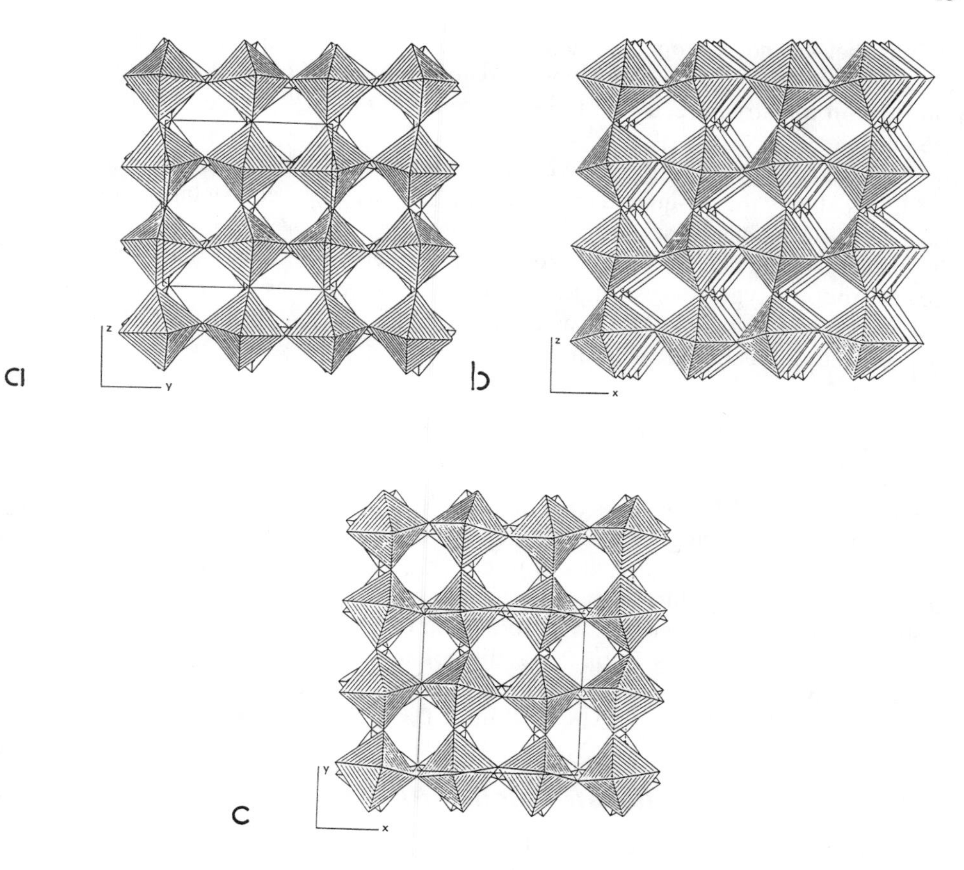

Figure 25. Structure of the monoclinic form of WO_3 (a) along $\vec{\mathbf{a}}$, (b) along $\vec{\mathbf{b}}$ and (c) along $\vec{\mathbf{c}}$: hatched squares represent the WO_6 octahedra that share their corners, forming diamond-shaped windows. (After Labbe.[44])

Figure 26. The perovskite $CaCu_3Mn_4O_{12}$ with diamond-shaped windows.

cations that are not surrounded by 12 oxygens but have a square-planar coordination. Such a structure can accommodate the host lattice of the hexagonal tungsten bronze structure, owing to the existence of 60°–120° angles, as we will see later.

Anion-excess perovskites have been mentioned in the literature,[45] $LaMnO_{3+\delta}$ being the oft-quoted example. $LaMnO_3$, as prepared, contains around 12% Mn^{4+}. This has been attributed to anion excess, but it is not clear where the excess oxygen can be located. Recent studies seem to indicate that Mn^{4+} is created not by oxygen excess, but because of the presence of vacancies in both A(La) and B(Mn) sites. $LaMnO_3$ with 33% or more of Mn^{4+} has been prepared by chemical or electrochemical means.

4.3 Oxygen-Deficient Perovskites

The formation of ordered anionic vacancies in the host lattice of perovskites leads to new mixed frameworks in which the BO_6 octahedra share their corners with polyhedra involving a smaller number of apices: pyramids, tetrahedra, and square-planar units. Several series of ternary oxides, involving copper, manganese, cobalt, and iron, illustrate these structural types derived from the perovskite motif (see, e.g., refs. 6 and 45). The formation of anionic vacancies in the perovskites, ABO_{3-x}, results from the ability of the B cations to adopt several coordinations (besides the octahedral one) and to exhibit mixed valence [e.g., Fe(III)/Fe(IV), Cu(II)/Cu(III), Mn(III)/Mn(IV)].

4.3.1 The Brownmillerite Family, $(ABO_3)_nABO_2$

The systems $Ca_2Fe_2O_5$-$ATiO_3$ (A = Ba, Sr, Ca) and $Ca_2Fe_2O_5$-$AFeO_3$ (A = Y, La, Gd) are characterized by the rich chemistry of the ABO_{3-x} oxygen-deficient perovskites. These phases, resulting from the creation of ordered anionic vacancies in the octahedral framework $[BO_3]_\infty$, are represented by the general formula $(ABO_3)_n \cdot ABO_2$. Their host lattice, $B_{n+1}O_{3n+2}$, is built up from multiple stoichiometric perovskite layers that are n octahedra thick, linked through single layers of BO_4 tetrahedra, as shown in Fig. 27.[46,47] The tetrahedral layers, BO_2, are derived from the perovskite framework by the creation of rows of vacancies along the $\langle 110 \rangle$ direction of the perovskite cubic cell, in such a way that one oxygen-deficient row alternates with one fully occupied row (Fig. 28). The cell parameters of the different n members of this series are related to those of the perovskite in a simple way. Thus, one can obtain an orthorhombic cell with the following relations:

$$a_n \approx b_n \approx a_c\sqrt{2} \qquad \text{and} \qquad c_n \approx (n+1)a_c \text{ or } 2(n+1)a_c$$

The doubling of the c parameter results generally from the shifting of the rows of oxygen vacancies between two successive tetrahedral BO_2 layers along $\vec{\mathbf{a}}$. This is the case in the first member of this series, $Ca_2Fe_2O_5$, well known as the brownmillerite structure. This structure is also observed in other oxides such as

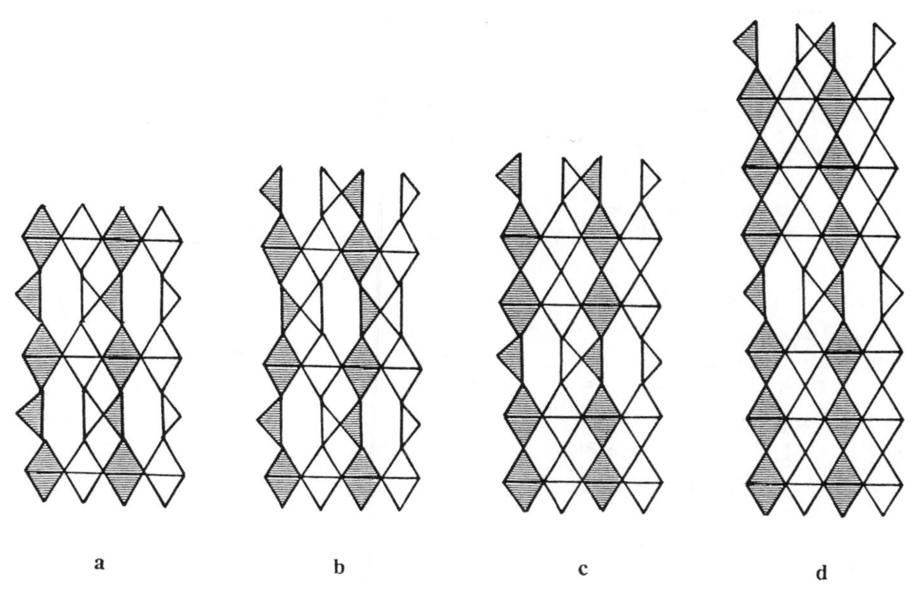

Figure 27. Schematic structure of the brownmillerite derived phases $(ABO_3)ABO_2$: (a) $n = 1$ hypothetical structure of $Ca_2Fe_2O_5$, (b) $n' = 1$ actual structure of $Ca_2Fe_2O_5$, (c) $n = 2$, and (d) $n = 3$. The BO_6 octahedra (lozenge shaped) and the BO_4 tetrahedra (triangles), which are viewed along the ⟨110⟩ direction of the perovskite, form rows at two levels alternately (hatched and open polyhedra). (After Raveau[46] and Grenier et al.[47])

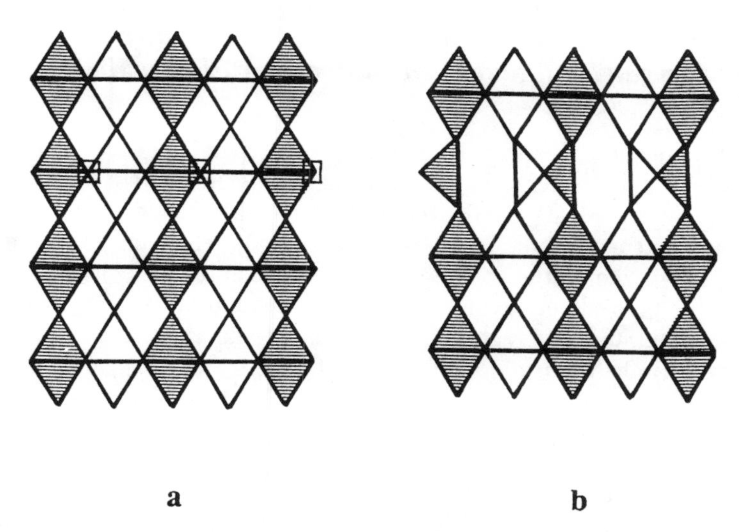

Figure 28. Creation of a BO_2 tetrahedral layer from the "BO_3" host lattice of the perovskite. Starting from the lattice of the perovskite (a) built up from BO_6 octahedra (hatched and open lozenges), by creation of oxygen vacancies (squares), one obtains the BO_2 lattice (b) built up from BO_4 tetrahedra (triangles). (After Raveau.[46])

Ca_2FeAlO_5, Ca_2FeCoO_5, and Sr_2CoO_5 (high temperature form). The hypothetical structure with the parameter $c \approx 2a_c$ (Fig. 27a) has not been observed, but is replaced by a superstructure with $c \approx 4a_c$ (Fig. 27b). In these oxides, the tetrahedral sites are occupied by the Fe^{3+} ions, which also take the octahedral coordination. Note that iron is trivalent and does not exhibit mixed valence here. An example of mixed valence for iron is given by the system $SrFeO_3/Sr_2Fe_2O_5$. The two limits, $SrFeO_3$ and $Sr_2Fe_2O_5$, correspond to the $n = \infty$ and $n = 1$ members, respectively. Mixed valent compounds, corresponding to the presence of Fe(III)/Fe(IV), have been observed in $SrFeO_{3-x}$.

A more complex distribution of the tetrahedral and octahedral layers is possible. This corresponds to nonintegral n values, which occur because of the intergrowth of two integral n members. This is the case in the oxides $Ca_4YFe_5O_{13}$, $Ca_5Fe_4O_{13}$ ($n = 1.5$), and $Ca_7Fe_6TiO_{18}$ ($n = 1.33$), which show an ordered intergrowth of the $n = 1$ and $n = 2$ members (Fig. 29), where single tetrahedral $[FeO_2]_\infty$ layers alternate with multiple $[BO_3]_\infty$ octahedral layers of different thicknesses. The possibility of partial replacement of iron by manganese in these phases has been shown. However, the distribution of oxygen vacancies is significantly modified. One observes the formation of vacancies in the octahedral layers, and an excess oxygen in the tetrahedral layer. This behavior is easily explained by the ability of Mn(III) to exhibit pyramidal coordination. Thus, in $Ca_2Fe_{2-x}Mn_xO_5$, Fe will have octahedral and tetrahedral coordinations while Mn will have pyramidal coordination. All these oxides possess interesting magnetic properties, some of them exhibiting spin-glass-like properties.

4.3.2 Three-Dimensional Mixed Frameworks Involving Pyramids, Octahedra, and Square-Planar Groups in Oxygen-Deficient Perovskites Containing Mn, Co, and Cu

Trivalent manganese, being a Jahn–Teller ion, generally occurs in pyramidal coordination. Mixed valent oxygen-deficient perovskites of Mn have been synthesized in which Mn(IV) and Mn(III) ions are in octahedral and pyramidal coordination, respectively. $Ca_2Mn_2O_5$ and $Ca_2Co_2O_5$, which have the same composition as brownmillerite, possess different structures. The latter is closely related to the perovskite, having an ordered formation of oxygen vacancies.[6,45] $Ca_2Mn_2O_5$ exhibits a different type of ordering of anionic vacancies (Fig. 30).[48] The major form has an orthorhombic cell ($a \approx a_p\sqrt{2} \approx 5.4$ Å, $b \approx 2a_p\sqrt{2}$ Å ≈ 10.2, and $c \approx a_p \approx 3.7$ Å), the host lattice, $[Mn_2O_5]_\infty$, being built up of corner-sharing MnO_5 pyramids (Fig. 30 a,b). This framework is deduced from the $[BO_3]_\infty$ lattice by creating oxygen vacancies in the basal plane of the BO_6 octahedra; rows of oxygen atoms are eliminated along $\langle 001 \rangle_p$, forming tunnels parallel to this direction where the Ca^{2+} cations are located. $Ca_2Co_2O_5$ also has an orthorhombic structure with a similar b parameter, but with the a and c parameters double of those of $Ca_2Mn_2O_5$. In this structure one oxygen out of two is eliminated along $\vec{c}$ so that we obtain twice more oxygen-deficient ions parallel to $\vec{c}$; this leads to a doubling of the a parameter.

Figure 29. Schematized structures of (a) $Ca_4YFe_5O_{13}$ (n = 1.5) and (b) $Ca_7Fe_6TiO_{18}$ (n = 1.33). For n = 1.5, the double layers of FeO_6 octahedra (lozenges) alternate with the single layers of FeO_4 tetrahedra (triangles). For n = 1.33, single layers of FeO_4 tetrahedra (triangles) are stacked with double and single layers of $Fe(Ti)O_6$ octahedra alternately (lozenges).

Besides the Mn(III) oxygen-deficient perovskites, several other oxygen-deficient perovskites, characterized by a mixed valence of manganese, Mn(III)/Mn(IV), have been isolated: $CaMnO_{2.8}$, $CaMnO_{2.75}$, $CaMnO_{2.67}$, and $CaMnO_{2.55}$. These are derived from the stoichiometric perovskite by an ordered arrangement of the anionic vacancies, leading to a framework built up of MnO_6 octahedra and MnO_5 pyramids. For instance, the structure of the tetragonal oxide $CaMnO_{2.8}$ ($a \approx a_p\sqrt{5} \approx 8.3$ Å and $c \approx 2a_p \approx 7.5$ Å) is deduced from the perovskite by an ordered elimination of rows of oxygen atoms along $\vec{c}$. These oxides are insulating and exhibit interesting magnetic characteristics.

Several copper oxides crystallize in structures closely related to those of the manganese and cobalt oxide perovskites but exhibit metallic or semimetallic properties as a result of the presence of the Cu(II)/Cu(III) couple. This mixed

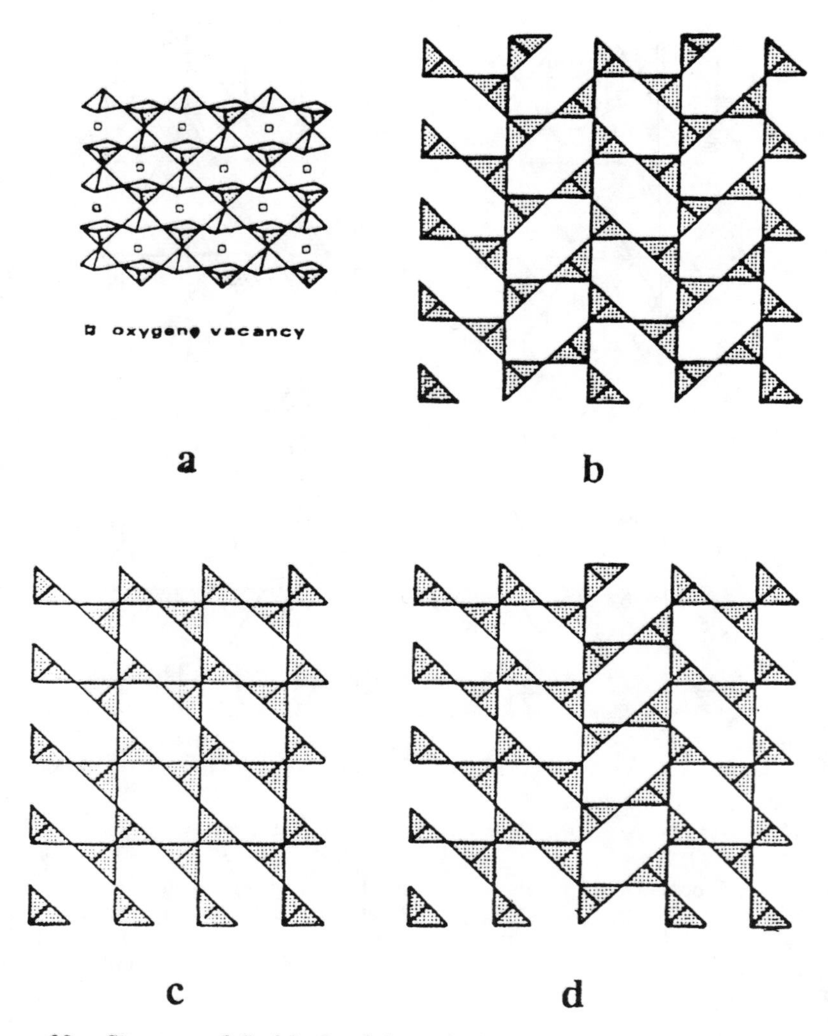

Figure 30. Structure of $Ca_2Mn_2O_5$. Schematic drawings corresponding to the major form (a-b) showing the rows of oxygen vacancies parallel to $[001]_p$ in the ReO_3-type framework. Structures of two minor forms (c-d) showing different orientations of the rows of tunnels. (After Reller et al.[48])

valence corresponds to a delocalization of holes over the copper–oxygen framework, leading to p-type conductivity. Because the framework is three-dimensional, there are strong magnetic interactions, and the compounds are not superconductors. $BaLa_4Cu_5O_{13+\delta}$ and $La_{8-x}Sr_xCu_8O_{20}$ are typical of such cuprates. The tetragonal structure of $BaLa_4Cu_5O_{13}$ ($a \approx 8.6$ Å $\approx a_p\sqrt{5}$ and $c \approx 3.8 \approx a_p$) consists of corner-sharing CuO_5 pyramids and CuO_6 octahedra (Fig. 31a)

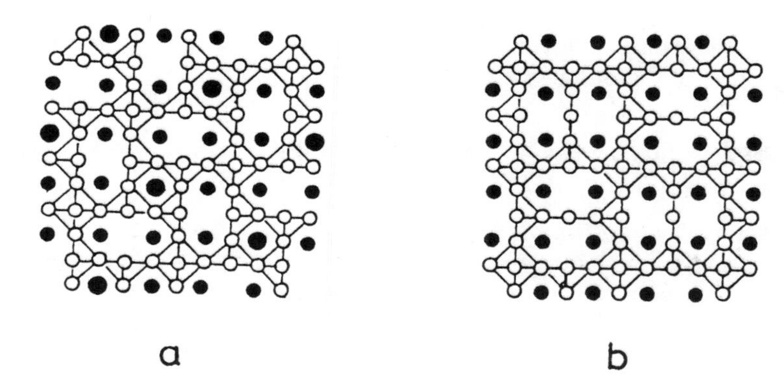

Figure 31. (a) Projection of the structure of $La_4BaCu_5O_{13+\delta}$ onto (001): open circles, corresponding to oxygens, form corner-sharing O_5 pyramids and O_6 octahedra where the copper atoms (not represented) are located; solid circles represent the La and Ba cations. (b) Projection of the structure of $La_{8-x}Sr_xCu_8O_{20}$ onto (001): open circles, corresponding to oxygen, form corner-sharing CuO_5 pyramids, CuO_6 octahedra, and CuO_4 square-planar groups, with copper (not shown); solid circles, La and Sr cations. (After Raveau et al.[49])

forming hexagonal tunnels and perovskite cages. The Ba^{2+} ions are located in the perovskite cages, whereas the La^{3+} ions occupy the hexagonal tunnels. One typical feature of the $[Cu_5O_{13}]_\infty$ host lattice is the geometry of the hexagonal tunnels, which is rather different from the ideal model derived from the perovskite. The O–O–O angles are close to 70° instead of 90°, and some oxygens are displaced toward the center of the tunnels. It is remarkable that the CuO_6 octahedra are slightly flattened, in contrast to the Cu(II) oxides, which generally exhibit elongated octahedra. The CuO_5 pyramids exhibit a large Cu—O apical distance compared to the four equatorial distances, which indicates the tendency of copper to adopt the square-planar coordination. A remarkable feature of such a structure is its ability to absorb additional oxygen ($\delta \approx 0.12$), leading to a double $BaLa_4Cu_5O_{13}$-type cell, $\sqrt{10a_p} \times \sqrt{10a_p} \times a_p$. The latter superstructure may result either from a distortion of the O_{13} cell or from an ordered introduction of oxygen into the tunnels.

The oxygen-deficient perovskite type phase $La_{8-x}Sr_xCu_8O_{20}$ has been isolated for $1.28 \leq x \leq 1.92$. It crystallizes in the tetragonal system ($a \approx 2\sqrt{2}a_p$ and $c \approx a_p$). The $[Cu_8O_{20}]_\infty$ framework (Fig. 31b) is very similar to that of $BaLa_4Cu_5O_{13}$. In both the structures, one can recognize Cu_5O_{22} groups of five corner-sharing polyhedra built up of one CuO_6 octahedron and four CuO_5 pyramids, forming hexagonal tunnels running along $\vec{\mathbf{c}}$. However, the Cu_5O_{22} groups are arranged differently in the two structures; there is corner sharing of the pyramids in $BaLa_4Cu_5O_{13}$, whereas they are linked through CuO_4 square-planar groups in $La_{8-x}Sr_xCu_8O_{20}$. One observes that the CuO_4 square-planar group exhibits two perpendicular directions, parallel to $\langle 110 \rangle$ and $\langle 1\bar{1}0 \rangle$ respectively. As a result, the hexagonal tunnels are differently oriented in the two structures (i.e., they share only their corners in $La_4BaCu_5O_{13}$, whereas they form pairs by sharing one CuO_4 group in the La-Sr cuprate).

4.3.3 *The Layered Cuprates $Ln_2BaCu_3O_{7-\delta}$, $La_3Ba_3Cu_6O_{14}$, and $YBaFeCuO_5$*

The structure of orthorhombic $YBa_2Cu_3O_7$ ($a \approx a_p$, $b \approx a_p$, $c \approx 3a_p$: Fig. 32a) derives from the stoichiometric perovskite by the elimination of rows of oxygen atoms parallel to the ⟨010⟩ direction at the levels $z = 0$ and $z = \frac{1}{2}$ in the orthorhombic cell and by an ordered distribution of Ba^{2+} and Y^{3+} ions in the A sites of the ABO_3 perovskite in such a way that two barium planes alternate with one yttrium plane along $\vec{\mathbf{c}}$. Thus, the structure consists of triple $[Cu_3O_7]_\infty$ layers of corner-sharing polyhedra, parallel to ⟨001⟩, whose cohesion is ensured by planes of yttrium cations. Each $[Cu_3O_7]_\infty$ layer consists of two sheets of corner-sharing CuO_5 pyramids connected to each other through $[CuO_2]_\infty$ rows of corner-sharing CuO_4 square-planar groups. The $[CuO_2]_\infty$ rows run along $\vec{\mathbf{b}}$. The structure is therefore often described as an association of Cu—O chains and pyramidal copper–oxygen layers. The Cu—O apical distance of the pyramid is close to 2.3 Å, significantly larger than the equatorial Cu—O distances (1.92–1.96 Å), showing the tendency of copper to adopt the square-planar coordination within the layers. The two-dimensional character of this structure, associated with the mixed valency Cu(II)/Cu(III) is at the origin of the high critical temperature of 92 K for this superconductor. Cuprates of many other lanthanides (Ln), $LnBa_2Cu_3O_7$, exhibit the same structure except those of cerium and terbium. The first one cannot be synthesized because of the formation of Ce(IV), which is a small cation.

Heating orthorhombic (superconducting) $YBa_2Cu_3O_7$ in air around 1070 K results in tetragonal (nonsuperconducting) $YBa_2Cu_3O_{6.4}$. The phase can also be synthesized by quenching a mixture of oxides (BaO, Y_2O_3, CuO) from 1270 K to room temperature. This tetragonal form designated T1 ($a \approx a_p$, $c \approx 3a_p$) differs from the orthorhombic one only in the positions of some of the oxygen atoms and in the oxygen content. One observes (Fig. 32b) triple layers of corner-sharing CuO_n polyhedra whose cohesion is ensured by a plane of yttrium cations. The ordering of the yttrium and barium planes, of the "1-2" type along $\vec{\mathbf{c}}$ remains unchanged. In the triple layers of polyhedra, one clearly recognizes two layers of corner-sharing CuO_5 pyramids. The main difference is that the rows of CuO_4 groups in the orthorhombic form are replaced by oxygen-deficient CuO_6 octahedra, whose basal corners are less than half-occupied, giving rise to copper with two kinds of coordination: twofold coordination for univalent copper and square-planar or pyramidal for Cu(II)/Cu(III).

The variation of oxygen stoichiometry in these cuprates can be carried out (without destroying the structure) further, leading to tetragonal $YBa_2Cu_3O_6$, designated T2 ($a = a_p$ and $c \sim 3a_p$). This oxide is obtained by annealing the orthorhombic form or the tetragonal form *T*1 to 870 K in vacuum. The structure of this oxide (Fig. 32c) derives from that of orthorhombic $YBa_2Cu_3O_7$ by remov-

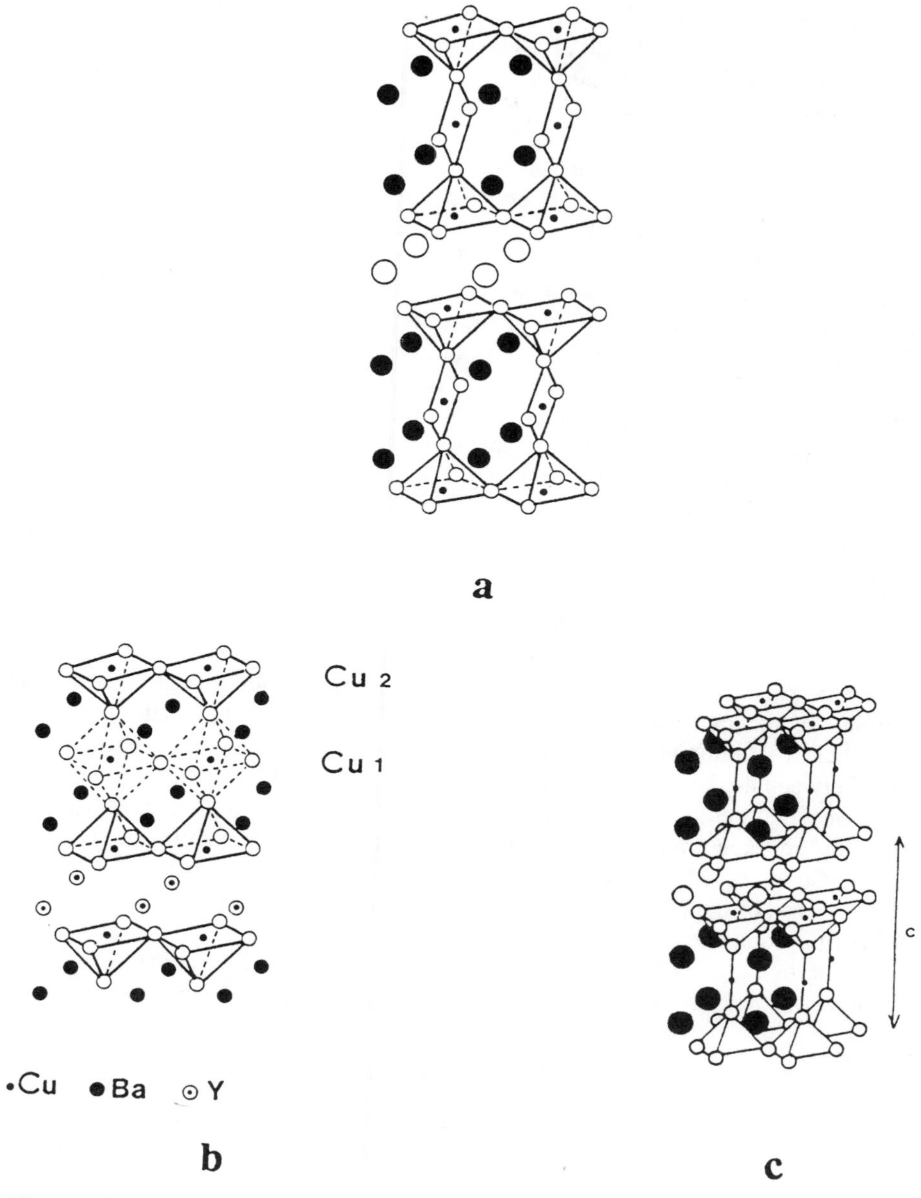

Figure 32. Structure of $LnBa_2Cu_3O_7$ showing the corner-sharing CuO_5 pyramids and CuO_4 square-planar groups as well as the CuO_2 sticks. (a) The 92 K orthorhombic superconductor $YBa_2Cu_3O_7$. (b) The tetragonal semiconductor $YBa_2Cu_3O_{6.3}$. (c) The insulating tetragonal cuprate $YBa_2Cu_3O_6$. (After Raveau et al.[49])

ing the oxygen atoms in the $[CuO_2]_\infty$ rows located at the same level as copper. This oxide can be formulated $YBa_2Cu_2^{II}Cu^{I}O_6$, with the divalent copper located in the CuO_5 pyramids forming the unchanged $[CuO_{2.5}]_\infty$ pyramidal layers and the univalent copper in twofold coordination ensuring the connection between these layers. This phase containing Cu(II) and Cu(I) is insulating.

Besides the T1 and T2 tetragonal forms, an oxygen-rich isotypic phase, $YBa_2Cu_3O_{6.8}$ with tetragonal symmetry, designated *T*3, has been isolated with $a \approx a_p$ and $c = 3a \approx 3a_p$. The average structure of this phase determined from neutron powder diffraction is similar to *T*1; it differs from the latter by a higher occupancy factor of the basal plane of the octahedra. High resolution electron microscopy shows that the *T*3 form is more complex than expected from neutron diffraction. One observes 90° oriented domains resulting from the lack of mismatch between a and c (c is rigorously equal to $3a$). Such domains can be interpreted as shown in Figure 33,[49] which gives two possible models. The presence of domains shows that at the scale of the grains, the structure corresponds to a three-dimensional connection of the CuO_n polyhedra. Interestingly, this cuprate is not a superconductor, in spite of the presence of mixed valence of copper.

The partial replacement of copper in $YBa_2Cu_3O_7$ by elements such as iron, cobalt, nickel, zinc, and palladium is possible in the orthorhombic as well as in the tetragonal phase. Such substitution leads to a dramatic change in properties (e.g., superconductivity). The partial replacement of barium by lanthanum is also possible. This leads to the oxide $La_3Ba_3Cu_6O_{14}$, whose tetragonal structure is similar to that of tetragonal $LaBa_2Cu_3O_{7-\delta}$ (Fig. 32b) except that there is a statistical distribution of La^{3+} and Ba^{2+} cations in the $[Cu_3O_7]_\infty$ layers, the cohesion between the layers being ensured by La^{3+} cations. The oxide can intercalate oxygen when annealed in an oxygen atmosphere at low temperatures, leading to a progressive transition from a semiconducting to a semimetallic state. $YBaFeCuO_5$ represents another ordered, oxygen-deficient perovskite. Its structure consists of double pyramidal $[FeCuO_5]_\infty$ layers, whose cohesion is ensured by yttrium ions, the barium ions being located inside the layers. This tetragonal oxide ($a \approx a_p$, $c \approx 2a_p$) derives from the stoichiometric perovskite by the removal of rows of oxygens at the level of yttrium cations in an ordered manner.

4.3.4 Other Oxygen-Deficient Perovskites

Several oxides derived from the perovskite structure by the creation of anionic vacancies, such as $SrTiO_{2.5}$ and $SrVO_{2.5}$, are of great interest although their structures have not been determined in detail.

4.4 Close-Packed Structures Derived from the Perovskites

Several oxides with the formula ABO_3 exhibit structures closely related to the perovskite, based on close-packed AO_3 layers that are stacked in different ways, leaving octahedral sites for the B cations. Such structures are obtained for large A cations (Ba or Sr) and for transition elements (e.g., Fe, Co, Ni, Nb). The

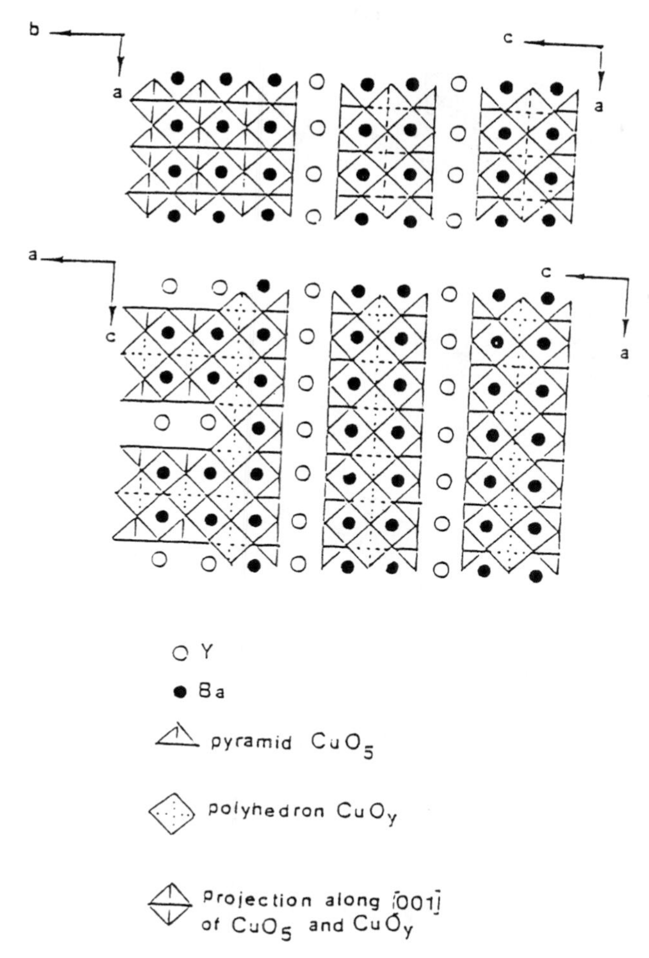

Figure 33. Two examples of idealized models of straight junctions for $YBa_2Cu_3O_{6.8}$ "T_3." The nature of the polyhedra is not specified, except for the CuO_5 pyramids along [100] or [010]. (After Raveau et al.[49])

analogy with the cubic perovskite structure depends on the nature of the stacking of the AO_3 layers. When the latter is entirely cubic (c) one obtains only corner-sharing BO_6 octahedra, leading to the cubic perovskite (Fig. 34a). On the other hand, for a pure hexagonal stacking of the AO_3 layers, the BO_6 octahedra can share only two opposite faces forming chains (Fig. 34b), as in $BaBiO_3$ (2H). The two structures can be intergrown with variable periodicities of the cubic and hexagonal close packings, resulting in the generation of many *polytypes,* closely related to the perovskite structure.[2,6] This is illustrated by the 6H polytype found in hexagonal $BaTiO_3$ (Fig. 34c), which has cch-type stacking sequence of the BaO_3 layers, giving rise to triple octahedral perovskite layers sharing octahedral faces. In the same way, the 4H polytype found in $BaMnO_3$ (Fig. 34d) consists

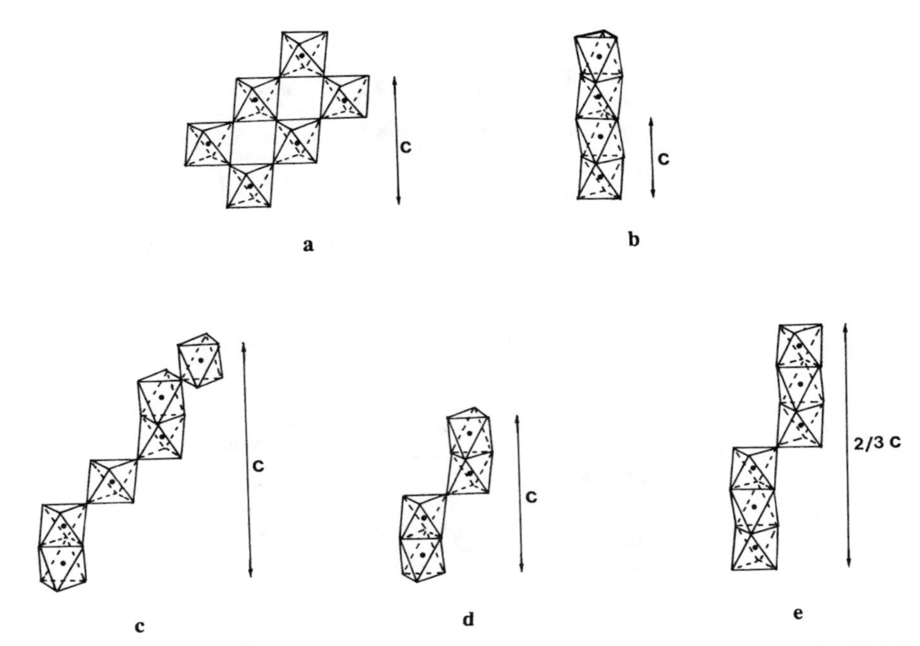

Figure 34. Linking of BO_6 octahedra in ABO_3 perovskite polytypes: (a) 3C, (b) 2H, (c) 6H, (d) 4H, and (e) 9R.

of double octahedral perovskite layers sharing the faces of the MnO_6 octahedra, with hc-type stacking of the BaO_3 layers. More complicated stackings such as the 9R polytype of $BaRuO_3$ (Fig. 34e) are observed, corresponding to double perovskite layers intergrown with triple layers of face-sharing octahedra, with chh-type stacking of the BaO_3 layers. These different kinds of stacking of the AO_3 layers allow the generation of close-packed structures that are deficient in the B cation. This is the case in $A_5Ta_4O_{15}$ (A = Sr, Ba: Fig. 35a) and $A_4Re_2MO_{12}$ (A = Sr, Ba; M = Mg, Zn, Co: Fig. 35b), whose AO_3 stackings correspond to the sequences hhccc and hhcc respectively. Note that in $A_5Ta_4O_{15}$, tantalum deficiency prevents the occurrence of face-sharing.

Several phases derived from close-packed structures exhibit significant oxygen deficiency. This is the case in $BaMnO_{3-x}$ ($0 \leq x \leq 0.50$), $BaFeO_{3-x}$ ($0 \leq x \leq 0.50$) and $BaCoO_{3-x}$ ($0 \leq x \leq 0.90$). Although the existence of anionic vacancies in these structures is well established, their complex distribution is not known with certainty.

4.5 Intergrowths of Perovskites with Other Structures

4.5.1 Intergrowth of Stoichiometric Perovskites with Rock Salt Structures

An examination of the structures of the cubic perovskite $SrTiO_3$ (Fig. 36a) and of the rock salt structure of SrO (Fig. 36b) shows that both are characterized by

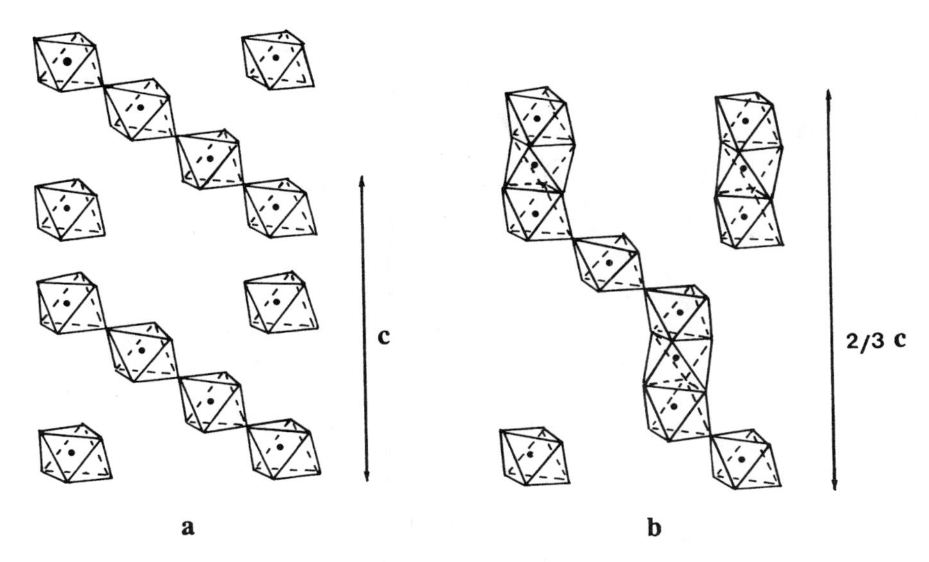

Figure 35. Linking of BO_6 octahedra in B-cation-deficient, close-packed structures (a) $Ba_5Ta_4O_{15}$ and (b) $A_4Re_2MO_{12}$. (After Wells.[4])

identical (001) planes with the composition "SrO." As a result, the two structures can be juxtaposed, leading to a series of intergrowth phases with the general formula $(SrTiO_3)_nSrO$ in which the multiple perovskite layers, which are n octahedra thick, alternate with single "SrO" rock salt layers as shown in Figure 37 for the three first members of the series Sr_2TiO_4 ($n = 1$), $Sr_3Ti_2O_7$ ($n = 2$), $Sr_4Ti_3O_{10}$ ($n = 3$). Theoretically, one can have different members of the series ranging from $n = 1$ to $n = \infty$ (the perovskite $SrTiO_3$). However, the disorder in the stacking increases as n increases, so that for high n values, the SrO layers appear as extended defects. The ideal cells of such intergrowths are related to that of the perovskite and of SrO in the following way:

$$a \approx a_{\mathrm{p}} \approx \frac{a_{\mathrm{SrO}}}{\sqrt{2}}$$

$$c_n \approx 2\left(n\mathbf{a}_{\mathrm{p}} + \frac{a_{\mathrm{SrO}}}{\sqrt{2}}\right) = a_{\mathrm{p}}(2n + \sqrt{2}) = a_{\mathrm{SrO}}(n\sqrt{2} + 1)$$

Nonintegral n members can also be expected; for example, $n = 1.5$ corresponds to an intergrowth of $n = 1$ and $n = 2$. This type of intergrowth is not limited to the titanates. Several series of phases, $(ABO_3)_nAO$, have been synthesized with A = Ca, Sr, Eu(II) and B = Mn, Ti. Numerous other oxides adopt these structures. This is the case of the oxides $La_{n+1}Ni_nO_{3n+1}$ which can be written as $(LaO)(LaNiO_3)_n$ where La_2NiO_4 ($n = 1$) has the K_2NiF_4 structure (for a review of oxides of this family, see ref. 50), and $LaNiO_3$ ($n = \infty$) is a perovskite. This structure of the La/Ni/O system is similar to that of the analogous Sr/Ti/O system

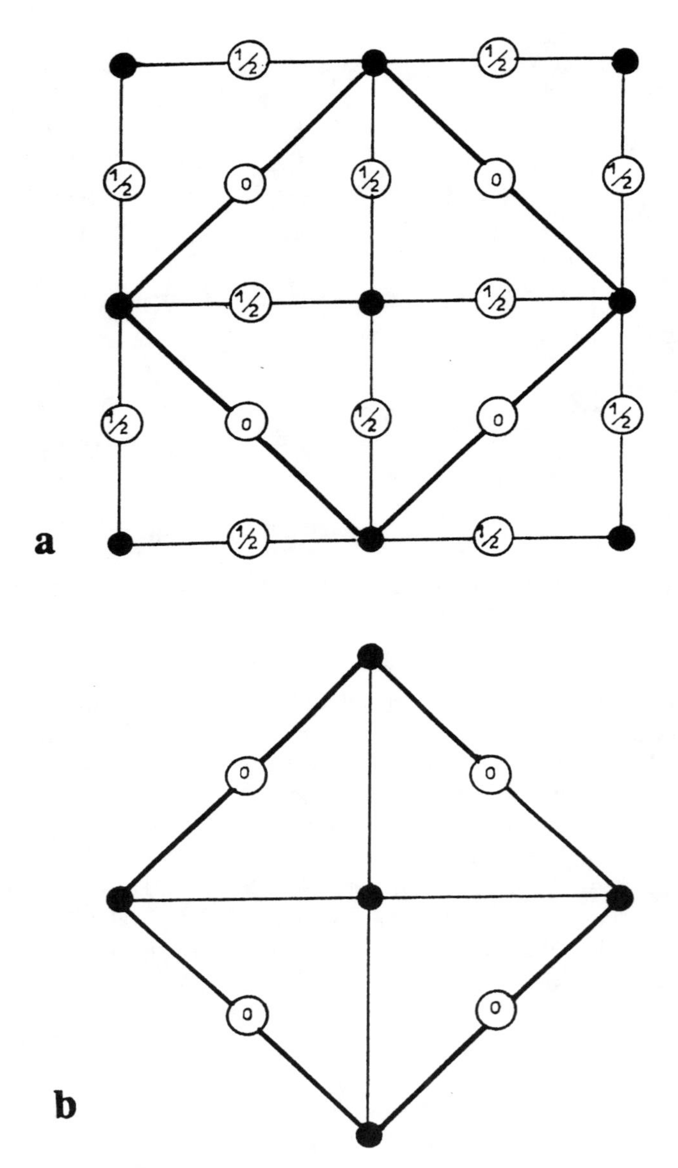

Figure 36. (a) Projection of the oxygen and strontium sublattice of the perovskite structures compared with (b) the $z = 0$ plane of the rock salt SrO structure onto (100); oxygen and strontium atoms represented by open and solid circles, respectively.

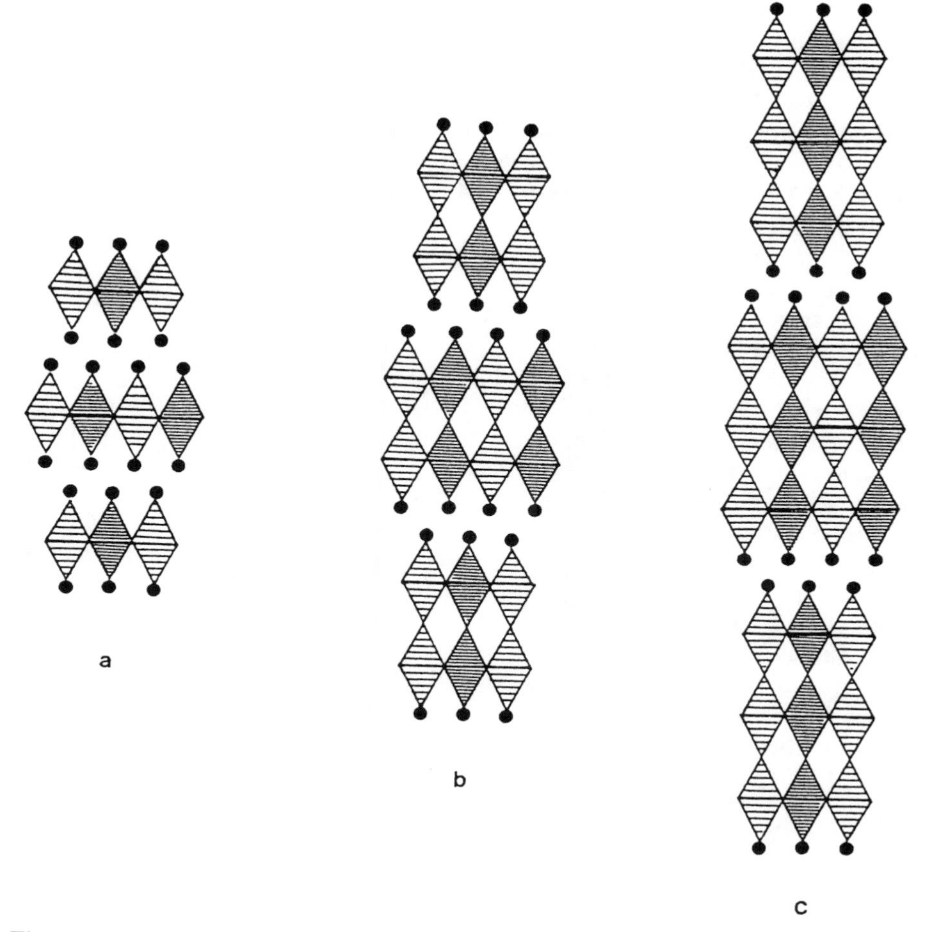

Figure 37. Structure of the intergrowth $(SrTiO_3)_nSrO$: (a) $n = 1$, (b) $n = 2$, (c) $n = 3$. The TiO_6 octahedra (hatched lozenges) from (a) single, (b) double, and (c) triple layers interconnect with strontium (solid circles) oxygen layers. (After Raveau.[46])

$Sr_{n+1}Ti_nO_{3n+1}$ described by Ruddlesden and Popper.[51] Another important example is the series of the cuprates $La_{2-x}A_xCuO_4$ (A = Ca, Sr, Ba) synthesized in 1980 by Raveau et al (see for a review ref. 49) in which high temperature superconductivity ($T_c \sim 35$–40 K) was observed for the first time in 1986 by Bednorz and Müller. La_2CuO_4 ($x = 0.0$) is the parent cuprate of this series that corresponds to the member n = s(K_2NiF_4) structure.

Although theoretically possible, SrO-rich intergrowths with the formulation $(SrO)_nSrTiO_3$, which would have multiple rock salt layers, have not been observed (for all $n > 1$). Multiple rock salt type layers have however been observed in the cuprates $Bi_2Sr_2CuO_6$ and $Tl_2Ba_2CuO_6$.[49] Both the structures can be described (Fig. 38a) as the intergrowth of single perovskite layers, $[SrCuO_3]_\infty$ and $[BaCuO_3]_\infty$, with triple rock salt layers, $[(BiO)_2(SrO)]$ and $[(TlO)_2(BaO)]_\infty$. The unit cell parameters of these oxides are therefore closely related to the perovskite. $Bi_2Sr_2CuO_6$ is orthorhombic with $a \approx a_p\sqrt{2}$, $b \approx 5a_p\sqrt{2}$, and $c \approx 2 \times 24.4$ Å, whereas $Tl_2Ba_2CuO_6$ is tetragonal with $a \approx a_p$ and $c \approx 23.2$ Å. Both the oxides are high T_c superconductors, with a maximum T_c of 23 K for the bismuth cuprate and a maximum T_c of 92 K for the thallium cuprate. In reality, the crystal chemistry of these phases is complex, the rock salt layers being strongly distorted with respect to the ideal structure. As a result, one observes a modulation of the structure in the bismuth cuprate, which can accept an excess of oxygen, according to the formulation $Bi_2Sr_2CuO_{6+\delta}$. In the same way, the thallium cuprate can exhibit a small deviation from oxygen stoichiometry, leading to the formulation $Tl_2Ba_2CuO_{6\pm\delta}$, but does not show incommensurate satellites in the electron diffraction patterns, in contrast to the bismuth cuprate. The variation in oxygen stoichiometry, although small, has a dramatic influence on the superconducting properties of these phases. For instance, the T_c of the cuprate $Tl_2Ba_2CuO_{6\pm\delta}$, can be increased progressively from 0 K up to 92 K, by successive annealing in a reducing atmosphere at low temperatures, without any detectable change in the structure. The oxides $TlBa_2CuO_{5\pm\delta}$ (nonsuperconducting) and $TlBa_{2-x}Ln_xCuO_{5-\delta}$ ($T_c \approx 30$–40 K) are also characterized by multiple layers of the rock salt type. Their structures (Fig. 38b) consist of an intergrowth of single perovskite layers, $[BaCuO_3]_\infty$, with double rock salt type layers, $[(TlO)(BaO)]_\infty$. In these tetragonal phases ($a \approx a_p \approx 3.8$ Å, $c \approx 9.5$ Å) also, the rock salt layers are distorted and there are possibly oxygen vacancies in the $[TlO]_\infty$ layers.

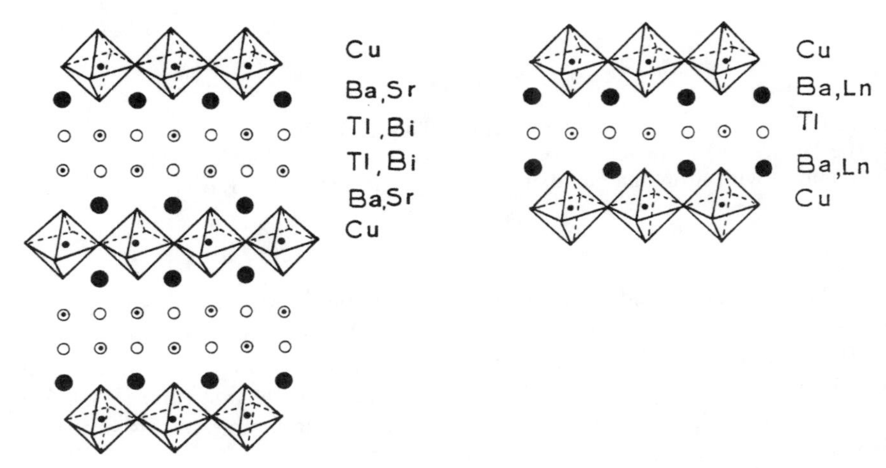

Figure 38. Schematic structures of (a) $Bi_2Sr_2CuO_6$ and $Tl_2Ba_2CuO_6$ and (b) $TlBa_2CuO_{5-\delta}$ and $TlBa_{2-x}Ln_xCuO_{5-\delta}$. (After Raveau et al.[49])

4.5.2 *Intergrowth of Oxygen-Deficient Perovskites with Rock Salt Layers in the Layer Cuprates, $[ACuO_{3-x}]_m[AO]_n$*

The ordered creation of oxygen vacancies in the perovskite layers of the $(ABO_3)_m(AO)_n$ intergrowths leads to a large family of cuprates known for their superconducting properties at high temperatures (see for a review ref. 49). Such cuprates, corresponding to the general formula $(ABO_{3-x})_m(AO)_n)$, can be synthesized because of the ability of copper to adopt pyramidal, square-planar, and other coordinations. They can be represented by the symbol $[m, n]$, where m and n represent the number of copper layers and the $[AO]_\infty$ layers in the perovskite and rock salt slabs, respectively. Different members of this series are listed in Table 8. The $m = 1$ members were described earlier; they correspond to single octahedral layers intergrown with single ($n = 1$), double ($n = 2$), or triple ($n = 3$) rock salt layers as in La_2CuO_4, $TlBa_2CuO_5$, and $Tl_2Ba_2CuO_6$, respectively. The $m = 2$ members are characterized by double pyramidal copper layers, deduced from the perovskite, by the elimination of rows of oxygen atoms along $\vec{a}$ (Fig. 39). They differ only in the thickness of the rock salt layer, which is single ($n = 1$) in $LaSrCaCu_2O_6$ (Fig. 39a), double ($n = 2$) in $TlBa_2CaCu_2O_7$ (Fig. 39b), and triple ($n = 3$) in $Tl_2CaCu_2O_8$ and $Bi_2Sr_2CaCu_2O_8$ (Fig. 39c). The elimination of rows of oxygen vacancies in the triple perovskite layers leads to triple copper layers, built up from one layer of corner-sharing, square-planar CuO_4 groups sandwiched between two pyramidal copper layers. This leads to the $m = 3$ members, such as $TlBa_2Ca_2Cu_3O_9$ (Fig. 40a) and $Tl_2Ba_2Ca_2Cu_3O_{10}$ or $Bi_2Sr_2Ca_2Cu_3O_{10}$ (Fig. 40b), which involve double ($n = 2$) and triple rock salt layers, respectively. Further increase in the number of copper layers ($n > 3$) leads to additional layers of square-planar groups between the pyramidal copper layers, as in $TlBa_2Ca_3Cu_4O_{11}$ and $Tl_2Ba_2Ca_3Cu_4O_{12}$ (both have two layers of CuO_4 groups along with two and three rock salt layers, respectively). Note that for $n = \infty$, we should obtain an oxygen-deficient perovskite built up of $[CuO_2]_\infty$ layers of CuO_4 square-planar groups only, interleaved with calcium ions. This is indeed the case in $Sr_{0.15}Ca_{0.85}CuO_2$ (Fig. 41). One can also subdivide this large family of intergrowths into series differing in the multiplicity of the rock salt type layers as follows:

$Tl_2Ba_2Ca_{m-1}Cu_mO_{2m+4}$ and $Bi_2Sr_2Ca_{m-1}Cu_mO_{2m+4}$, whose triple rock salt layers are formed of thallium or bismuth bilayers.

$TlBa_2Ca_{m-1}Cu_mO_{2m+3}$ and $Tl_2Sr_2Ca_{m-1}Cu_mO_{2m+3}$, whose double rock salt layers are formed of thallium monolayers.

$A_2(Ca, Y, Sr)_{m-1}Cu_mO_{2m+2}$, which possess single rock salt layers [e.g., La_2CuO_4 ($m = 1$), $La_2SrCu_2O_6$ ($m = 2$), $PbBaSrYCu_3O_8$ ($m = 3$)].

Besides these layered cuprate superconductors, there are more complex intergrowths of rock salt and oxygen-deficient perovskite structures. A good example

Table 8 Superconducting Layered Cuprates $[m,n]$

	m			
n	1	2	3	4
1	[1,1] $La_{2-x}A_xCuO_4$ La_2CuO_4 (A = Ca, Sr, Ba) $T_c \approx 35–40$ K	[2,1] $La_{2-x}A_{1+x}Cu_2O_6$ A = Ca, Sr $T_c = 60$ K	[3,1] $PbBaYSrCu_3O_8$ $T_c = 50$ K	[4,1]
2	[1,2] $Tl_{0.5}Pb_{0.5}Sr_2CuO_{5-\delta}$ $T_c = 40$ K $Tl_{1-x}Pr_xSr_{2-y}Pr_yCuO_{5-\delta}$ $T_c = 40$ K $HgBa_2CuO_{4+\delta}$ $T_c = 94$ K	[2,2] $TlBa_2CaCu_2O_{7-\delta}$ $T_c \approx 60$ K $TlSr_2CaCu_2O_{7-\delta}$ $T_c \approx 50$ K $Pb_{1-y}M_ySr_2Ca_{1-x}Y_xCu_2O_{7-\delta}$ (M = Sr, Ca, Cd, Gd, Cu, . . .) $T_c = 20–80$ K $Tl_{0.5}Pb_{0.5}Sr_2CaCu_2O_{7-\delta}$ $T_c \approx 70–80$ K	[3,2] $TlBa_2Ca_2Cu_3O_9$ $T_c \approx 120$ K $Tl_{0.5}Pb_{0.5}Sr_2Ca_2Cu_3O_9$ $T_c \approx 120$ K $HgBa_2Ca_2Cu_3O_8$ $T_c \approx 133$ K	[4,2] $TlBa_2Ca_3Cu_4O_{11}$ $T_c \approx 108$ K
3	[1,3] $Tl_2Ba_2CuO_{6\pm\delta}$ $T_c = 30–92$ K $Bi_2Sr_2CuO_{6+\delta}$ $T_c \approx 10–22$ K	[2,3] $Tl_2Ba_2CaCu_2O_8$ $T_c \approx 105$ K $Bi_{2-x}Sr_{2-x}CaCu_2O_{8+\delta}$ $T_c \approx 85$ K $Bi_{2-x}Pb_xSr_2Ca_{1-x}Y_xCu_2O_8$ $T_c \approx 85$ K $\rightarrow$ N.S.	[3,3] $Tl_2Ba_2Ca_2Cu_3O_{10}$ $T_c \approx 125–130$ K $Bi_{2-x}Pb_xSr_2Ca_2Cu_3O_{10}$ $T_c \approx 110$ K	[4,3] $Tl_2Ba_2Ca_3Cu_4O_{12}$ $T_c \approx 115$ K

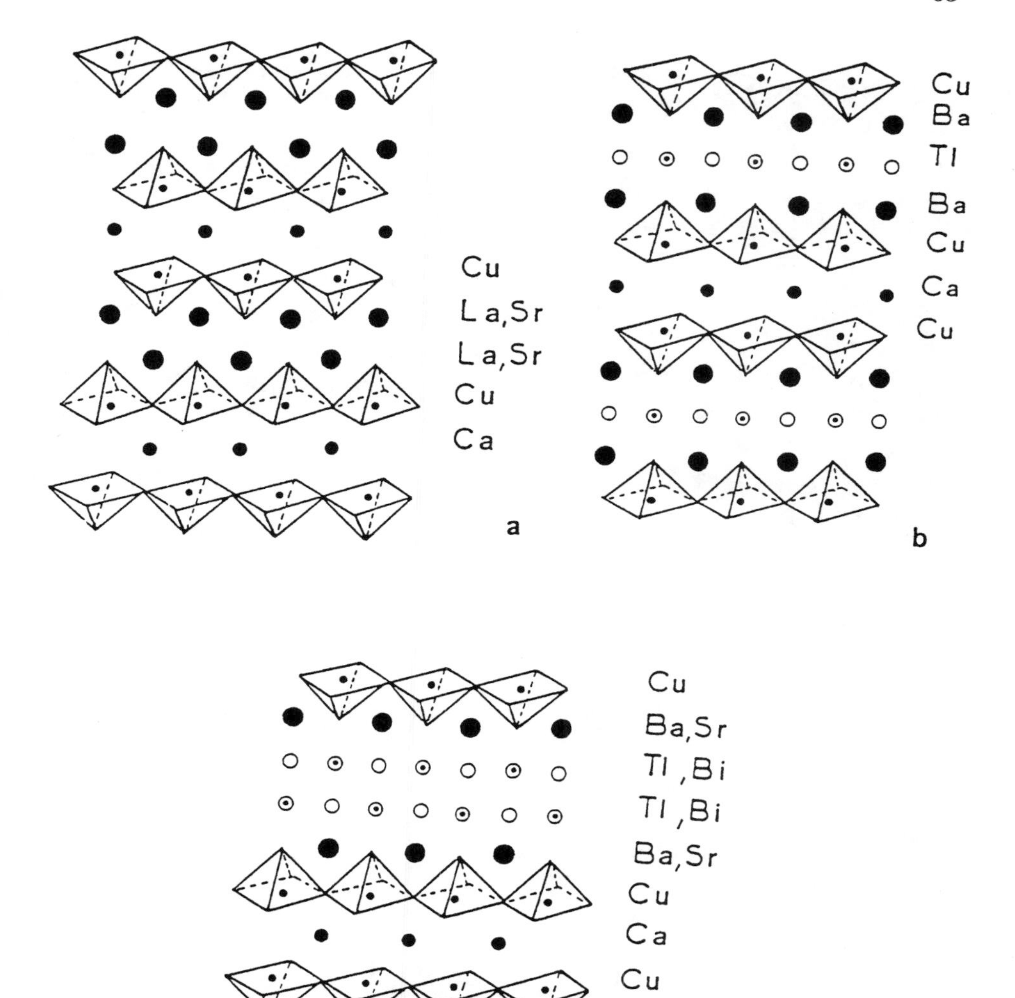

Figure 39. Schematic structure of the $m = 2$ cuprates $(ACuO_{3-x})_m(AO)_n$: (a) $LaSrCaCu_2O_6$, (b) $TlBa_2CaCu_2O_7$, and (c) $Tl_2Ba_2CaCu_2O_8$ or $Bi_2Sr_2CaCu_2O_8$. (After Raveau et al.[49])

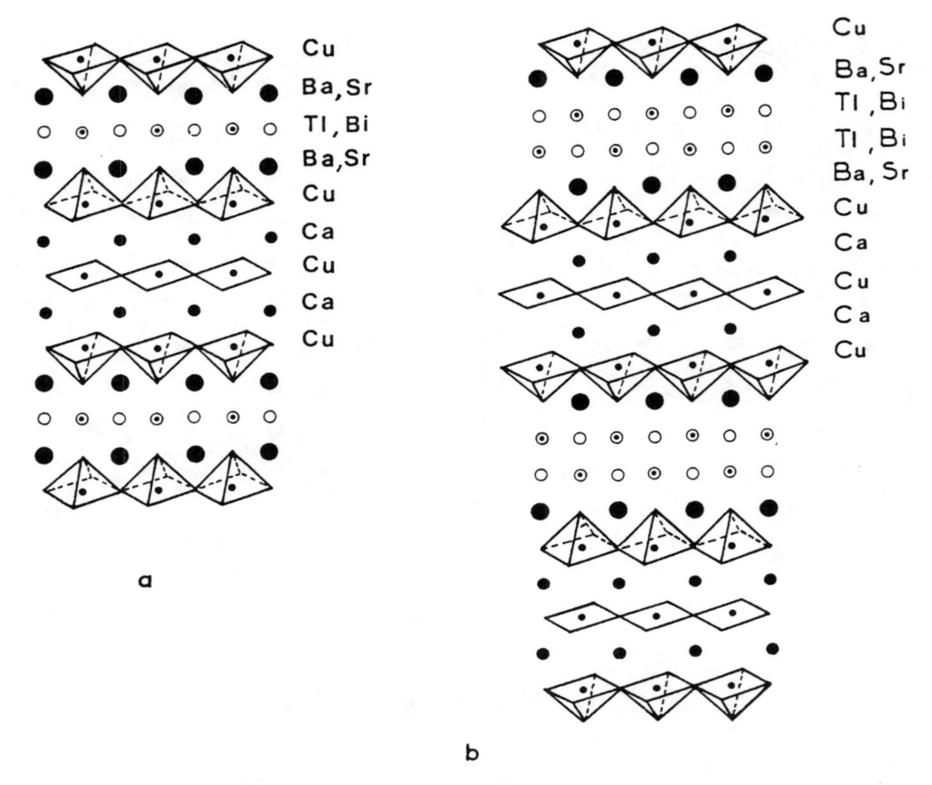

Figure 40. Schematized structure of the $m = 3$ cuprates $(ACuO_{3-x})_m(AO)_n$: (a) $TlBa_2Ca_2Cu_3O_9$ and (b) $Tl_2Ba_2Ca_2Cu_3O_{10}$ or $Bi_2Sr_2Ca_2Cu_3O_{10}$. (After Raveau et al.[49])

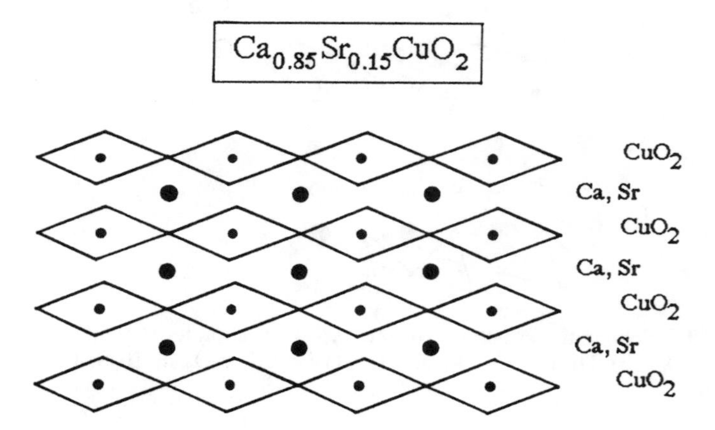

Figure 41. Schematized structure of $Sr_{0.15}Ca_{0.85}CuO_2$. (After Raveau et al.[49])

is provided by the layered cuprate $Sr_6Nd_3Cu_6O_{17}$. This oxide, which can be formulated as $Sr_2NdCuO_{5.66}$, has an orthorhombic cell ($a \approx a_p$, $b \approx 3a_p$, $c \approx$ 20 Å). It is characterized by a similar arrangement of the metallic atoms, and to a great extent, of the oxygen atoms as in $La_2SrCu_2O_6$ (Fig. 42a). Several of the oxygen atoms and anionic vacancies are however distributed differently. This structure is actually intermediate between that of $YBa_2Cu_3O_7$ and $La_2SrCu_2O_6$. It can be described as a regular intergrowth of $YBa_2Cu_3O_7$-type ribbons, $[Sr_{1.16}Nd_{0.84}Cu_2O_{4.66}]_\infty$, running along $\vec{\mathbf{b}}$ with the rock salt type layers, $[Sr_{1.16}Nd_{0.16}O]_\infty$. The distribution of cations in the different layers is novel. The smaller cations, Nd^{3+}, tend to be located preferentially in eightfold coordination in the 123 ribbons, between the CuO_5 pyramids, whereas the larger ones, Sr^{2+}, are preferentially distributed in the rock salt layers. Another structure, closely related to $Sr_6Nd_3Cu_6O_{17}$, is that of $Sr_2NdCu_2O_6$, whose ideal model (Fig. 42b) consists of six-sided tunnels built up from rings of corner-sharing CuO_5 pyramids. Such tubes are arranged in such a way that they form fluorite-type cages and rock salt layers where Nd and Sr are located. Neither of these oxides is superconducting.

Finally, Sr_2CuO_3 (or Ca_2CuO_3) (Fig. 43a), with an orthorhombic structure ($a \approx a_p$, $b \approx 3.5$ Å, $c \approx 12.7$ Å), can be considered as an intergrowth of the oxygen-deficient perovskite layer with the SrO rock salt layer. This structure is closely related to that of La_2CuO_4 (Fig. 43b) and is derived from the latter by eliminating alternate rows of oxygen atoms along $\vec{\mathbf{b}}$ in the perovskite layer. This results in the formation of $[CuO_2]_\infty$ rows of corner-sharing CuO_4 square-planar groups running along $\vec{\mathbf{a}}$; these rows form layers parallel to (001). Thus, the structure of these cuprates can be described as an oxygen-deficient intergrowth of the K_2NiF_4 type.

4.5.3 Intergrowths of Perovskite with "Bi_2O_2" Layers Giving Rise to Aurivillius Phases

Oxides with the general formula, $(Bi_2O_2)^{2+}(A_{n-1}B_nO_{3n+1})^{2-}$ have been synthesized (A = Ca, Sr, Ba, Pb, K, Na, Bi; B = Ta, Nb, Ti, W, Mo). Extensively studied because of their ferroelectric properties, they are characterized by layers of two kinds: pure octahedral layers $[B_nO_{3n+1}]_\infty$, and alternating pyramidal $[Bi_2O_2]_\infty$ layers. The $[Bi_2O_2]_\infty$ layers are built up of edge-sharing BiO_4 pyramids and are parallel to (001) (Fig. 44). This particular coordination of Bi(III) is due to its $6s^2$ lone pair, which can be assimilated to an anion and is oriented along $\vec{\mathbf{c}}$. The octahedral $[B_nO_{3n+1}]_\infty$ layers consist of distorted ReO_3-type sheets where the A cations are incorporated. The thickness of these layers (along $\vec{\mathbf{c}}$) depends on the n value characterizing the number of ReO_3-type layers stacked between two successive $[Bi_2O_2]_\infty$ layers. Koecklinite, Bi_2MoO_6 (Fig. 44a), known for its catalytic properties, and the ferroelectric oxides $Bi_2SrNb_2O_9$ (Fig. 44b) and $Bi_4Ti_3O_{12}$ (Fig. 44c), represent, respectively, the $n = 1$, 2, and 3 members of

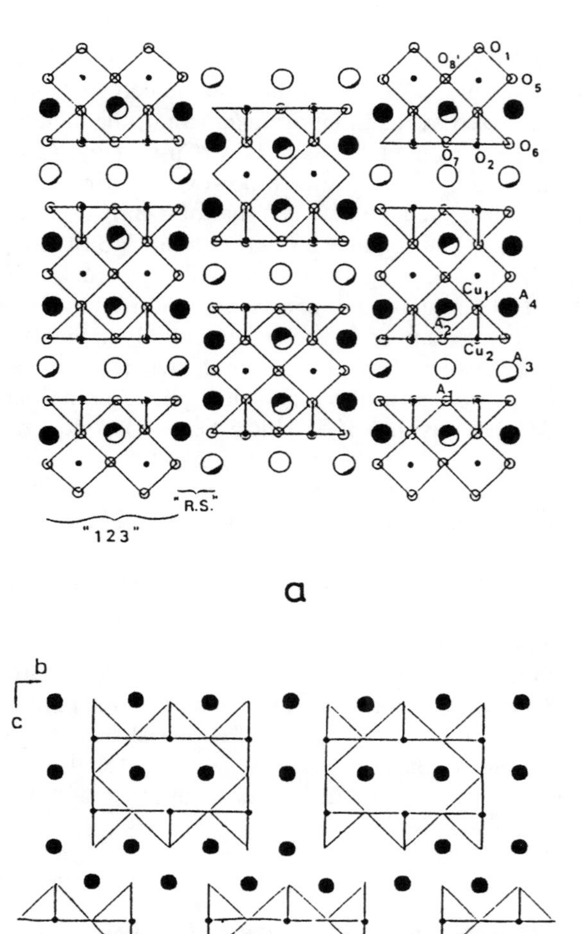

Figure 42. (a) Crystal structure of $Sr_6Nd_3Cu_6O_{17}$. Copper (small solid circles) and oxygen (small open circles) form CuO_5 pyramids and CuO_4 square-planar groups. Large solid circles correspond to strontium and large open circles to neodymium, whereas half-shaded circles correspond to a mixed occupancy Nd-Sr. (b) Schematic drawing of the limit compound $Sr_2NdCu_2O_6$. Small solid circles represent copper and large solid circles neodymium and strontium. Note the CuO_5 pyramids forming six-sided tunnels. (After Raveau et al.[49])

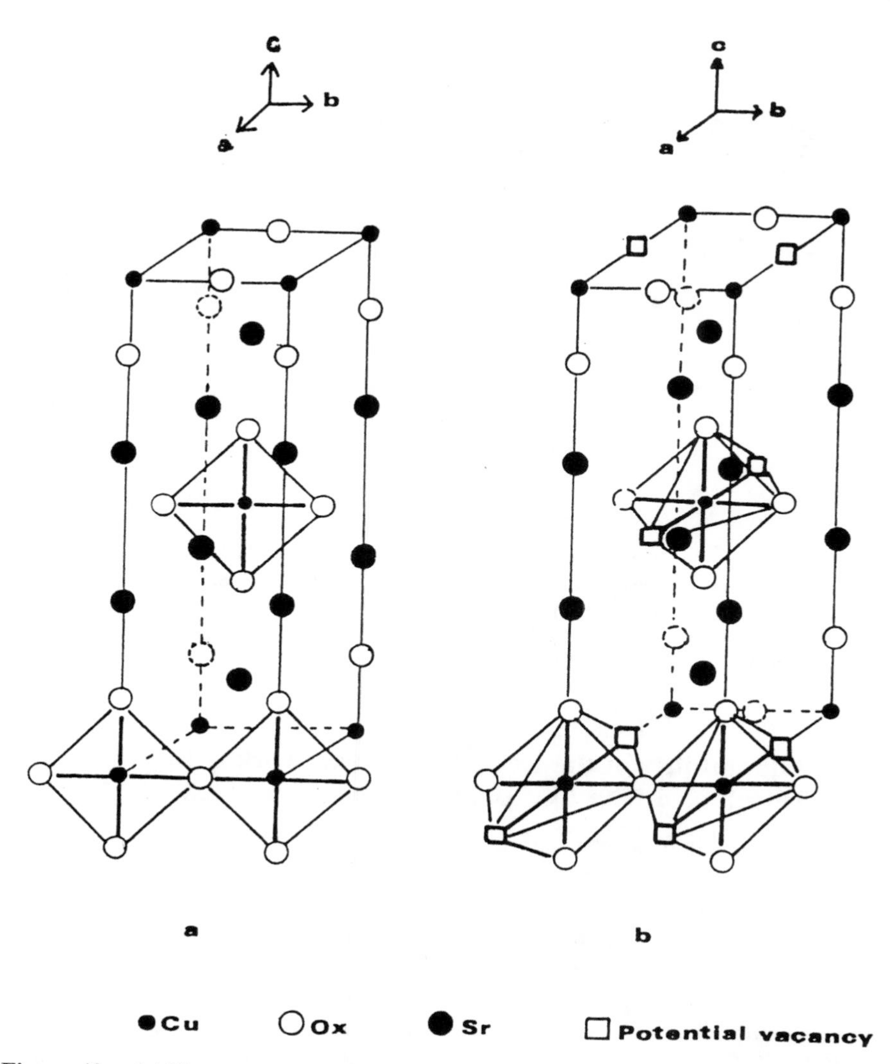

Figure 43. (a) The structure of Sr_2CuO_3, which can be deduced from (b) the La_2CuO_4-type structure, by eliminating alternate rows of oxygen atoms along $\vec{\mathbf{b}}$ (After Raveau.[46])

this family. Relationships with the perovskite structure leads to a tetragonal or an orthorhombic (pseudotetragonal) cell with $a \approx b \approx a_p$ or $a_p\sqrt{2}$ and $c \approx 8.5 + 2na_p$. Numerous other members can be synthesized by compensated substitutions of the A and B ions, and of the oxygens by fluorine. Microphases involving nonintegral n values can also be synthesized. They correspond to a sequence of perovskite layers of different thicknesses (along $\vec{\mathbf{c}}$) and can be considered as intergrowths of integral n members.

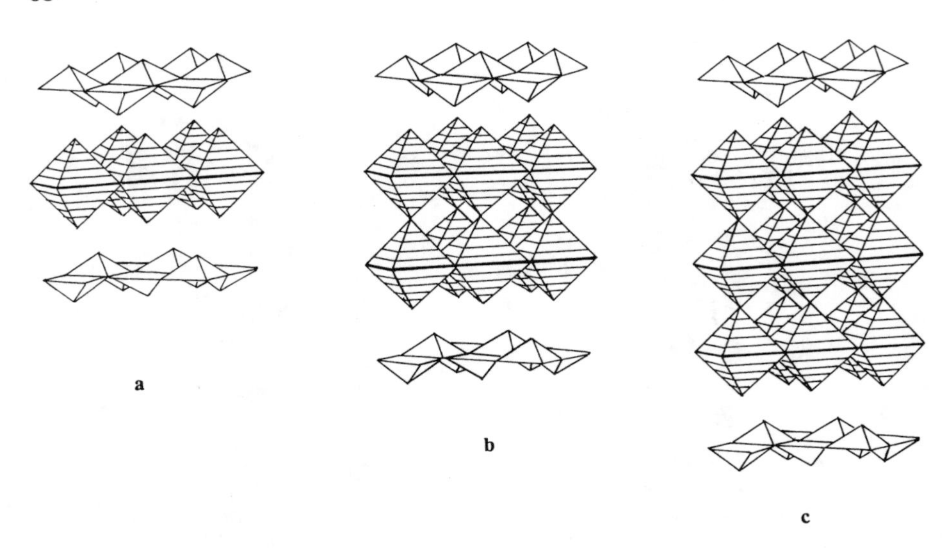

Figure 44. Structure of the Aurivillius phases $(Bi_2O_2)^{2+}(A_{n-1}B_nO_{3n+1})^{2-}$. In the layer of edge-sharing BiO_4 pyramids, the oxygen atoms sit in the basal plane, whereas the apical site is occupied by bismuth. The BiO_6 octahedra (hatched) form (a) single layers ($n = 1$) as in Bi_2MoO_6, (b) double layers ($n = 2$) as in $Bi_2SrNb_2O_9$, or (c) triple layers ($n = 3$) as in $Bi_4Ti_3O_{12}$. (After Wadsley.[37])

4.6 Adaptability of the ReO_3-Type Framework with PO_4 Tetrahedra: The Phosphate Tungsten Bronzes

The phosphate tungsten bronzes form a large family whose members demonstrate the ability of the monophosphate group, PO_4, or the diphosphate group, P_2O_7, to accommodate the octahedral ReO_3-type framework.[46] These oxides result from there replacement of one BO_6 octahedron by one PO_4 or P_2O_7 group in the octahedral B_4O_{20} structural unit (Fig. 45). The first family, called the diphosphate tungsten bronzes with hexagonal tunnels ($DPTB_H$), have a monoclinic structure. They form a series of microphases having the general formula $A_x(P_2O_4)_2(WO_3)_{2m}$, with A = K, Rb, Tl, Ba, $x \approx 1$, and m an integer. The host lattice (Fig. 46) is built up of corner-sharing WO_6 octahedra and P_2O_7 groups, and the structure can be described as the stacking of ReO_3-type slices connected through single layers of diphosphate groups. At the junction between successive ReO_3-type slices, WO_6 octahedra and P_2O_7 groups form six-sided tunnels running along $\vec{\mathbf{b}}$, where the A cations are located. The ReO_3-type slices have a variable width characterized by the number of octahedra, m, forming strings between two diphosphate planes, that is, parallel to [101]. The even-m members (Fig. 46a) differ from the odd-m members (Fig. 46b) by a relative translation of the diphosphate planes. As a result, in the even-m members, all the octahedral strings parallel to [100] are identical and consist of $m/2$ WO_6 octahedra, whereas in the odd-m members, the corresponding strings are built of $(m + 1)/2$ and $(m -$

Figure 45. (a) Replacement of a BO_6 octahedron by a PO_4 tetrahedron in an octahedral B_4O_{20} unit. (b) Accommodation of a diphosphate group.

1)/2 octahedra. In the different members of the series, the parameters of the monoclinic cell are related to those of the perovskite in the following manner:

$$a_m \approx \frac{m}{2}\, a_{\mathrm{p}} + k,\ b_{\mathrm{m}} \approx 2a_{\mathrm{p}},\ c_{\mathrm{m}} \approx 2a_{\mathrm{p}}\sqrt{5} \qquad \text{for even-}m\text{ members}$$

$$a_{\mathrm{m}} \approx a_{\mathrm{p}} \left[\frac{(m-1)^2}{2}\right]^{1/2} 1 + k,\ b_{\mathrm{m}} \approx 2a_{\mathrm{p}},\ c_{\mathrm{m}} \approx 2a_{\mathrm{p}}\sqrt{5}$$

for odd-m members

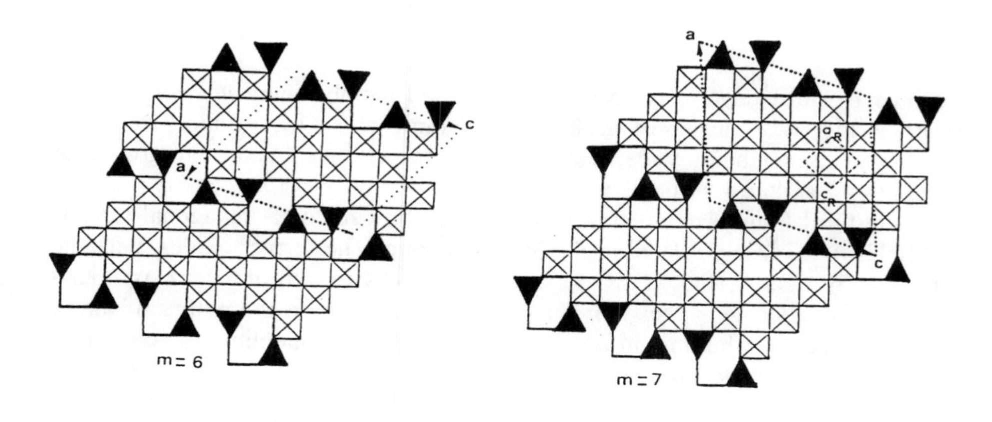

Figure 46. Host lattice of the diphosphate tungsten bronzes $A_xP_2O_4\ (WO_3)_{2m}$ with hexagonal tunnels: (a) an even-m member, $m = 6$ and (b) an odd-m member, $m = 7$. The WO_6 octahedra (crossed squares) and the P_2O_7 groups (solid triangles) form six-sided tunnels where the A cations (not represented) sit (after Raveau[46]).

where $k \approx 2.7$ Å (close to $a_p\sqrt{2}$) and $\beta \approx 117°$ for even-*m* members and $\beta \approx 117 - \omega$ for odd-*m* members [$\tan \omega \approx 2/(m - 1)$]. Note the tendency to form regular tunnels—that is, with a geometry involving 90° and 120° angles for the six-sided tunnels, and almost 90° for perovskite tunnels.

The monophosphate tungsten bronzes with pentagonal tunnels ($MPTB_P$) represent the second family of the PTB series. These oxides with an orthorhombic structure have the general formula $(PO_2)_4(WO_3)_{2m}$. Their host lattice (Fig. 47) is composed of corner-sharing WO_6 octahedra and PO_4 tetrahedra, each monophosphate group (i.e., each PO_4 tetrahedron) sharing its four corners with WO_6 octahedra. This mixed framework is characterized by strong anisotropy. It can also be described as the stacking of ReO_3-type slices with a variable width, connected to each other through single layers of PO_4 tetrahedra parallel to (001). At the junction between two ReO_3-type slices, the PO_4 tetrahedra and WO_6 octahedra form pentagonal tunnels which are empty. Note that the geometry of these tunnels is also characterized by 120° and 90° angles. The members of this family differ in the thickness of the ReO_3-type slices and by the parity of the number *m* of octahedra forming the octahedral string between two successive tetrahedral planes. In the even-*m* members (Fig. 47a), the octahedral slices have a constant width corresponding to strings of $m/2$ octahedra along the $\langle 100 \rangle$ direction. In the odd-*m* members (Fig. 47b), the strings of octahedra parallel to the $\langle 100 \rangle_{ReO_3}$ direction are built up of $(m + 1)/2$ and $(m - 1)/2$ octahedra alternately. The parameters of the orthorhombic cell of the different members of this family are related to that of the perovskite in the following way:

$$a_m \approx a_p\sqrt{2},\ b_m \approx a_p\sqrt{3},\ c_m \approx 2\left(k + \frac{m}{2} - a_p \cos 35°\right), \quad k \approx 2.45 \text{ Å}$$

However, an exception is observed in the odd member with $m = 5$. Its host lattice (Fig. 47c) consists of stacking of two kinds of ReO_3-type slice, which are four and six octahedra wide; the structure is to be treated as an intergrowth of the $m = 4$ and $m = 6$ members.

The third family, monophosphate tungsten bronzes with hexagonal tunnels ($MPTB_H$), is closely related to the $MPTB_P$ group. The former oxides, with the general formula $A_x(PO_2)_4(WO_3)_{2m}$, are stabilized by cations of the size of potassium or smaller (A = Na, K). The structure depends on the nature of the A cation; it is orthorhombic in potassium bronzes, and monoclinic or sometimes triclinic in sodium bronzes. In spite of these distortions, one observes ReO_3-type slices identical to those obtained in $MPTB_P$s, connected through monophosphate planes, but forming hexagonal tunnels at the junction of the two slices where the A cations are located. In reality, the frameworks of $MPTB_P$s and $MPTB_H$s are built up of identical octahedral layers bordered with PO_4 tetrahedra (Fig. 48). The $MPTB_H$ structure differs from the $MPTB_H$ structure by the relative orientation of two successive layers; one obtains the $MPTB_H$ framework by a rotation of 180° of one layer out of two in the $MPTB_P$ structure (Fig. 49). The

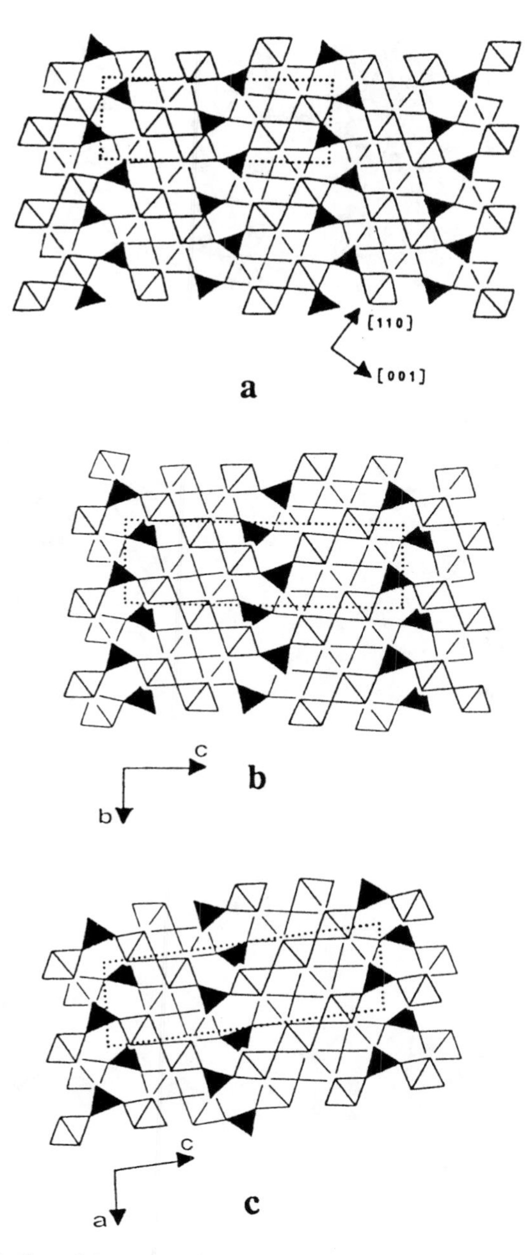

Figure 47. Host lattice of the monophosphate bronzes $(PO_2)_4(WO_3)_{2m}$ with pentagonal tunnels; the WO_6 octahedra (lozenges) form ReO_3-type layers interconnected with PO_4 tetrahedra (solid triangles). (a) the even-*m* member ($m = 4$) $P_4W_8O_{32}$ consists of strings of four WO_6 octahedra forming a fish-bone array. The odd-*m* member, $m = 5$, consists either of (b) strings of five octahedra or (c) of an intergrowth of the $m = 4$ and $m = 6$ members. (After Raveau.[46])

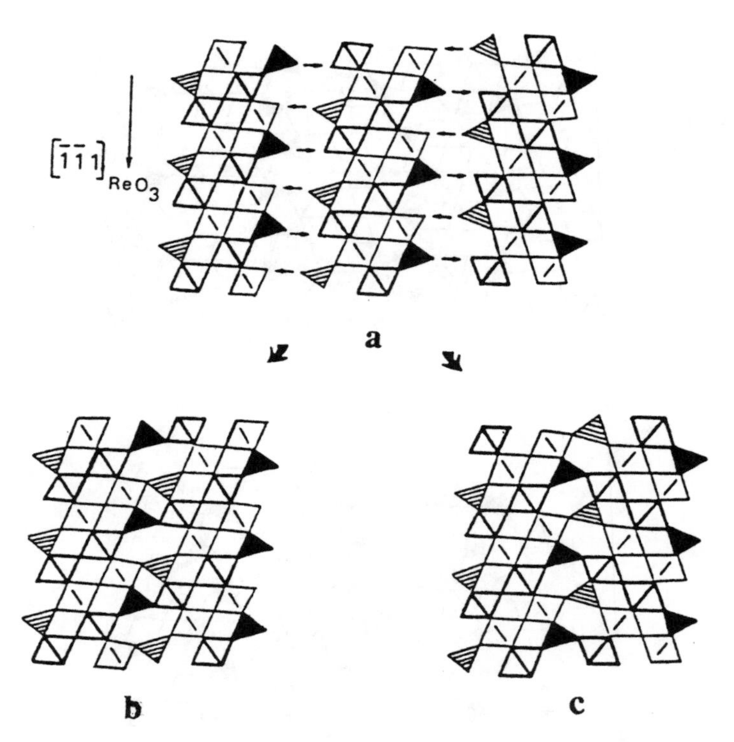

Figure 48. Structural relationships between monophosphate tungsten bronzes with pentagonal tunnels ($MPTB_P$) and those with hexagonal tunnels ($MBTB_H$). (a) The ReO_3-type layers bordered with PO_4 tetrahedra observed in both structures. The assemblage of these layers forms (b) $MPTB_H$ and (c) $MPTB_P$. (After Raveau.[46])

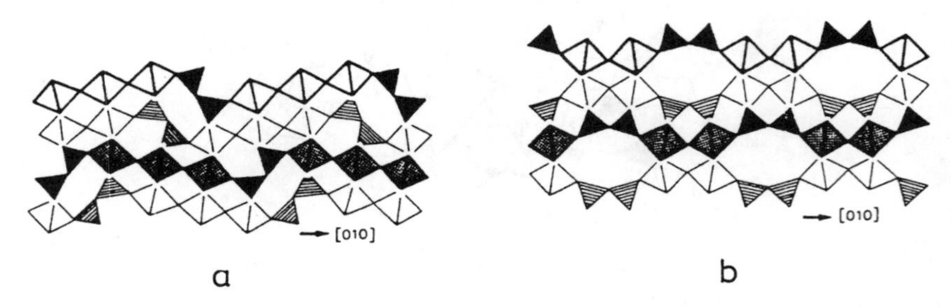

Figure 49. (a) $P_8W_{12}O_{52}$: projection of the structure onto the (010) plane. The strings of WO_6 octahedra (lozenges) and diphosphate groups (triangles) form pentagonal and rectangular tunnels that are empty. (b) The phosphate bronze $CsP_8W_8O_{40}$: view of the structure along $\vec{c}$. The WO_6 octahedra from ReO_3-type columns interconnected through P_2O_7 groups (solid and hatched triangles). Note the octagonal tunnels occupied by cesium. (After Raveau.[46])

octahedral chains form a fish-bone array from one layer to the other as in $MPTB_{P}s$ (Fig. 49c), but are parallel to one another in the $MPTB_{H}s$ (Fig. 48b). The odd-*m* members are often characterized by a translation of one phosphate plane out of two with respect to the even-*m* members. Sometimes, one observes an intergrowth of two even-*m* members. The parameters of the monoclinic cell of these bronzes are related to that of the ideal perovskite in the following way:

$$a \approx a_p\sqrt{3}, \qquad b \approx a_p\sqrt{2}, \qquad c \approx \frac{ma_p \cos 35° + 2k \cos 17°}{\sin \beta}$$

$$\tan \beta \approx \frac{ma_p \cos 35° + 2k \cos 17°}{2(a_p\sqrt{3} - k \sin 17°) - ma_p \sin 35°} \qquad \text{with } a_p \approx 3.8 \text{ Å}$$

$$k \approx 2.6 \text{ Å}$$

Note that geometry of the tunnels is characterized by angles close to 90° and 120°.

Besides the three major families already discussed, with *m* ranging from 4 to 16, a fourth family has been observed by high resolution electron microscopy. In contrast to the other three, however, it has not been isolated as a pure phase. Bronzes of this family correspond to the formula $(P_2O_4)(WO_3)_{2m}$. Their host lattice is composed of ReO_3-type slices connected through diphosphate groups just as in $DPTB_{H}s$. They differ from the $DPTB_{H}s$ in the nature of the tunnels at the junction between two ReO_3-type slices, which are pentagonal and seem to be empty. These oxides represent a series of potential diphosphate tungsten bronzes with pentagonal tunnels ($DPTB_{P}$), which had not been isolated till recently.

The ability of the PO_4 tetrahedra to accommodate the ReO_3-type framework is also observed in the oxides $P_8W_{12}O_{52}$ and $CsP_8W_8O_{40}$. The first host lattice consists of corner-sharing WO_6 octahedra and P_2O_7 groups (Fig. 49a) and exhibits different sorts of tunnels. In this structure one recognizes ReO_3-type columns, built up of 2 × 3 octahedra, linked together by their corners through P_2O_7 groups. The bronze $CsP_8W_8O_{40}$ is also built up of corner-sharing WO_6 octahedra and P_2O_7 groups (Fig. 49b), forming octagonal tunnels where the cesium cations are located. In spite of such specificity, this phase is also related to the perovskite; one observes octahedral B_4O_{20} blocks, forming ReO_3-type columns running along $\vec{\mathbf{c}}$. These columns of 2 × 2 octahedra are linked through P_2O_7 groups. Note that the angles of the B_4O_{20} units are not 90° in the cubic perovskite; rather, they exhibit a diamond-shaped configuration (60–120° angles) as in WO_3.

The W(VI)/W(V) mixed valence observed in all the foregoing bronzes causes a delocalization of the electrons over the $[WO_3]_\infty$ framework. Consequently, these materials exhibit electrical properties similar to Na_xWO_3; that is, they are excellent metallic conductors (and for this reason are called bronzes). They differ, however, from the A_xWO_3 bronzes in their anisotropy of conductivity. The disposition of the phosphate planes in the three families would be expected to give rise to two-dimensional conductivity. These materials also exhibit charge-density waves. Furthermore, the presence of empty cages or tunnels affords the possibility of intercalation of cations like lithium or sodium. These structures also offer the possibility of formation of extended defects.

Frameworks of the kind just discussed are not limited to the P/W/O and A/P/W/O systems alone. There are two forms of Mo_4O_{11} (monoclinic and orthorhombic) that represent the $m = 6$ members of the $MPTB_H$ and $MPTB_P$ series, respectively; in these host lattices the PO_4 tetrahedra are replaced by MoO_4 tetrahedra. No other member of this family has yet been found. Bronzoids $NbPO_5$ and $NbTaO_5$, isotypic with the bronze PWO_5, are known. These phases are insulating owing to the d^0 configuration of Nb(V) and Ta(V).

4.7 Introduction of Carbonate and Other Oxyanions in the Perovskite Structure

Carbonate groups can be accommodated in the perovskite structure of cuprates, giving rise to a layered structure in which single perovskite layers are connected through layers or rows of CO_3 groups. This is the case in the oxycarbonate $Sr_2CuO_2CO_3$, whose structure (Fig. 50) consists of single perovskite layers connected through single layers of CO_3 groups.[53] Several superconducting copper oxycarbonates are derived from this structure. For example, in $Tl_{0.5}Pb_{0.5}Sr_4Cu_2O_7CO_3$[54] and $Bi_2Sr_4Cu_2CO_3O_8$[55] (Fig. 51), single octahedral perovskite layers are connected through layers of CO_3 groups. These structures can

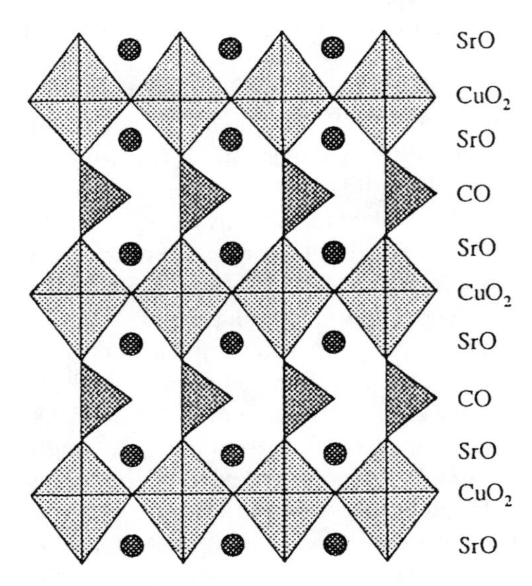

Figure 50. Structure of the oxycarbonate $Sr_2CuO_2CO_3$. Layers of CuO_6 octahedra are interconnected through rows of triangular CO_3 groups. (After Fornichev et al.[53])

Figure 51. Structures of (a) $Tl_{0.5}Pb_{0.5}Sr_4Cu_2CO_3O_7$ (after Huvé et al.[54]) and (b) $Bi_2Sr_4Cu_2CO_3O_8$ (after Pelloquin et al.[55]) Both structures are built up from single layers of CuO_6 octahedra interconnected with rows of CO_3 groups intergrown with double rock salt layers in the Tl phase and triple rock salt layers in the Bi phase.

be described as intergrowths of $SrCuO_2CO_3$ with $Tl_{0.5}Pb_{0.5}Sr_2CuO_5$ and $Bi_2Sr_2CuO_6$, respectively. In the oxygen-deficient perovskite $YBa_2Cu_3O_7$, square-planar CuO_4 groups are replaced in an ordered manner by CO_3 groups. The result is a series of oxycarbonates of the general formula $(Y,Ca)_n(BaSr)_{2n}Cu_{3n-1}CO_3O_{7n-3}$ with closely related structures (Fig. 52).[56] Other oxyanion derivatives of cuprates in which the CuO_4 units are replaced with NO_3^-, PO_4^{3-}, SO_4^{2-}, and BO_3^{3-} have been described in the recent literature (see, e.g., refs. 57–59).

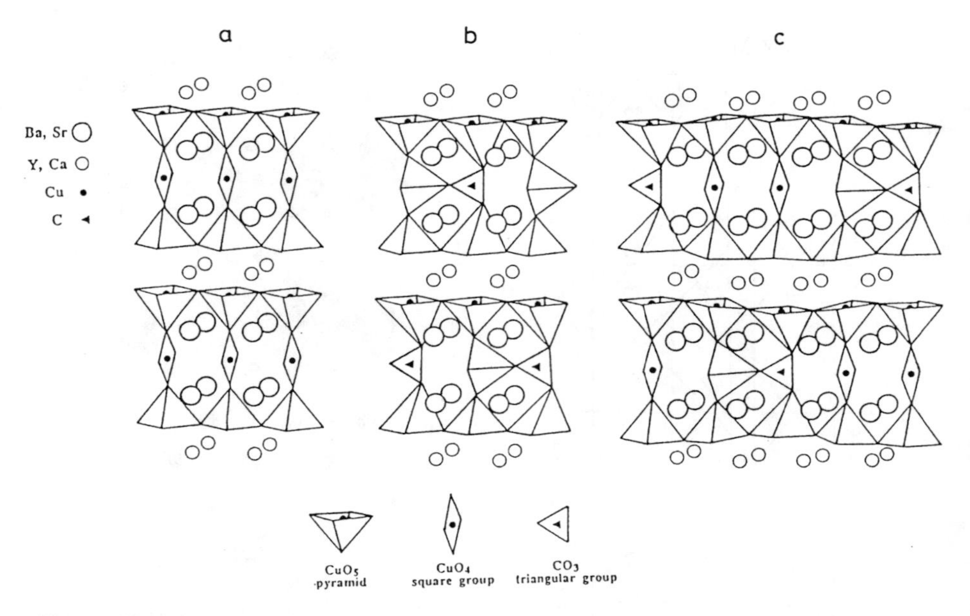

Figure 52. Structure of the oxycarbonates $(Y,Ca)_n(Ba,Sr)_{2n}Cu_{3n-1}CO_3O_{7n-\delta}$ derived from $YBa_2Cu_3O_7$: (a) n = $YBa_2Cu_3O_7$, (*b*) $n = 1$ involving rows of CO_3 groups and CuO_5 pyramids, (*c*) $n = 2$ involving rows of CO_3 groups, CuO_5 pyramids and CuO_4 square-planar groups. (After Raveau et al.[56])

5 Octahedral Tunnel Structures: Bronzes and Bronzoids

Metal–oxygen octahedra, BO_6, form host lattices characterized by large tunnels where cations can be located.[37] In this respect, perovskites form a large family described by four-sided tunnels. Besides the perovskites, there are other tunnel structures that can be classified according to the size and the shape of the tunnels. The number of such oxides is quite large, and we will examine a few representative ones belonging to the main structural types.

5.1 Tunnel Structures with Angles of 90°

The bronze Na_xTiO_2 exhibits tunnels with a square section similar to those observed in the perovskite. The host lattice, $[TiO_2]_\infty$ is similar to that of a perovskite but differs in that the TiO_6 octahedra share their edges instead of their corners along two directions (Fig. 53), whereas along the third direction, they share corners as in a perovskite. The sodium content that can be incorporated in the tunnels is not well defined ($x = 0.20$), and the tunnels are never fully occupied. The $[TiO_2]_\infty$ framework is distorted because of electrostatic repulsions introduced by edge-sharing octahedra. As a result, Na^+ has only four near neighbors in the perovskite edge. In Na_xTiO_2, Ti exhibits mixed valence, Ti(III)/

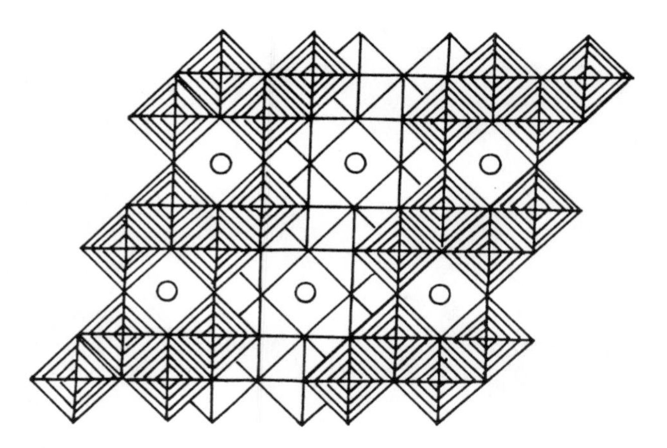

Figure 53. Projection of the structure of the bronze Na_xTiO_2 onto the (110) plane. The TiO_6 octahedra (crossed squares) share edges forming square tunnels where the sodium ions (open circles) are located. (After Wadsley.[37])

Ti(IV), and metallic conductivity. The isotypic oxide $AlNbO_4$ has all its tunnels empty and does not exhibit electron delocalization, Nb being in the +5 oxidation state (d^0). It is consequently a bronzoid. Note that the $[TiO_2]_\infty$ framework results from the ReO_3-type framework by shearing the structure to form edge-sharing octahedra in place of corner-sharing octahedra. Consequently, Na_xTiO_2 belongs to a large family of oxides called shear structures (see Section 11).

Perovskite tunnels can become associated by their faces to form rectangular tunnels. This is found in the titanates $A_2Ti_6O_{13}$ (A = Na, K, Rb), which exhibit a monoclinic cell ($a \approx 15.5$ Å, $b \approx 3.8$ Å, $c \approx 9.1$ Å). Their host lattice, $[Ti_6O_{13}]_\infty$, consists of structural units of 2 × 3 edge-sharing octahedra (Fig. 54a). Stacking of these units, along $\vec{\mathbf{b}}$ creates along this direction infinite octahedral walls that are three octahedra wide (Fig. 54b). In the (010) plane, the units share their corners (Fig. 54c), resulting in rectangular tunnels that consist of three face-sharing perovskite tunnels, called "3P." In these tunnels, two perovskite cages out of three are occupied by the A cations, which are displaced toward each other with respect to the ideal positions (Fig. 54c).[60] A "4P" structure has been prepared by soft chemistry; the host lattice of this phase, $K_2Ti_8O_{17}$, consists of units of 2 × 4 edge-sharing octahedra. All these oxides are insulators, owing to the d^0 configuration of Ti(IV).

Based on the B_4O_{20} units present in the rutile structure, one can generate new structures with larger tunnels. This is the situation in hollandites (A_xMnO_2) and in substituted titanates $A_x(Ti_{1-y}B_y)O_2$, where B is a divalent or a trivalent ion and A = Ba, K, Rb, Tl, Pb. The hollandite structure (Fig. 55) is tetragonal ($a \approx 9.8$ Å, $c \approx 2.8$ Å) and can be derived from the rutile structure by edge-sharing of the octahedra in the B_4O_{20} units. This results in large square tunnels that generally are less than half-occupied, x being close to 0.25 in the mineral. The

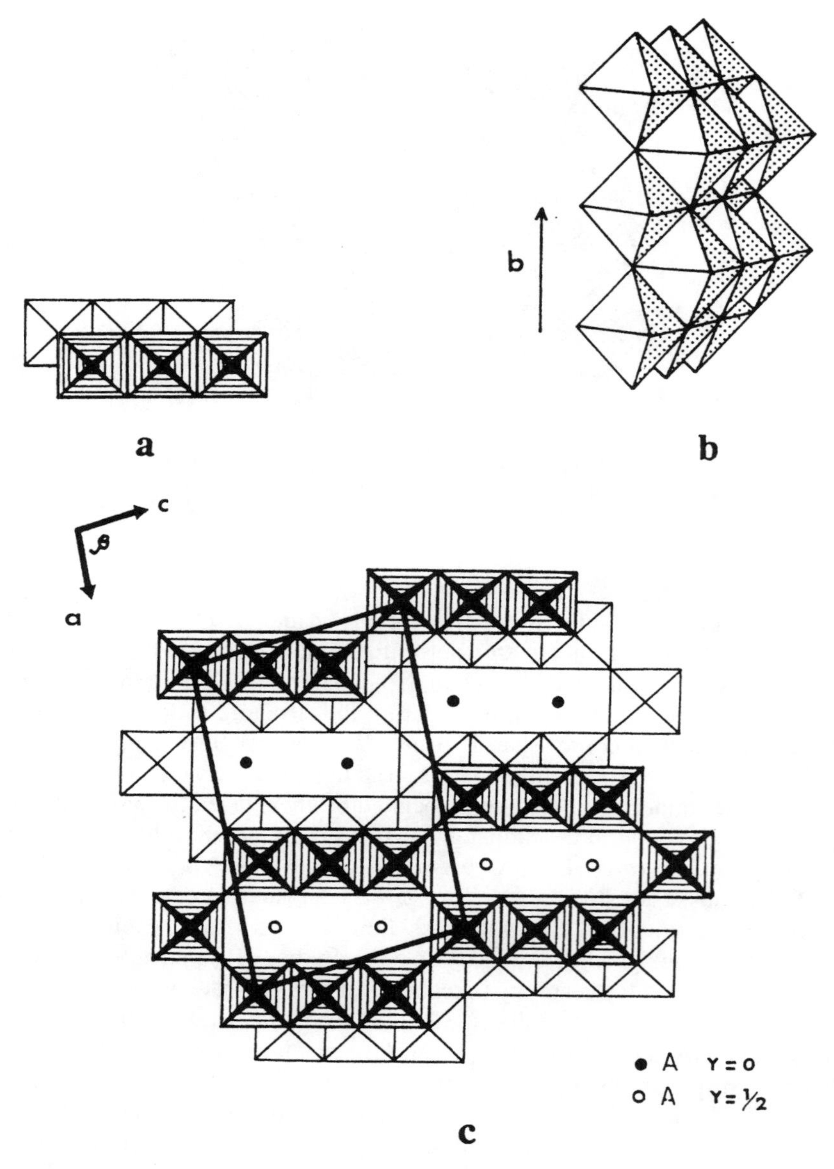

Figure 54. Structure of the $K_2Ti_6O_{13}$-type oxides. (a) structural units of 2 × 3 edge-sharing octahedra, (b) infinite walls of edge-sharing octahedra three octahedra wide, and (c) projection onto (010) showing the rectangular tunnels. (After Andersson and Wadsley.[60])

Figure 55. Projection of the hollandite structure A_xBO_2 onto the (001) plane. The BO_6 octahedra (lozenges) form B_4O_{20} units that share edges. It results in large square tunnels where the A cations (circles) are located. (After Wadsley.[37])

tunnels can be empty or partially occupied by water molecules, as in α-MnO_2. Some oxyhydroxides adopt this structure, an example being FeOOH; its composition is not well defined and could involve the presence of water or even of chlorine in the tunnels.

The structures of ramsdellite and psilomelane are closely related to those of rutile and hollandite. The ramsdellite structure is characterized by an orthorhombic cell ($a \approx 4.5$ Å, $b \approx 10$ Å, $c \approx 2.9$ Å). Its host lattice results from that of rutile or of hollandite, by replacing the B_4O_{20} unit by the B_8O_{34} unit. Such units (Fig. 56a) are obtained by replacing each octahedron of the B_4O_{20} unit by two edge-sharing octahedra. The $[BO_2]_\infty$ host lattice of the ramsdellite (Fig. 56b) is obtained by the association of the B_8O_{34} units by their corners along two directions in the (001) plane, and by their edges along $\vec{\mathbf{c}}$, forming rectangular tunnels (Fig. 56b). This structure, first observed in γ-MnO_2 with empty tunnels, is stabilized by the incorporation of small cations such as lithium. Nonstoichiometric ramsdellites, where lithium is distributed in the octahedral sites and in the tunnels simultaneously, have been synthesized. This is the situation in complex oxides such as in one of the forms of $LiFeSnO_4$, $Li_{1+x}(Li_{2x/3}Fe_{1-x}Sn_{1+x/3})O_4$ ($0 \leq x \leq 0.25$), $Li_{0.8}(Li_{0.29}Ti_{1.4})O_4$, and $Li_3FeSb_2O_8$. In these oxides, the lithium ions that sit in the tunnels exhibit a tetrahedral coordination. There is a statistical distribution of lithium in the two kinds of octahedral site in the first two oxides, and an ordered distribution in $Li_3FeSb_2O_8$ (leading to a tripling of the c parameter, hence the name triramsdellite). The structure of psilomelane, $Ba_xMnO_2(0.4 - x)H_2O$, with $0.1 < x < 0.15$; is also characterized by B_4O_{20} units. The monoclinic cell ($a \approx 9.6$ Å, $b \approx 2.9$ Å, $c \approx 13.9$ Å, $\beta \approx 92°$) is related to the structures of hollandite and ramsdellite. However, the angles of the tunnels differ from 90°.

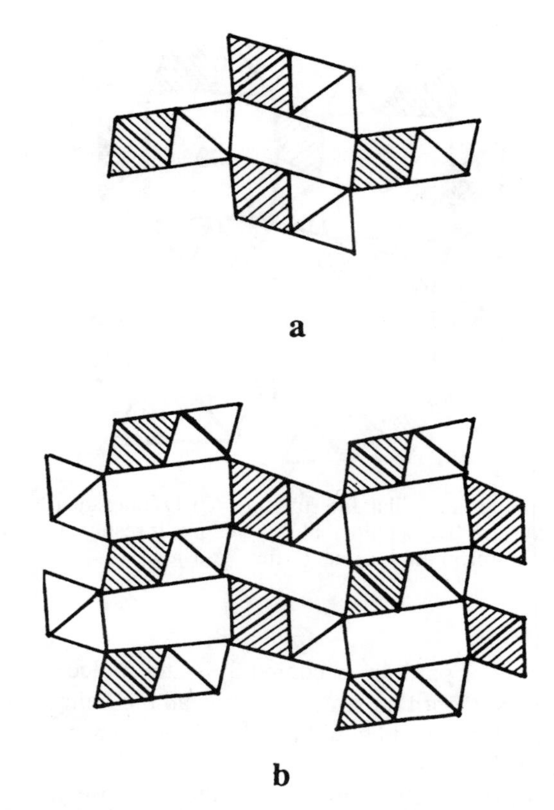

Figure 56. Structure of ramsdellite. (a) B_8O_{34} units and (b) projection of the structure onto (001). (After Wells.[4])

The host lattice involving B_4O_{20} units shares the edges along $\vec{\mathbf{a}}$ and $\vec{\mathbf{b}}$, and the edges are linked through an additional octahedron, sharing its edges with two units along $\vec{\mathbf{c}}$. As a result one obtains rectangular tunnels where the Ba^{2+} cations and H_2O molecules are intercalated.

5.2 Tunnel Structures with Angles of 60°–120°

The structure of the hexagonal tungsten bronze (HTB) A_xWO_3 (A = K, Rb, Ti, Cs, In), discovered by Magnéli,[61] is a typical host lattice involving tunnels with 120° angles. The host lattice $[WO_3]_\infty$ (Fig. 57) consists of corner-sharing WO_6 octahedra; the latter share all their corners in the three directions of the hexagonal cell ($a \approx 7.4$ Å, $c \approx 7.6$ Å), resulting in the formation of hexagonal tunnels running along $\vec{\mathbf{c}}$, where the A cations are located. The large size of these tunnels shows that such a structure is readily obtained under normal pressure, for large A cations (with a size equal to or higher than that of potassium). For smaller A

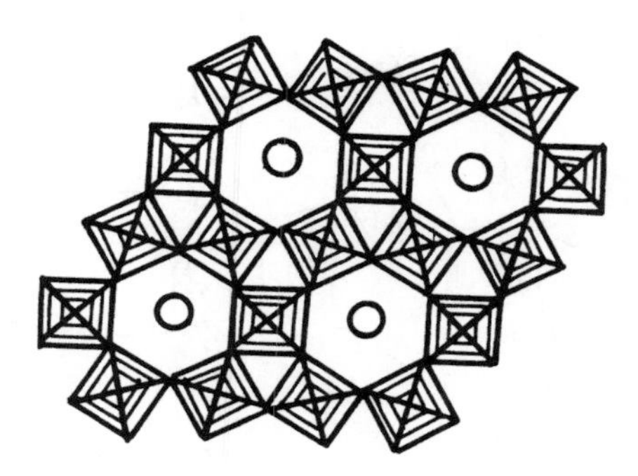

Figure 57. The hexagonal tungsten bronze (HTB)$K_{0.33}WO_3$: projection onto the (001) plane. (After Magnéli.[61])

cations, the structure can be stabilized by application of high pressure ($\sim$ 60 kbar). The A cation is located at the same level as the apical oxygen of the WO_6 octahedra along $\vec{\mathbf{c}}$ and the occupancy factor of the tunnels is variable ($0.20 \leq x \leq 0.33$). For $x = 0.33$, the tunnels are fully occupied, whereas for $x < 0.33$, the A cations are distributed statistically in the tunnels.

Hexagonal bronzes of the type K_xMoO_3 can be synthesized only under high pressures ($\sim$ 65 kbar). Isotypic oxides, but with insulating properties, are obtained by replacing the d^1 element W(V) by a d^0 element such as Nb(V) and Ta(V). This is the case in the bronzoid $A_x(M_{1-x}W_x)O_3$, with M = Ta, Nb and A = K, Rb, Ti, Cs ($0.20 < x \leq 0.33$). An examination of the (001) projection of the structure (Fig. 57) shows the presence of tricapped trigonal prismatic cages, which can accommodate small cations. Making use of this feature, HTB bronzes of the type $Li_yA_xWO_3$, and the bronzoid $K_2Ta_{3.4}O_9$ with HTB structure have been synthesized. The presence of A cations is not absolutely necessary for the stability of the HTB framework. Thus, $(MoW_{11})O_{30}$ and $(MoW_{14})O_{45}$ with similar structures have been obtained in the form of whiskers. Although it is not a tunnel structure, UNb_3O_{10} is closely related to the HTBs. Its orthorhombic cell ($a \approx a_H$, $b \approx a_H\sqrt{3}$, $c \approx 2c_H$) results from the stacking of the HTB octahedral layers, shifted by $a/2$ with respect to each other along the three directions of the hexagonal cell, $\langle 100\rangle$, $\langle 010\rangle$, and $\langle 110\rangle$. Consequently, the hexagonal tunnels are suppressed, one NbO_6 octahedron being located on either side of a hexagonal window. In this structure, the uranium cation is located in the plane of the hexagonal window, sandwiched between two oxygen atoms along $\vec{\mathbf{c}}$, so that it forms an hexagonal bipyramid UO_8. Along the $\vec{\mathbf{c}}$ direction, successive NbO_6 octahedra are turned by 60°, explaining the quadrupling of the c parameter.

By shearing the HTB structure along $\vec{\mathbf{c}}$ at the level of the apical corners of the WO_6 octahedra, one obtains HTB slices shifted with respect to each other, as shown for the tungstate $Tl_2W_4O_{13}$ (Fig. 58) with an orthorhombic cell ($a \approx 7.3$ Å, $b \approx 37.9$ Å, $c \approx 3.8$ Å).[62] This structure consists of HTB layers shifted by $c/2$ alternately, the cohesion between such layers being ensured by Tl^+ cat-

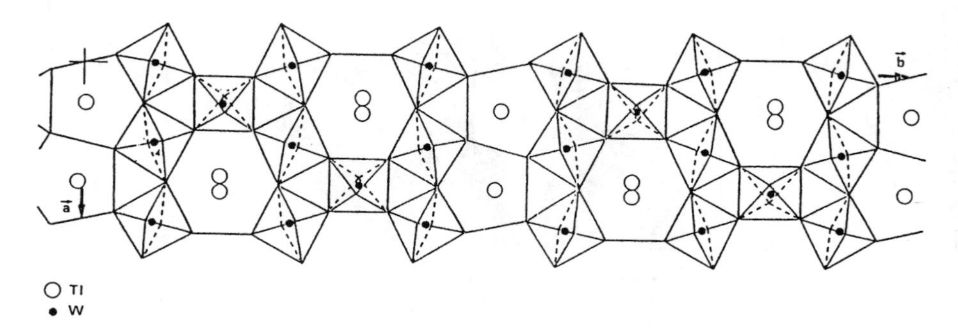

Figure 58. Projection of the structure of $Tl_2W_4O_{13}$ onto the (001) plane. The tungsten atoms (small solid circles) form octahedra that share corners, resulting in HTB slabs interleaved with Tl^+ cations (open circles). (After Goreaud et al.[62])

ions. This arrangement has the character of tunnel as well as of layer structures. The HTB $[WO_3]_\infty$ framework can be associated with a diamond-shaped WO_3 framework, leading to the formation of structures called intergrowth tungsten bronzes (ITBs). These bronzes can be described as the intergrowth of HTB slices with distorted ReO_3-type slices (Fig. 59).[63] Numerous oxides belong to this family, including A_xWO_3 bronzes ($x < 0.10$), $ACu_3M_7O_{21}$, $Ca_2TlNb_5O_{15}$, $UMo_{20}O_{64}$, and UMo_2O_8.

The oxides $A_xV_xMo_{1-x}O_3$ (A = K, Rb, Cs; $0 < x \leq 0.166$) are closely related to the HTBs. Their hexagonal host lattice ($a \approx 10.5$ Å, $c \approx 30.7$ Å) consists of triple rows of corner-sharing octahedra running along $\vec{\mathbf{c}}$ just as in the HTBs (Fig. 60a).[64] Unlike the HTBs, these rows are shifted by $c/2$ alternately and share edges laterally. The tunnels are consequently less numerous than in the HTBs; accordingly, $x = 0.166$ corresponds to the full occupation of the tunnels. $KMo_5O_{15}(OH)\cdot 2H_2O$ is isotypic with $A_xV_xMo_{1-x}LO_3$, but the oxygen atoms are partly replaced with OH groups and by H_2O molecules and the vacancies at the molybdenum sites are distributed in an ordered manner. Also related to the

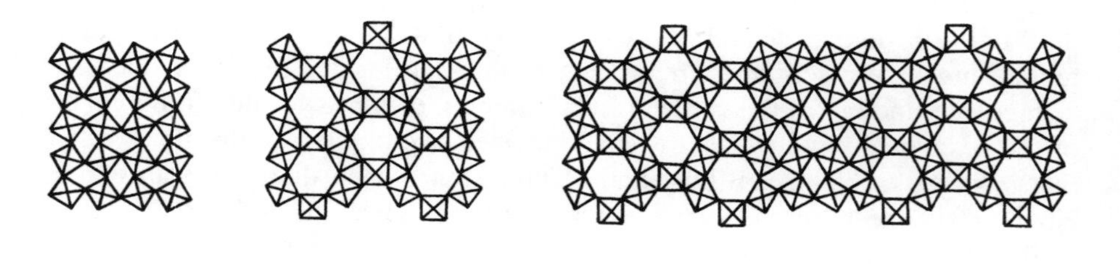

Figure 59. Association of (a) diamond-shaped ReO_3-type slices and (b) HTB slices leading to (c) the ITB structure. (After Kihlborg and Sharma.[63])

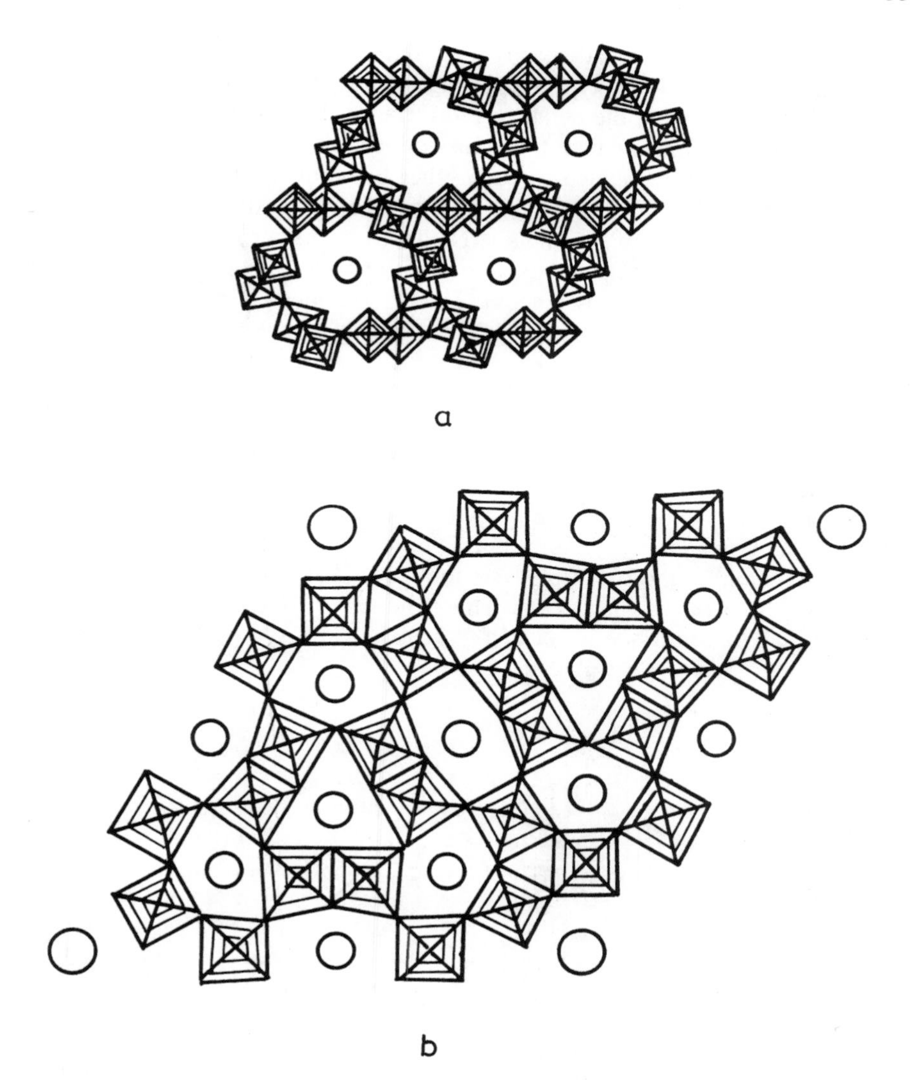

Figure 60. (a) Hexagonal $A_x(V_xMo_{1-x})O_3$: view along $\vec{\mathbf{c}}$ (After Galy and Darriet.[64]) (b) Structure of $BaTa_2O_6$: view along $\vec{\mathbf{c}}$. (After Layden.[65])

HTB's is $BaTa_2O_6$, with a hexagonal cell ($a \approx 21.1$ Å, $c \approx 3.9$ Å). Its host lattice (Fig. 60b)[65] has hexagonal rings of corner-sharing TaO_6 octahedra as in HTBs, but the rings share edges instead of octahedral corners. As a result, one observes besides the HTB hexagonal tunnels, a second kind of tunnel with a triangular shape. Both types of tunnel are fully occupied by Ba^{2+} ions. No deviation from stoichiometry is evidenced in this insulating phase.

The tantalate KTa_5O_{13} with an orthorhombic structure ($a \approx 5.6$ Å, $b \approx 10.7$

Å, $c \approx 16.8$ Å) has a host lattice (Fig. 61a) built up of corner- and edge-sharing octahedra forming α-PbO_2-type slices (Fig. 61b).[66] The α-PbO_2 type slices are linked by the corners of the octahedra to form (at the junction of two adjacent slices), six-sided tunnels where the potassium cations are located. Two successive α-PbO_2 slices are related to each other by a mirror plane, and thus this structure can be described as a chemical twin of the α-PbO_2 structure. Also characterized by hexagonal tunnels is $Na_2Ti_9O_{19}$, with a monoclinic cell ($a \approx$ 12.2 Å, $b \approx 3.8$ Å, $c \approx 15.3$ Å). Its host lattice consists of Na_xTiO_2 layers linked to each other through TiO_6 octahedra. The Na^+ cations are present in the hexagonal tunnels, at the junction between two Na_xTiO_2-type layers.

Although they do not form large tunnels, diaspore and aeschynite can be described as having tunnels with 60–120° angles. The structure of diaspore, α-AlO(OH) (Fig. 62a),[4] with an orthorhombic structure ($a \approx 4.4$ Å, $b \approx 10$ Å, $c \approx 2.8$ Å), is derived from that of ramsdellite (see Fig. 56) by a compression along one direction, leading to a flattening of the tunnels, which are significantly smaller but large enough to accept hydrogen. Such a structure has also been observed in α-FeOOH (geoethite) but has not been found in synthetic oxides. An examination of the (010) planes shows that diaspore has a nearly perfect close packing of the oxygen atoms. This is however not the case of aeschynite, $CaTa_2O_6$, whose orthorhombic cell ($a \approx 11$ Å,$b \approx 7.5$ Å, $c \approx 5.4$ Å) has tunnels

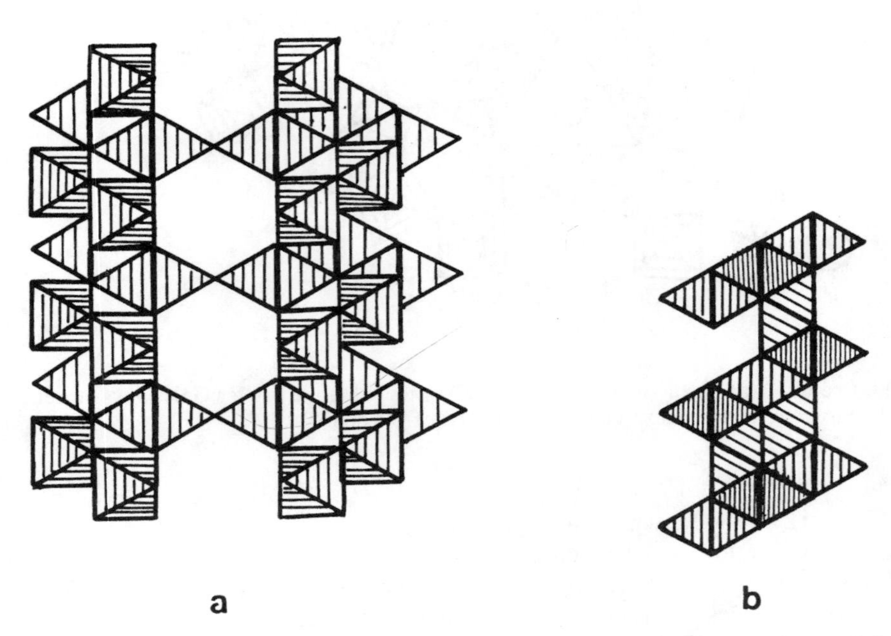

Figure 61. (a) Projection of the structure of KTa_5O_{13} onto (010) and (b) the α-PbO_2 slices viewed along $\vec{\mathbf{a}}$. (After Awadalla and Gatehouse.[66])

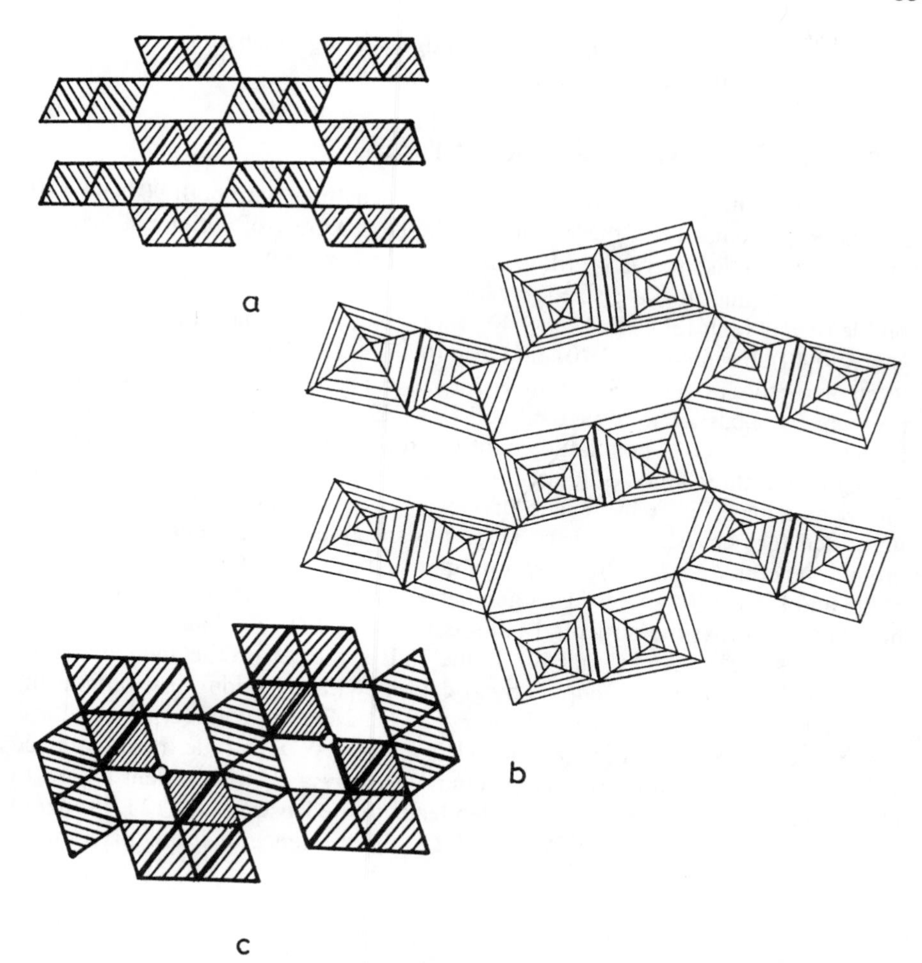

Figure 62. (a) Diaspore, α-AlOOH: projection of the structure onto (010). (After Wells.[4]) (b) Projection of the structure of aeschynite $CaTa_2O_6$ along $\vec{b}$ showing the edge-sharing TaO_6 octahedra forming tunnels where the Ca^{2+} ions are located. (After Jahnberg.[67]) (c) Projection onto the (100) plane of the hollandite-distorted framework of $Ba_2Ti_9O_{20}$. One recognizes the distorted hollandite tunnels built up from edge-sharing TiO_6 octahedra (hatched lozenges) that are obstructed by two octahedra (darker lozenges). (After Fallon and Gatehouse.[68])

that are similar but larger (Fig. 62b),[67] where the Ca^{2+} cations are located with a bicapped trigonal-prismatic coordination.

The distorted hollandite type structure of $Ba_2Ti_9O_{20}$ is not a tunnel structure, but it should be considered here. It has a $[Ti_9O_{18}]_\infty$ framework similar to that of hollandite (see Fig. 55) but forms tunnels with angles of 60–120° instead of 90°

(Fig. 62c).[68] The presence of additional edge-sharing octahedra in the tunnels leads to a cage structure of this oxide.

5.3 Structures with Pentagonal Tunnels

Most oxides with pentagonal tunnels involve structures with both 90° and 120° angles. Representative of these oxides is the large family of tetragonal tungsten bronzes (TTB) discovered by Magnéli. These bronzes with the formula A_xWO_3 are found for intermediate-sized A cations (A = K, In, Na, Ba, Pb). The magnitude of x is high for K and In ($0.40 < x \leq 0.60$), but small for sodium ($0.30 < x < 0.40$), barium ($x \approx 0.20$), and lead ($0.17 < x < 0.35$). The corresponding tetragonal bronzes of molybdenum, K_xMoO_3, can be synthesized only under high pressure (65 kbar). The tetragonal cell ($a \approx 12$ Å, $c \approx 3.8$ Å $\approx a_p$) of these bronzes has a $[WO_3]_\infty$ host lattice built up from corner-sharing WO_6 octahedra, giving rise to three kinds of tunnel: perovskite with a square section, triangular, and pentagonal (Fig. 63a).[69] The first two types of tunnel have already been described for perovskites and the HTB's, respectively. The pentagonal tunnels are not generally characterized by a regular section, but instead exhibit two 90° angle and three 120° angles. This geometry results from the composition of the host lattice: perovskite structural units B_4O_{20} (Fig. 63b) (90° angles) and HTB B_3O_{15} units (60° angles) (Fig. 63c). In the TTB's, the perovskite and pentagonal tunnels can be partially occupied by A cations with a random distribution; the maximum occupancy is obtained for $x = 0.60$. Metallic conductivity of these compounds is explained, just as in HTB's and Na_xWO_3, by the mixed valence of tungsten. In the same way, replacing W(V) by a d^0 pentavalent cation to eliminate the mixed valence of tungsten leads to the isostructural TTB bronzoid $K_xB^V_xW^{VI}_{1-x}O_3$ (B = Nb, Ta). As in HTB's, the existence of triangular tunnels

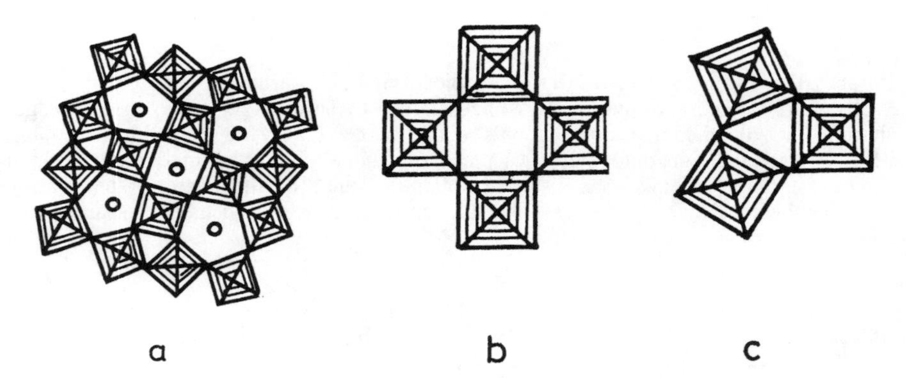

Figure 63. (a) Structure of the tetragonal tungsten bronze (TTB), K_xWO_3: projection onto the (001) plane. (After Magnéli.[69]) (b) The octahedral perovskite units B_4O_{20} and (c) HTB units (B_3O_{15}) in the TTBs.

allows small cations with a tricapped trigonal prismatic coordination to be intercalated. This leads to $K_xLi_yWO_3$ bronzes and the $K_6Ta_{10.80}O_{30}$ bronzoid, where the additional lithium and tantalum atoms occupy prismatic cages.

Numerous tantalates and niobates of divalent cations adopt the TTB structure. This is the case in bronzoids, AB_2O_6 (A = Sr, Ba, Pb; B = Nb, Ta), in which the perovskite and the pentagonal tunnels are partially occupied by the A cations. The simultaneous presence of monovalent and divalent A cations allows one to synthesize TTB bronzoids with fully occupied tunnels according to the formula $A_4A'_2B_{10}O_{30}$ (B = Nb, Ta; A = Ba, Sr; A′ = Na, K). A number of these niobates exhibit a tetragonal or an orthorhombic cell derived from TTBs ($a \approx b \approx a_{TTB}\sqrt{2}$, $c \approx c_{TTB}$). Small atomic displacements with respect to the ideal TTB structure are found, giving rise to interesting phenomena and superstructures. The niobates are important compounds owing to ferroelectric properties resulting from a rattling of niobium inside the NbO_6 octahedra. This is the case in the famous "banana," $Ba_4Na_2Nb_{10}O_{30}$, extensively studied for its ferroelectric and nonlinear optical properties. The TTB structure can be stabilized by an oxygen or a fluorine atom instead of an A cation. Such phases are obtained in Nb_2O_5/WO_3 and Ta_2O_5/WO_3, the most representative oxides being $Nb_{16}W_{18}O_{94}$ and $Ta_4W_7O_{31}$. $Nb_{16}W_{18}O_{94}$ has an orthorhombic cell ($a \approx a_{TTB}$, $b \approx 3a_{TTB}$, $c \approx c_{TTB}$), and its host lattice $[B_{30}O_{90}]$ is the distorted TTB framework (Fig. 64),[70] with flattened perovskite cages. It differs from the TTBs by having no cations in the perovskite tunnels and by exhibiting an ordered distribution of the Nb—O—Nb chains in the pentagonal tunnels leading to the formulation $(NbO)_4(B_{30}O_{90})$ with B = Nb, W. Indeed 4 pentagonal tunnels out of 12 are occupied by the Nb—O chains in an ordered manner. The niobium atoms of these chains are located at the same level as the niobium and tungsten atoms of the BO_6 octahedra, whereas the oxygen atoms take the place of A cations, so that one obtains NbO_7 pentagonal bipyramids in the tunnels. A similar structure

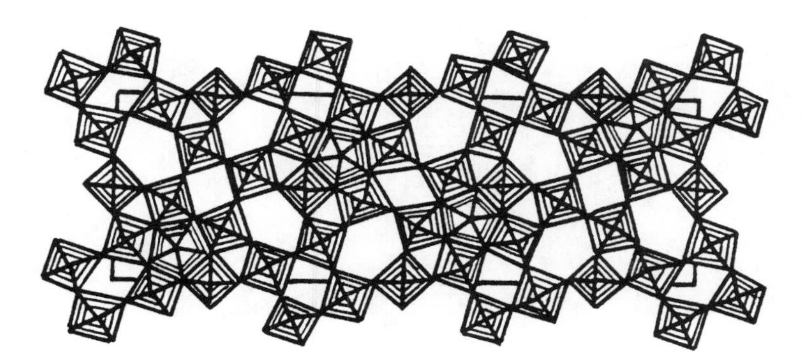

Figure 64. Projection of the structure of the bronzoid $Nb_{16}W_{18}O_{94}$ onto the (001) plane. The $Nb(W)O_6$ octahedra (hatched squares) form pentagonal tunnels that are partly filled with Nb—O—Nb chains forming NbO_7 bipyramids (hatched). (After Sleight.[70])

is found in $Ta_4W_7O_{31}$ with an ordered occupancy of 4 pentagonal tunnels out of 16 in the tetragonal cell ($a \approx 2a_{TTB}$, $c \approx c_{TTB}$) by Ta—O chains, leading to the formula $(TaO)_4(B_{40}O_{120})$. The simultaneous presence of A cations in the perovskite tunnels and in the pentagonal tunnels, and of B—O chains in the pentagonal tunnels, has been observed in several structures. This is the case in the bismuth niobate $Bi_6Nb_{34}O_{94}$, whose orthorhombic structure derives from that of $Nb_{16}W_{18}O_{94}$ by partially replacing the Nb-O chains by bismuth. In this respect, the pseudoternary system $WO_3/B_2O_5/AB_2O_6$ (A = Ba, Sr, Pb; B = Ta, Nb) is very rich, having a wide homogeneity range $A_\alpha(BO)_\beta BO_3$ (Fig. 65),[71] with various possible superstructures and defects.

The bronze $Ba_{0.15}WO_3$, having an orthorhombic structure ($a \approx 8.9$ Å, $b \approx 10$ Å, $c \approx 3.8$ Å), is closely related to TTBs in that it has the $[WO_3]_\infty$ host lattice of corner-sharing WO_6 octahedra forming similar pentagonal tunnels (Fig. 66a),[72] characterized by 90° and 120° angles. This bronze is also related to the TTB's by the existence of 60°-type W_3O_{15} units that share their corners in the (001) plane. The niobate and the tantalate having the formula $KCuB_3O_9$ (B = Nb, Ta) are also orthorhombic but differ from $Ba_{0.15}WO_3$ in that there is a doubling of the c parameter as a result of the tilting of the BO_6 octahedra with respect to $\vec{\mathbf{c}}$ (which allows a square-planar coordination for Cu). The BO_3 host lattice (Fig. 66b)[73] is similar to that of $Ba_{0.15}WO_3$, except for the tilting of the octahedra. In both the structures, the cations (Ba^{2+} and K^+) are in pentagonal tunnels, but the pentagonal tunnels are less than half-occupied in the case of barium and fully occupied in the case of potassium. The Cu^{2+} cations are located in the diamond-shaped perovskite cages (this difference from $Ba_{0.15}WO_3$ arises from Jahn–Teller distortion in the case of copper), exhibiting considerable similarity to the $CaCu_3B_4O_{12}$ perovskites.

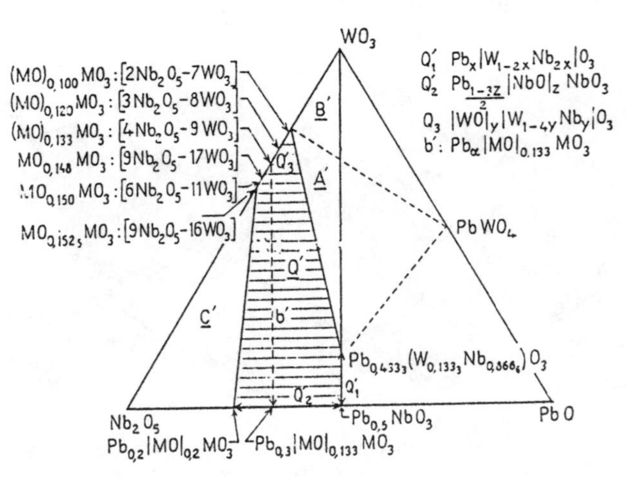

Figure 65. Homogeneity range of the nonstoichiometric bronzoids, $A_\alpha(BO)_\beta BO_3$: example of the system $WO_3/Nb_2O_5/PbO$. (After Le Parmentier.[71])

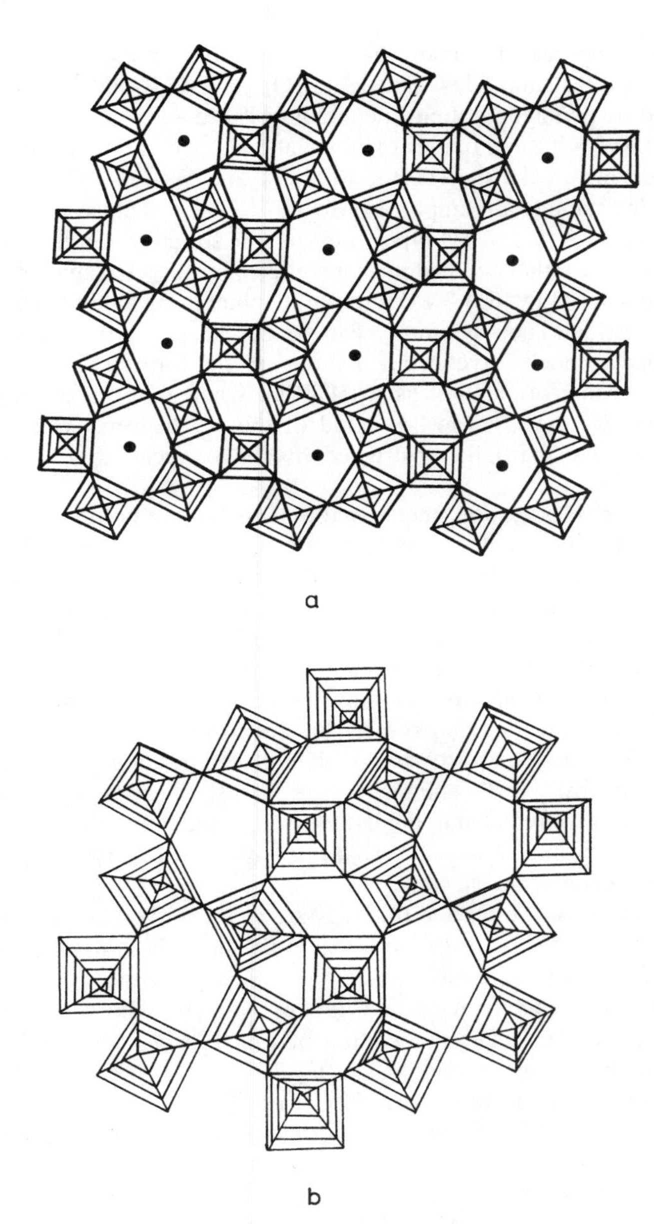

Figure 66. (a) Projection of the structure of the bronze $Ba_{0.15}WO_3$ along $\vec{\mathbf{c}}$. The WO_6 octahedra (hatched) form pentagonal tunnels occupied by barium cations (circles). (After Michel et al.[72]) (b) Structure of $KCuNb_3O_9$: view along $\vec{\mathbf{c}}$. The NbO_6 octahedra are tilted with respect to $Ba_{0.15}WO_3$. The K^+ ions in pentagonal tunnels and copper in square-planar coordination are not represented. (After Groult et al.[73])

Pentagonal bipyramidal coordination of molybdenum and tungsten is observed in several suboxides. Thus, $Mo_{10}O_{28}$, $W_{18}O_{49}$, and $Mo_{17}O_{47}$ are all characterized by pentagonal tunnels containing "Mo—O" or "W—O" chains (see Fig. 67).[74,75] All three oxides exhibit flattened perovskite tunnels, just like the TTB bronzoids. Like the HTB's, $Mo_{10}O_{28}$ and $W_{18}O_{49}$ exhibit hexagonal tunnels, but the tunnels are empty. In $Mo_{10}O_{28}$, which is closely related to the TTB's, one observes wide TTB-type regions in the structure. Unlike other tunnel structures, some of the octahedra share edges in $W_{18}O_{49}$ and $Mo_{17}O_{47}$. In addition, $Mo_{17}O_{47}$ (Fig. 68)[76] exhibits large tunnels of a kind not observed in any other oxide. Their geometry can be described on the basis of HTB-type hexagonal tunnels: five octahedra out of six form the hexagonal ring (i.e., keep the same orientation as in HTB's), whereas the sixth edge of the ring is formed by the O—Mo—O bond of the octahedron. As a result, the hexagonal tunnel is partially obstructed by the oxygen atom of the MoO_6 octahedron.

Some of the oxides exhibit pentagonal tunnels but are not closely related to the TTB's or the HTB's. Two examples are the titanoniobates and tantalates, ATi_3BO_9 (A = K, Rb, Tl; B = Nb, Ta), and oxides of the $CaFe_2O_4$ and $CaTi_2O_4$ series. The substituted titanates, ATi_3MO_9, which have an orthorhombic cell ($a \approx 6.5$ Å, $b \approx 3.9$ Å, $c \approx 15$ Å) are built up of units of 2 × 2 edge-sharing octahedra (Fig. 69a). Along $\vec{\mathbf{b}}$, such units share their edges, forming infinite columns, whereas they share octahedral corners in the (010) plane (Fig. 69b).[77] Although the structure is stabilized by the full occupancy of the pentagonal tunnels by the A cations, an isotypic oxide with empty tunnels, $Ti_2Nb_2O_9$, has been synthesized by "soft" chemistry (see Part III, Section 1). Lithium and sodium have been intercalated in the latter oxide, using butyllithium and sodium naphthalide. The bronzes $Na_xTi_2Nb_2O_9$ ($x \approx 0.30$) and $Li_xTi_2Nb_2O_9$ ($0.15 \leq x < 1$) have thus been synthesized.

The structures of $CaFe_2O_4$ and $CaTi_2O_4$ are characterized by infinite ribbons of edge-sharing octahedra similar to those in rutile.[78] These rows of octahedra are associated two by two, just as ramsdellite and diaspore (Fig. 70a). The double octahedral ribbons share their corners to form tunnels whose section can be described as pentagonal. The association of four blocks of two edge-sharing octahedra can be made in two ways (Fig. 70b,c), affording $CaTi_2O_4$ and $CaFe_2O_4$. Both structures are orthorhombic ($a \approx 9$ Å, $b \approx 3$ Å, $c \approx 10$–11 Å) but differ from each other in the mode of linking of the double octahedral ribbons. In $CaFe_2O_4$, two successive double rows share their corners along $\vec{\mathbf{c}}$ (Fig. 70d) whereas in $CaTi_2O_4$, two successive double rows share their edges along $\vec{\mathbf{b}}$ (Fig. 70e). Numerous oxides exhibit the $CaFe_2O_4$ structure: $SrFe_2O_4$, β-$CaCr_2O_4$, CaV_2O_4, ASc_2O_4 (A = Ca, Mg, Sr), Eu_2SrO_4, AIn_2O_4, (A = Ca, Sr), $CaYb_2O_4$, $NaBB''O_4$ (B = Sc, Fe; B″ = Ti, Sn, Zr, Hf), BaB_2O_4 (B = Ln: Nd–Lu), and SrB_2O_4 (B = Y, La–Ho). Compounds having this structure are stabilized under high pressure (e.g., $NaAlGeO_4$) or by hydrothermal synthesis (e.g., $Na_{0.56}Fe_{0.28}Ti_{1.72}O_4$). In both structural types, the A cation exhibits the same coordination forming a bicapped or tricapped trigonal prism.

a

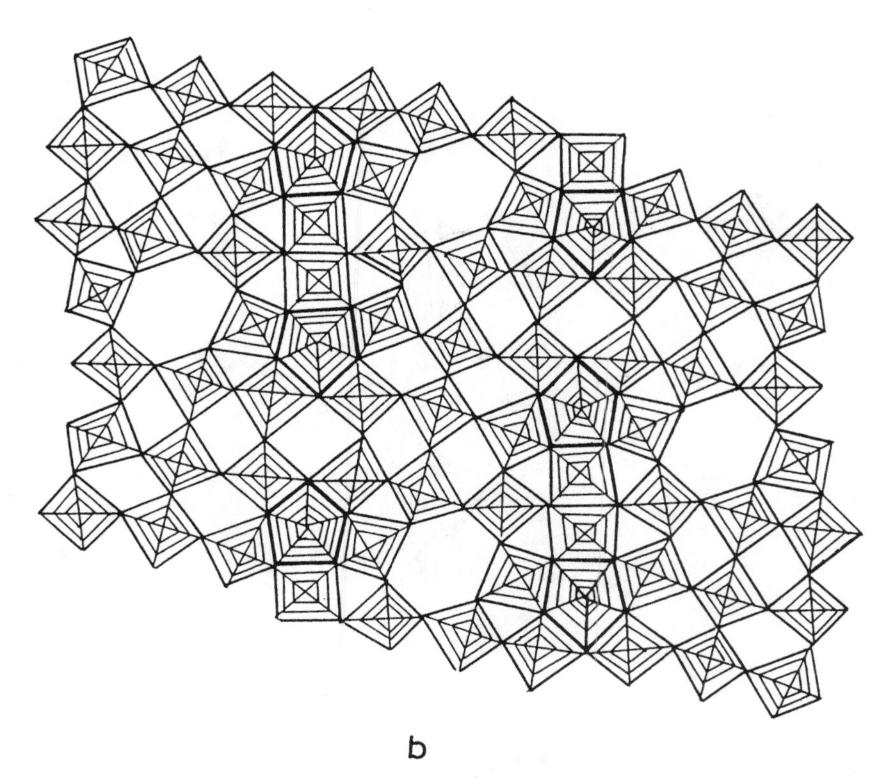

b

Figure 67. (a) Structure of $Mo_{10}O_{28}$: view along $\vec{c}$. (After Wadsley and Kihlborg.[74]) (b) Structure of the oxide $W_{18}O_{49}$: view along $\vec{b}$. (After Magnéli.[75])

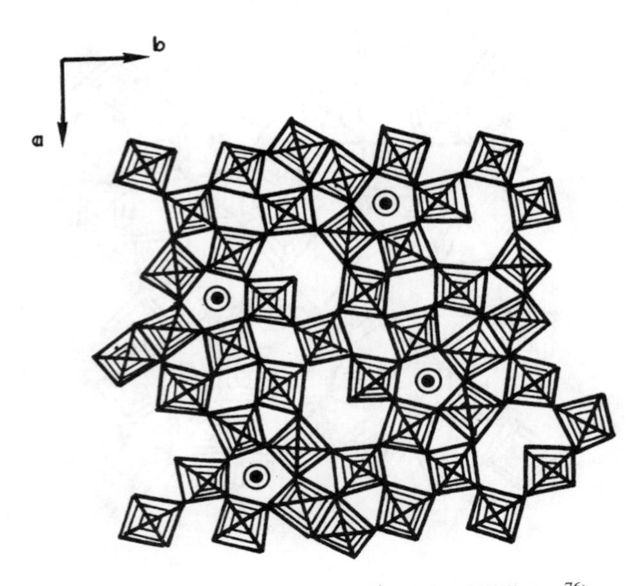

Figure 68. Structure of $Mo_{17}O_{47}$: view along $\vec{c}$. (After Kihlborg.[76])

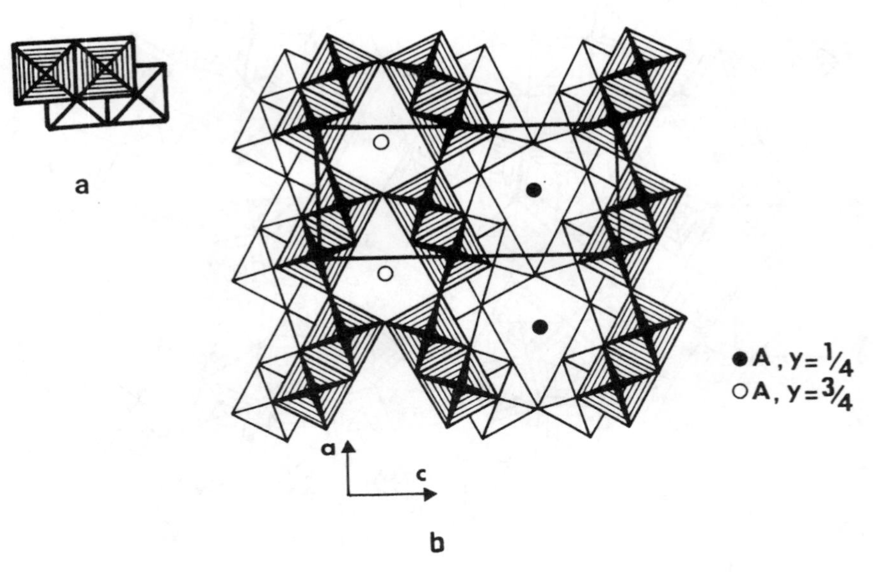

Figure 69. KTi_3NbO_9 structure: (a) units of 2 × 2 edge-sharing octahedra and (b) projection of the structure onto the (010) plane. (After Wadsley.[77])

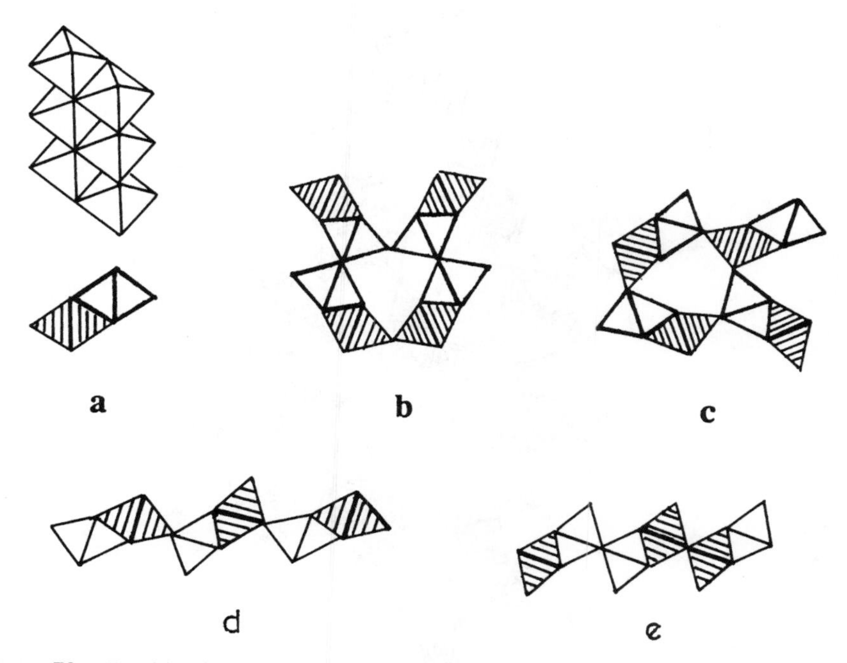

Figure 70. Double octahedral ribbons (a) forming $CaFe_2O_4$ and $CaTi_2O_4$ structures and their association leading to two different pentagonal tunnels in $CaTi_2O_4$ (b) and in $CaFe_2O_4$ (c). Association of the double octahedral ribbons by corner-sharing along $\vec{\mathbf{c}}$ in $CaFe_2O_4$ (d) and by edge-sharing along $\vec{\mathbf{b}}$ in $CaTi_2O_4$ (e). (After Reid et al.[78])

5.4 Other Examples of Octahedral Structures with Large Tunnels

It is difficult to present systematically all the different structure types characterized by a complex geometry of the tunnels. It should nevertheless be noted that most of the octahedral structures will tend to form tunnels with 90° or 60–120° angles. The tungstates $K_2W_3O_{10}$ and $K_2W_4O_{13}$ are two examples of frameworks forming large tunnels. In $K_2W_3O_{10}$, with a monoclinic structure ($a \approx 10.9$ Å, $b \approx 3.8$ Å, $c \approx 32$ Å, $\beta \approx 108°$), one observes (Fig. 71)[79] that the tunnels result from the association of several pentagonal and hexagonal tunnels. In the same way, $K_2W_4O_{13}$[80] with a hexagonal structure ($a = 15.53$ Å, $c = 3.85$ Å) exhibits, in addition to HTB-tunnels, large tunnels corresponding to the association of three pentagonal tunnels.

$Na_xTi_{2-x}Fe_xO_4$ ($0.15 < x < 0.90$), with an orthorhombic structure ($a \approx 9.2$ Å, $b \approx 3$ Å, $c \approx 11.3$ Å), has a host lattice close to that of $CaTi_2O_4$ or $CaFe_2O_4$. Its tunnels also form angles close to 60° and 120°, but the octahedra are more condensed, forming units of four edge-sharing octahedra. $ATi_2B_5O_{17}$ (A = Na,

Figure 71. $K_2W_3O_{10}$: projection of the structure onto (010). (After Okada et al.[79])

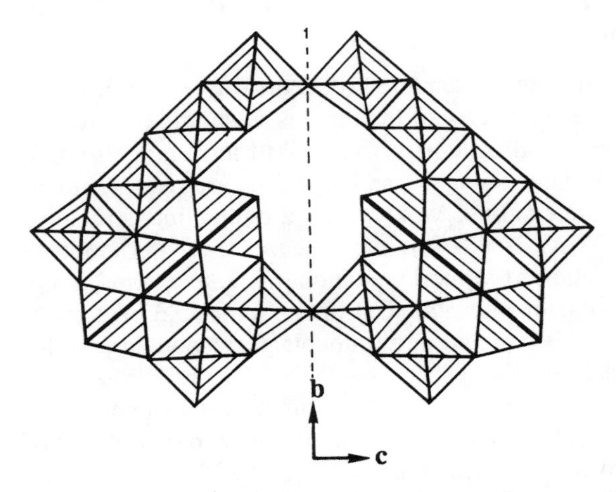

Figure 72. $KTi_2Ta_5O_{17}$ structure: view along $\vec{a}$. (After Gatehouse and Nesbit.[81])

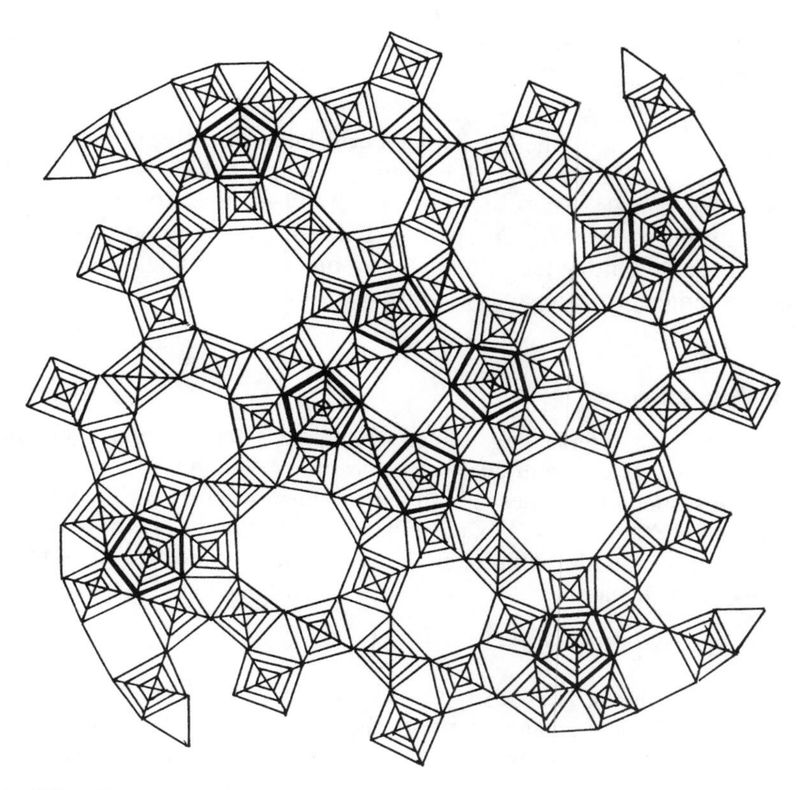

Figure 73. Structure of the GTB, $Rb_3Nb_{54}O_{146}$: projection onto (001). (After Gatehouse.[81])

K; B = Nb, Ta), with an orthorhombic structure ($a \approx 6.6$ Å, $b \approx 8.9$ Å, $c \approx 21.4$ Å), has tunnels of unusual shape but with 60° and 90° angles and a pseudohexagonal character, as shown on Figure 72.[81] The host lattice consists of rutile slices which are derived from each other by a mirror plane. These structures can be described as chemical twins of the rutile structure.

$Rb_3Nb_{54}O_{146}$ and $TlNb_7O_{18}$ are known as Gatehouse tungsten bronzes (GTBs). These compounds with a tetragonal cell ($a \approx 27.5$ Å, $c \approx 3.9$ Å) are the only octahedral oxides with very large tunnels. Their host lattice (Fig. 73)[81] is also related to the HTBs and TTBs. One recognizes the B_3O_{15} and B_4O_{20} octahedral units, forming HTB tunnels (where Rb^+ and Tl^+ ions are located), pentagonal tunnels occupied by Nb—O chains (NbO_7 pentagonal bipyramids), and empty perovskite tunnels. The NbO_6 octahedra are distorted, however, and the angles between the octahedral units deviate from 60° to 120° in order to form heptagonal tunnels containing monovalent cations.

6 Octahedral Intersecting Tunnel Structures: Pyrochlores and Relatives

In the pyrochlore oxides, the host lattice forms large cages resulting from intersecting tunnels; there are large deviations in the stoichiometry of the cations present in the tunnels. The cations are generally mobile, giving rise to ionic conductivity and ion exchange properties. These properties are somewhat similar to those of zeolites, although the tunnels are smaller; reversible hydration occurs in these materials just as in zeolites. (For a review see ref 82–83).

6.1 Pyrochlores

A large family of intersecting tunnel structures is that of nonstoichiometric pyrochlores, $A_{1+x}B_2O_6$, with cubic structure ($a \approx 10.5$ Å). They are prepared from the corresponding oxides at high temperatures when A is K, Rb, Cs, or T1 and B is (mainly) Nb, Ti, W, Mo, or Sb. Ion exchange properties of these oxides, especially of those with thallium, allow hydronium pyrochlores to be prepared according to the equation:

$$Tl_x(M_{1+x}W_{1-x})O_6 + xH^+_{aq} \rightarrow H_x(M_{1+x}W_{1-x})O_6{\cdot}H_2O + xTl^+_{aq}$$

Hydrated pyrochlores, $ABWO_6{\cdot}H_2O$ (A = Li, Na, Ag; B = Ta, Nb, Sb) are prepared by cationic exchange between the pyrochlore, $HBWO_6{\cdot}H_2O$, and the solution of the corresponding salt. Ion exchange with bivalent cations is more difficult but total exchange is achieved starting from $H_2Ta_2O_6{\cdot}H_2O$, leading to the $ATa_2O_6{\cdot}H_2O$ pyrochlores when A is Ca, Cd, or Pb. Only a partial exchange is obtained when A is Sr or Ba. Solid state ion exchange reactions allow sodium- or potassium-rich pyrochlores to be prepared starting from the thallium pyrochlores (making use of the volatility of thallium chloride):

$$\begin{array}{lcl} TI_{1+x}(Ta_{1+x}W_{1-x})O_6 + (1+x)ACl & \xrightarrow{770\ K} & A_{1+x}(Ta_{1+x}W_{1-x})O_6 + (1+x)TlCl \\ & & \quad\downarrow H_2O \\ & & A_{1+x}(Ta_{1+x}W_{1-x})O_6{\cdot}H_2O \end{array}$$

where A = Na^+, K^+, $0 \leq x \leq 1$.

Stoichiometric pyrochlores $NaTaO_3$ and $KTaO_3$ are stable at room temperature (but are very hygroscopic). They transform to a perovskite structure at high temperatures. Hydronium pyrochlores behave like Brønsted acids and permit the synthesis of $N_2H_5TaWO_6$ and NH_4TaWO_6 pyrochlores by reaction with hydrazine and ammonia, respectively. The $[B_2O_6]$ host lattice of these oxides (Fig. 74) is built up of corner-sharing BO_6 octahedra.[42,83]. One recognizes the B_3O_{15} octahedral units, linked along the $\langle 111 \rangle$ direction through another octahedron (whose ternary axis is parallel to this direction forming B_4O_{18} units), whereas laterally in the (111) plane they share corners forming HTB layers (Fig. 74b).

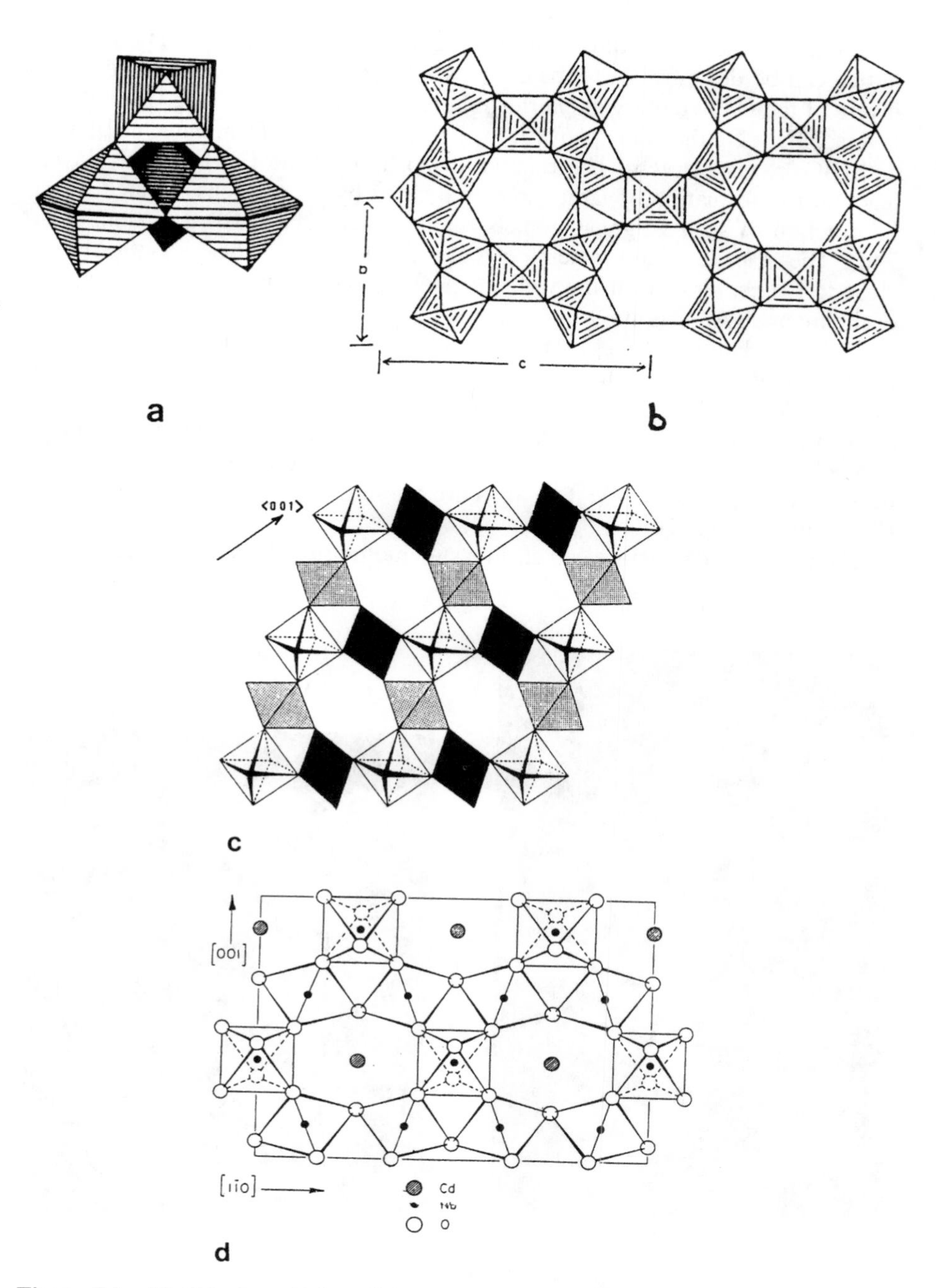

Figure 74. $[B_2O_6]_\infty$ host lattice of the pyrochlores $A_{1+x}B_2O_6$: (a) the octahedral unit B_4O_{18}, (b) HTB $[B_3O_9]_\infty$ layers parallel to $(111)_c$, (c) stacking of HTB $[B_3O_9]_\infty$ and $[BO_3]_\infty$ layers along $\langle 111 \rangle_c$, and (d) projection of the structure along $\langle 110 \rangle$ showing the hexagonal tunnels. (After Raveau[83] and Jona and Shirane.[42])

The latter octahedra are tilted with respect to the HTB structure. The $[B_2O_6]_\infty$ lattice can be described in terms of the stacking of the HTB layers and of the layers of the ⟨111⟩ ternary axis oriented octahedra alternately (Fig. 74c). These layers, parallel to the (111) plane of the cubic cell, can be formulated as $[B_3O_9]_n$ and $[BO_3]_n$, respectively. The relative disposition of the HTB $[B_3O_9]_\infty$ layers leads to the formation of large cavities defined by 18 oxygens, located at the intersection of six hexagonal tunnels running along the ⟨110⟩ directions (Fig. 74d). The A cations and water molecules are located in these cavities either close to the center or near the walls, depending on the size and the magnitude of the intercalated cation.

The host lattice $[B_2O_6]_\infty$ can be itself nonstoichiometric. This has been observed in $Rb_{20+x}(W_4O_6)_{1/3-x/12}W_{32}O_{108}$, which exhibits a tetragonal symmetry ($\mathbf{a} \approx 16$ Å, $\mathbf{c} \approx 10.5$ Å). In this oxide, the $W_{32}O_{108}$ framework forms tunnels (similar to those in the $A_{1+x}B_2O_6$ pyrochlores) that intersect with larger tunnels parallel to $\vec{\mathbf{c}}$. The latter result from the elimination (along $\vec{\mathbf{c}}$) of blocks of four octahedra W_4O_{18}, so that W_3O_{15} units are missing in an ordered fashion in the HTB region (Fig. 75).[84]

The pyrochlore structure can also be stoichiometric. The stoichiometric com-

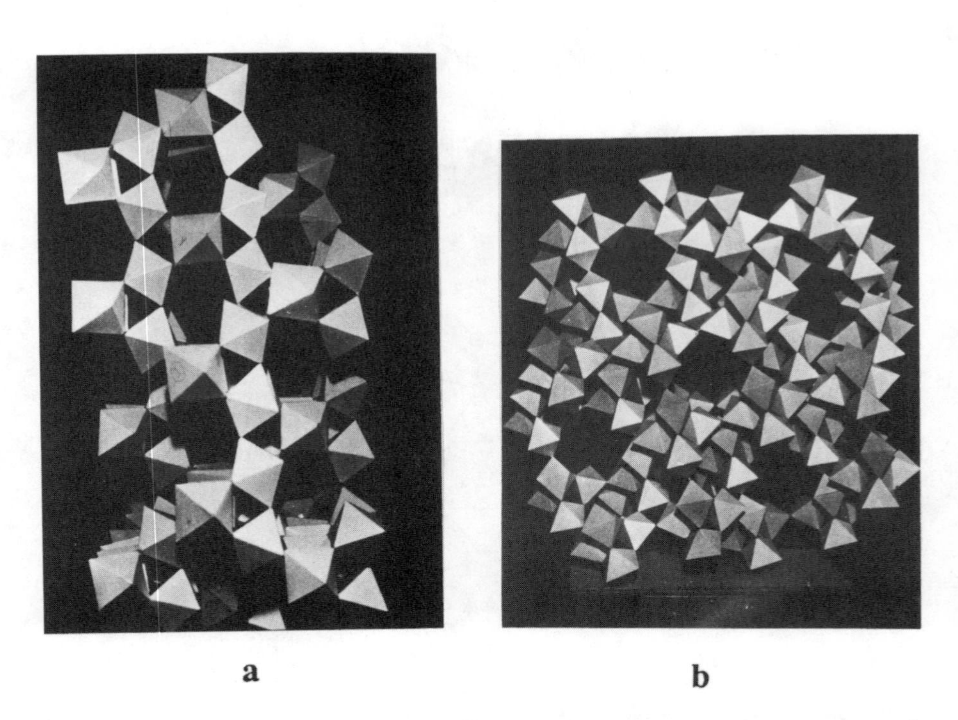

Figure 75. $Rb_{20+x}(W_4O_6)_{1/3-x/12}W_{32}O_{108}$: (a) Projection along ⟨110⟩ showing pyrochlore hexagonal tunnels; (b) the HTB layers parallel to (001) where the W_3O_{15} units are missing. (After Goreaud et al.[84])

position is $Ln_2B_2O_7$ (Ln = lanthanide; B = Ti, Sn, Zr) or $Cd_2B_2O_7$ (B = Ta, Nb, Sb). These oxides can be formulated as $(A_2O)(B_2O_6)$, and the host lattice is $[B_2O_6]_\infty$. The additional oxygens of the A_2O sublattice are at the center of the O_{18} cage, whereas the A cations are in the hexagonal HTB rings at the boundary between two cages. As a result, these oxides have a close-packed structure and do not exhibit cationic mobility (owing to the obstruction of the tunnels by oxygen). Nonstoichiometry on the A_2O sublattice (which has the Cu_2O-type structure) has however been observed. Indeed, the pyrochlores $Pb_{2-\alpha}(Ti,Ta)O_{7-\beta}$ and $Pb_{2-\alpha}(Ti,Nb)O_{7-\beta}$ show wide homogeneity ranges for the A cations as well as oxygen.

Structures of the minerals alunite, $KAl_3(SO_4)_2(OH)_6$, and crandallite, $CaAl_3(OH)_6[PO_3O_{0.5}(OH)_{0.5}]_2$, are closely related to the pyrochlore structure. The host lattices of these hexagonal oxohydroxides ($a \approx 7$ Å, $c \approx 17.3$ Å) are of interest and correspond to the formulas $KAl_3(OH)_6O_3$ and $CaAl_3(OH)_6O_3$, consisting of the HTB layers; $Al_3(OH)_6O_3$ is similar to the B_3O_9 layer observed in pyrochlores, connected through KO_6 or CaO_6 antiprisms (instead of BO_6 octahedra). The stacking of the layers is different from that of pyrochlores, forming much larger cavities where the "SO" and "PO" groups are located, and forming SO_4 or PO_4 tetrahedra with the other oxygens of the framework.

6.2 $A_2B_7O_{18}$ Oxides

Association of the two B_3O_{15} octahedral units through octahedral edges leads to a new octahedral unit, "B_6O_{24}" (Fig. 76). Such a unit of 2 × 3 edge-sharing octahedra is characterized by 60° and 120° angles. The oxides $A_2Ta_6TiO_{18}$ (A = Rb, Cs), which crystallize in the hexagonal system ($a \approx 7.5$ Å, $c \approx 8.2$ Å) have the $[B_7O_{18}]_\infty$ host lattice, built up of B_6O_{24} units. These units share their

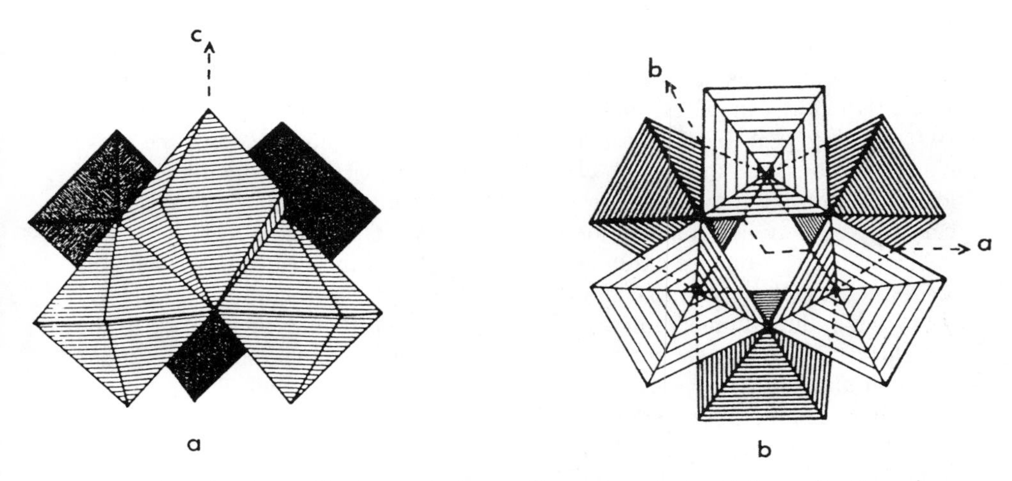

Figure 76. Octahedral B_6O_{24} unit built up of two edge-sharing B_3O_{15} units. Two views are given in (a) and (b).

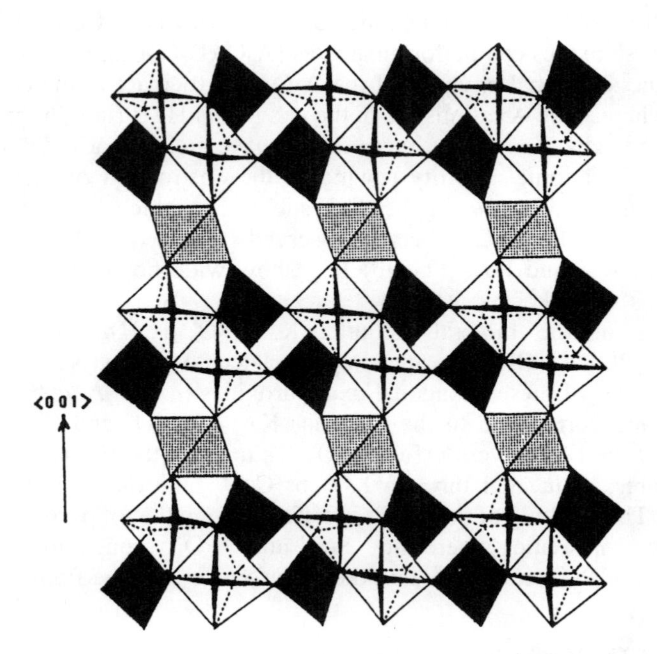

Figure 77. $Rb_2Nb_6TiO_{18}$ showing the stacking of the $[B_6O_{15}]_\infty$ and $[BO_3]_\infty$ layers along $\vec{c}$. (After Desgardin et al.[85])

corners laterally in the (001) plane forming $[B_6O_{15}]_\infty$ layers, which are derived from the HTB$[B_3O_9]_\infty$ layers in the following way. Each B_6O_{15} layer can be considered as an association of two B_3O_9 layers through the edges of the octahedra. Along $\vec{c}$, the B_6O_{24} units are linked through BO_6 octahedra whose ternary axis is parallel to this direction. Consequently, the octahedral three-dimensional lattice $[B_7O_{18}]_\infty$ can be attributed to the stacking of $[B_6O_{15}]_\infty$ layers and of $[BO_3]_\infty$ layers (Fig. 77).[85] Comparison of Figures 77 and 74 shows the close relationship between the $A_{1+x}B_2O_6$ pyrochlore and $A_2B_7O_{18}$, which have identical HTB and $[BO_3]_\infty$ layers. The $A_2B_7O_{18}$ structure is, in fact, derived from that of pyrochlore by replacing the B_3O_{15} unit by a B_6O_{24} unit.

The $A_2B_7O_{18}$ structure exhibits cavities similar to those in the pyrochlores, but formed by 21 oxygen atoms instead of 18 as in the pyrochlore. At the level of one cage, three tunnels intersect in the $A_2B_7O_{18}$ structure against six in the pyrochlore. These intersecting tunnels allow two-dimensional mobility of the A cations in the $[BO_3]_\infty$ layers (i.e., between two $[B_6O_{15}]_\infty$ layers). This results in ion exchange properties that allow the preparation of several hydrates and ammonium derivatives, including $A_2B_6TiO_{18}{\cdot}H_2O$ (B = Ta, Nb; A = Li, Na, Ag, K), $(H_3O)_2B_6TiO_{18}$, and $(NH_4)_2B_6TiO_{18}$.

6.3 Pyrochlore Intergrowths: HTB-$A_2B_7O_{18}$

The analogy between the pyrochlore and the $A_2B_7O_{18}$ structures due to the existence of similar HTB-$[B_3O_9]_\infty$ layers suggests the possibility of intergrowths of these structures according to the general formula $(A_2B_7O_{18})_n \cdot A_{2+x}B_4O_{12}$. In this formula, $A_{2+x}B_4O_{12}$ corresponds to the composition of the pyrochlore slice $(A_{1+x}B_2O_6)$. The hexagonal cell parameters of such intergrowths are:

$$a_n \approx a_{HTB} \approx a_{A_2B_7O_{18}} \approx a_{py}\sqrt{2} \approx 7.5 \text{ Å}$$

$$c_n \approx a_{py}\sqrt{3} + 3nc_{A_2B_7O_{18}} \approx 18 \text{ Å} + 3n \times 8.2 \text{ Å}$$

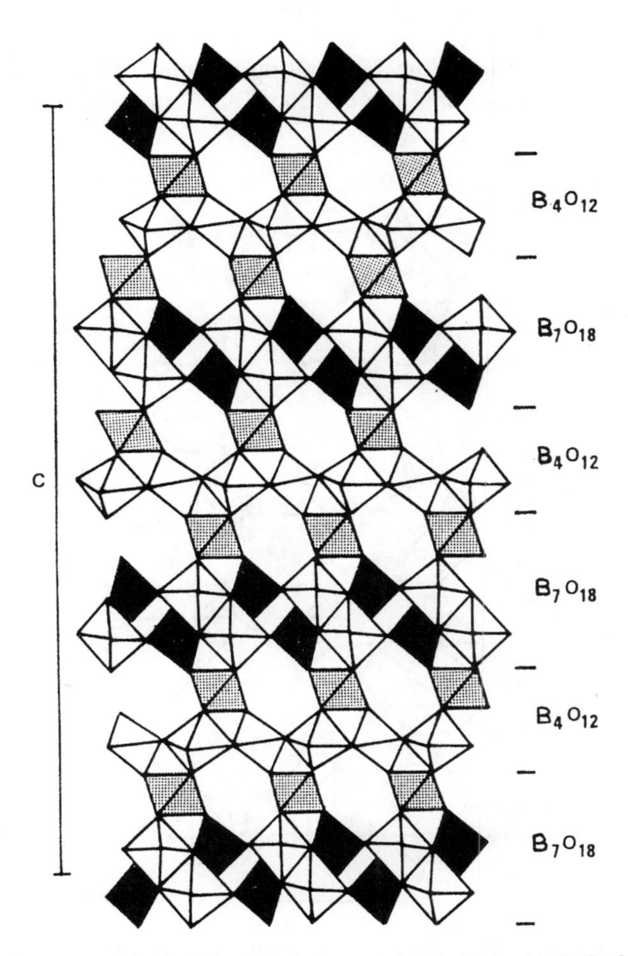

Figure 78. Structure of $Rb_{15}Nb_{33}O_{90}$, intergrowth of single $[A_2B_7O_{18}]_\infty$ layers with single $[A_2B_4O_{12}]_\infty$ pyrochlore layers. (After Michel et al.[86])

A series of oxides of the formula $A_4B'_{10}B''O_{30}$ corresponding to the first member of this series has been isolated for A = Rb, Tl, Cs, B′ = Ta, Nb, and B″ = W, Mo. Thus, $Rb_{15}Nb_{33}O_{90}$ and $Tl_{12}Nb_{33}O_{88.5}$ exhibit the same host lattice, with the a and c parameters close to 7.5 and 43 Å, respectively. This framework (Fig. 78)[86] consists of the intergrowth of single $A_2B_7O_{18}$ layers with single pyrochlore $A_2B_4O_{12}$ layers. In each layer, one recognizes intersecting tunnels where the A cations are located. Like the pyrochlores and $A_2B_7O_{18}$ oxides, these intergrowths also exhibit ion exchange properties. Hydronium, ammonium, and anhydrous as well as hydrated potassium, sodium, and silver $n = 1$ members have been prepared. Curiously, the other members of the series have not yet been characterized. The similarity between these two structures and HTB's suggests the possibility of intergrowth with the latter. This is the case in $A_{10}B_{29.2}O_{78}$ (A = K, Rb, Cs; B = Ta, Nb). These hexagonal oxides ($a \approx 7.5$ Å, $c \approx 36.5$ Å) have their $[B_{28}O_{78}]_\infty$ host lattice (Fig. 79) built up of double HTB$[(B_3O_9)_2]_\infty$ layers and $[B_6O_{15}]_\infty$ layers connected through the $[BO_3]_\infty$ layers of single BO_6 octa-

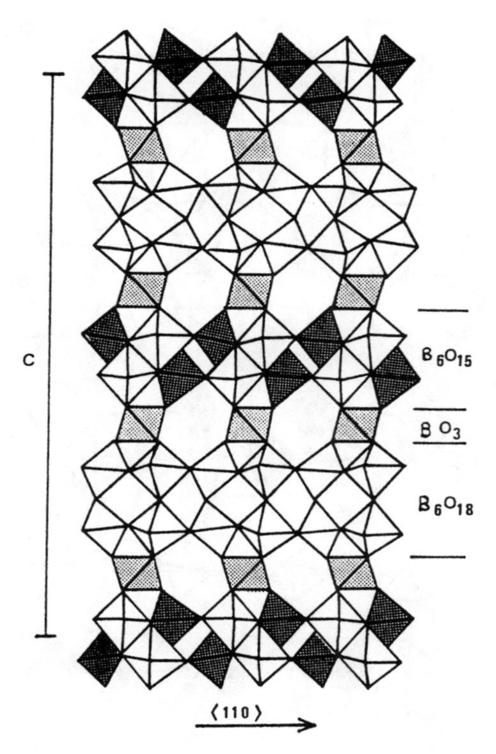

Figure 79. Structure of $Rb_{10}Nb_{29.2}O_{78}$, intergrowth of single $[Rb_2Nb_7O_{18}]_\infty$ layers with single $[Rb_2Nb_4O_{12}]_\infty$ pyrochlore layers and with double HTB $[(RbNb_3O_9)_2]_\infty$ layers. (After Michel et al.[86])

hedra.[86] Consequently, they can be described as an intergrowth of HTB, B_7O_{18}, and pyrochlore structures according to the formula $(A_2B_7O_{18})_2(A_2B_4O_{12})_2(AB_3O_9)_2$. In this host lattice, the A cations are located in the intersecting tunnels, whereas additional B atoms (Ta or Nb) are distributed in the trigonal prismatic cavities formed by two B_3O_{15} units within the double HTB layers. These oxides also exhibit ion exchange properties. Hydrates $A_{10}B_{29.2}O_{78.10}{\cdot}H_2O$ (A = H, Na, Ag, K) have been prepared.

6.4 $K_2Sb_4O_{11}$ and $K_3Sb_5O_{14}$

Although they are not transition metal oxides, two potassium antimonates deserve to be mentioned here: $K_2Sb_4O_{11}$,[87] with a monoclinic structure ($a \approx 19.5$ Å, $b \approx 7.5$ Å, $c \approx 7.2$ Å, $\beta \approx 91.8°$), and $K_3Sb_5O_{14}$,[87] with an orthorhombic structure ($a \approx 24.2$ Å, $b \approx 7.2$ Å, $c \approx 7.3$ Å) exhibit different host lattices built up of edge- and corner-sharing SbO_6 octahedra. The $K_2Sb_4O_{11}$ host lattice can be described on the basis of pairs of octahedra, Sb_2O_{10}, also observed in the cubic form of $KSbO_3$. Association of these units through the edges forms infinite rows of edge-sharing octahedra running along $\vec{\mathbf{a}}$, whereas association through their corners forms infinite ribbons showing a similarity to the rutile structure. Along $\vec{\mathbf{b}}$, some of these pairs share their corners, forming ReO_3-type rows sharing their edges. This framework allows large tunnels along $\mathbf{b}$ and $\vec{\mathbf{c}}$; the K^+ ions are located at the intersection of these tunnels, which allow a two-dimensional mobility.

In $K_3Sb_5O_{14}$, also, there are Sb_2O_{10} octahedral units. Association of such units through their corners leads to infinite rutile-type ribbons running along $\vec{\mathbf{b}}$. The lateral connection of these ribbons in the (001) plane is more complex than in $K_2Sb_4O_{11}$. Large tunnels are present along $\vec{\mathbf{c}}$ and $\vec{\mathbf{b}}$, where the K^+ ions are located; part of the K^+ ions are located at the intersection of these tunnels. These materials exhibit ion exchange properties similar to those of pyrochlores. Such reactions allow potassium to be replaced by some species (e.g., Na, Ag, Rb, Tl, Cs) using fused salts, and also the preparation of hydrated antimonates such as $A_2Sb_4O_{11}{\cdot}xH_2O$ (A = Na, Tl, K/Ag, K/Li), and $Li_3/Sb_5/O_{14}{\cdot}(H_2O)_{2.5}$ from aqueous solutions. All these oxides exhibit ionic conductivity; $(H_3O)_2Sb_4O_{11}(H_2O)_{1.1}$ is a good proton conductor.

7 Octahedral Lamellar Oxides

Octahedral lamellar oxides have layers that are electrically neutral or charged, and cohesion of the layers occurs through van der Waals interaction, hydrogen bonding, or ionic bonding. Two cases must be considered in this category of oxides. When the octahedra share only their corners, the layers are sufficiently thick to avoid distortion. Thin layers are also formed, but in such cases the octahedra share both edges and corners.

7.1 Niobates and Titanates of the Formula $A_nB_nO_{3n+2}$ and Molybdates of the Type $Cs_2Mo_nO_{3n+1}$ Derived from the Perovskite Structure

Oxides of the formula $A_nB_nO_{3n+2}$ form a large family in which A is an alkaline earth and/or a lanthanide ion and B is titanium and/or niobium. The different members of this series are found in $A_2Nb_2O_7/CaTiO_3$ (A = Ca, Sr) and $Ln_2Ti_2O_7/CaTiO_3$ (Ln = La, Nd) systems. The ideal structure (Fig. 80a) is orthorhombic, but monoclinic distortion often occurs. Accordingly, $Ca_2Nb_2O_7$ (n = 4), is found in both orthorhombic ($a \approx 26.5$ Å, $b \approx 5.5$ Å,$c \approx 7.7$ Å) and monoclinic ($a \approx 13.4$ Å, $b \approx 5.5$ Å, $c \approx 7.7$ Å, $\beta \approx 98°$) forms, which differ with respect to the distortions and tilting of the NbO_6 octaheda. The other members of the family generally exhibit an orthorhombic cell with the a and c parameters close to those of $Ca_2Nb_2O_7$, although the b parameter varies with the thick-

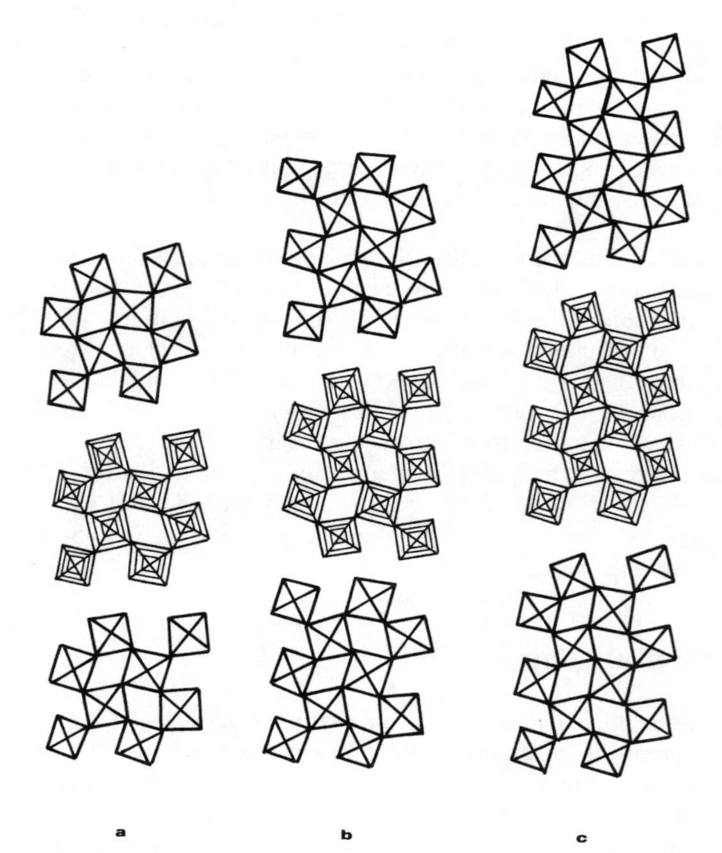

Figure 80. Layered perovskites, $A_nB_nO_{3n+2}$: (a) $n = 4$, (b) $n = 5$, and (c) $n = 6$. (After Nanot et al.[88] and Raveau.[46])

ness of the octahedral layer. The structure of these oxides (Fig. 80) consists of the anionic layers, $[B_nO_{3n+2}]^{n'-}$ $(2n < n' < 3n)$, whose cohesion occurs through the A^{2+} and A^{3+} cations.[46,88] These octahedral layers have a distorted perovskite configuration, leading to diamond-shaped tunnels running along $\vec{\mathbf{a}}$, where the A cations are located. The integral number n corresponds to the number of BO_6 octahedra, which determines the thickness of the perovskite slab. The successive slabs are shifted by $a/4$ with respect to each other, and the orthorhombic cell is related to the perovskite cell (a_p) as follows: $a \approx 2a_p$, $b \approx$?, $c \approx a_p\sqrt{2}$. The A cations have an irregular coordination ranging from 7 to 12, with a smaller coordination for the cations located between the layers than for those located in the perovskite tunnels. Nonintegral n values lead to the formation of intergrowths between members of integral n.

The cesium molybdates, $Cs_2Mo_nO_{3n+1}$,[81] are exemplified by two members of monoclinic structure: for $n = 5$, $Cs_2Mo_5O_{16}$ ($a \approx 21.4$ Å, $b \approx 5.6$ Å, $c \approx 14.3$ Å, $\beta \approx 122.7°$), and for $n = 7$, $Cs_2Mo_7O_{32}$ ($a \approx 21.5$ Å, $b \approx 5.6$ Å, $c \approx 19.5$ Å, $\beta \approx 127.4°$). The structure of these phases consists of octahedral $[Mo_nO_{3n+1}]^{2-}$ infinite layers whose cohesion occurs through the presence of Cs^+ cations. The octahedral slabs consists of ReO_3-type ribbons n octahedra wide. Laterally, in the (010) plane, the ReO_3-type ribbons share octahedral edges forming rock salt type ribbons.

A related series of oxides studied by Tournoux, Jacobson and others[52] is $A'[A_{n-1}B_nO_{3n+1}]$.

7.2 Lamellar Structures Built up of Edge-Sharing Octahedra

To build thin octahedral layers, it would be necessary to have a rigid chain of octahedra along one dimension. This implies that to ensure the presentation of a tilting around the mean direction of the chain, the octahedra should share their edges. Two ways of forming such chains can be considered. In the first, each octahedron shares two adjacent edges with its neighbors, forming infinite chains of edge-sharing octahedra. These chains can be described as double ReO_3-type chains (DRC) sharing the octahedral edges (Fig. 81a–c). In the second way, which is more simple, each octahedron shares two opposite edges with its neighbors, forming so-called rutile chains, i.e. infinite chains of edge-sharing octahedra (Fig. 81d, e).[89] It is easy to visualize how thin layers result from the association of such identical chains of octahedra.

7.2.1 Layered Oxides Resulting from the Association of DRC's

Structures of many of the lamellar oxides result from the association of double ReO_3-type chains. Isolated chains have been observed in $NaMoO_3F$. Association of such chains through octahedral corners leads to the formation of two ReO_3-type layers sharing their edges (Fig. 82a–c), which are in fact the (001) crystallographic shear planes encountered in many of the B_nO_{3n-1} oxides. The structure of MoO_3 (Fig. 82d) involves such layers held together by van der Waals forces.

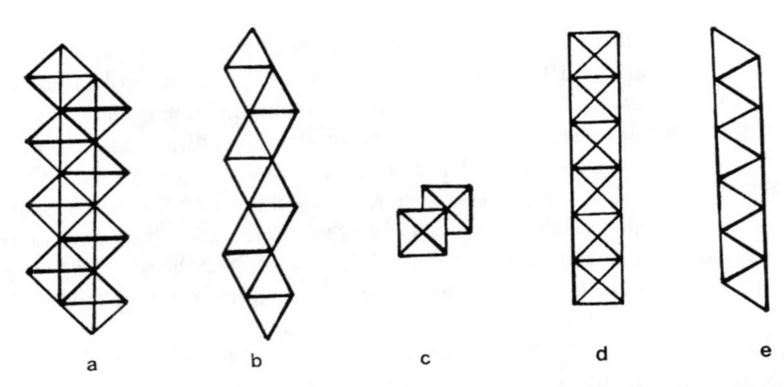

Figure 81. Double ReO_3-type chains resulting from the association of two ReO_3-type chains through the edges: (a) corner-on view of the octahedral chains, (b) edge-on view of the octahedral chains, (c) view along the chain direction, and (d) corner view of the rutile chains along with (e) edge-on view. (After Raveau.[89])

Association of the DRC's through the edges of octahedra can occur in different ways. Figure 83 shows two ways that lead to layered structures observed in several families of oxides. When two adjacent DRC's share one edge per unit of four octahedra according to Fig. 83a (mode I), the structure of brannerite ($ThTi_2O_6$: Fig. 84) is obtained[90]; this structure is found in various oxides, including $A_xV_xMo_{1-x}O_3$ and $A_xV_xW_{1-x}O_3$ (A = Li, Na, K, Rb, Cs). This structure tends to be close packed as the size of the A ion decreases (e.g., lithium compounds in which the Li^+ ions exhibit an almost perfect tetrahedral coordination). This structure is also derived from anatase by a shearing phenomenon.

When two successive DRC's share three edges per unit of four octahedra according to Figure 83b (mode II), one obtains a structure characterized by identical layers built up of infinite double ribbons of edge-sharing octahedra (Fig. 85a), found in several families of oxides. Four different structural families, which differ from one another only by the relative positions of the octahedral layers, are known:

1. The titanates, $A_xTi_{2-y}B_yO_4$ (A = Rb, Tl, Cs; B = Mn, Sc, Al, Mg, Ni, Zn, Fe), exemplified by $Rb_xB_2O_4$ (Fig. 85b).
2. $K_xTi_{2-x}B_yO_4$ (B = Mg, Ni, Cu, Fe, Zn) shown as $K_xB_2O_4$ (Fig. 85c).
3. γ-FeOOH, also called lepidocrocite (Fig. 85d).
4. β-$NaMnO_2$ and $LiMnO_2$ (Fig. 85e).

Examination of Figure 85 clearly indicates that the structure of $K_xB_2O_4$ (Fig. 85c) can be deduced from that of the $Rb_xB_2O_4$ (Fig. 85b) by a simple translation of one layer out of two by half an octahedral edge along the length of the octahedral ribbons, changing the pseudocubic coordination of the A ions (rubidium) into the prismatic trigonal coordination (potassium). The lepidocrocite structure (Fig. 85d) can similarly be derived from the $Rb_xB_2O_4$ structure by a

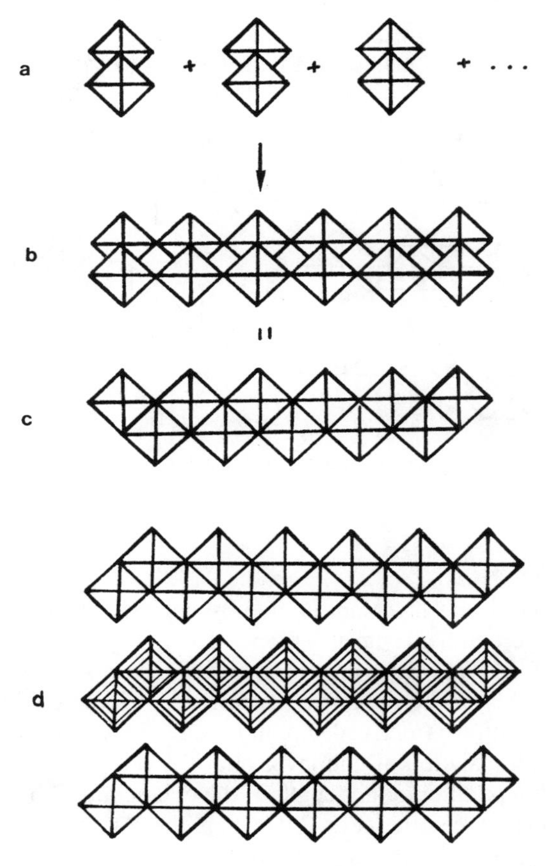

Figure 82. Association of an infinite number of DRCs through the corners (a) leads to the formation of (001) crystallographic shear planes, which can be represented in two ways (b and c), leading to (d) the MoO_3 structure. (After Raveau.[89])

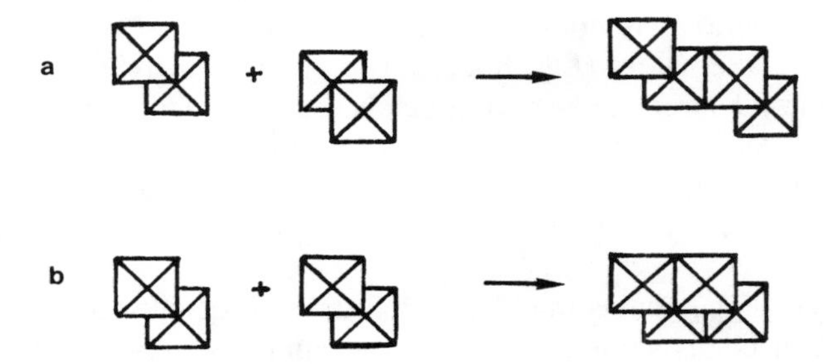

Figure 83. Two ways of connecting the DRCs through their edges: (a) mode I and (b) mode II. (After Raveau.[89])

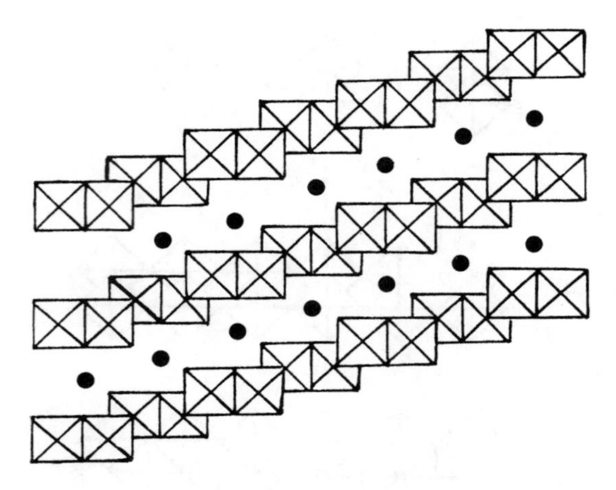

Figure 84. Layered structure of brannerite built up of TiO_6 edge-sharing octahedra interleaved with Th ions (dots). (After Ruh and Wadsley.[90])

simple translation of one layer out of two by half the height of an octahedron in the direction perpendicular to the plane of the projection. The structures of β-$NaMnO_2$ and $LiMnO_2$ (which are close packed) are closely related to the three other structures and can be described as formed of identical layers (Fig. 85e) whose relative position is derived directly from the γ-FeOOH structure (Fig. 85d) by a simple translation of one layer out of two by half an octahedral edge along the direction parallel to the length of the octahedral ribbons. The β-MnO_2 layers form octahedral cavities (Fig. 85f) where the Na^+ or Li^+ ions are located.

Layers can be generated by using simultaneously the two different modes of association described in Figure 83. This is the situation in the vanadium bronze δ-$Ag_xV_2O_5$. The idealized structure of this bronze can be described as due to the occurrence of association by modes I and II alternately, leading into infinite ribbons of edge-sharing octahedra (Fig. 86a–c).

Other forms of association by the simultaneous occurrence of modes I and II are possible, and these would give rise to layers of different types not identified hitherto. Nevertheless, attention must be drawn to another way of associating modes I and II alternately (Fig. 86d,e). The result is a hypothetical structure formed of units of three edge-sharing octahedra (Fig. 87a), which can be considered as the idealized structure of the vanadium bronzes $A_{1+x}V_3O_8$ (A = Li, Na, Ag; $x \approx 0.1$–0.50). Layers of these bronzes are generally described as built up of units of 2×2 edge-sharing octahedra linked through VO_5 pyramids (Fig. 87b).

Double ReO_3-type chains may be linked through octahedral corners as well as through their edges simultaneously in the same layer by mode II (Fig. 88). Titanates, titanoniobates, and tantalates having the formula $A_nB_{2n}O_{4n+2}$, and known for their ion exchange properties, correspond to this mode of association

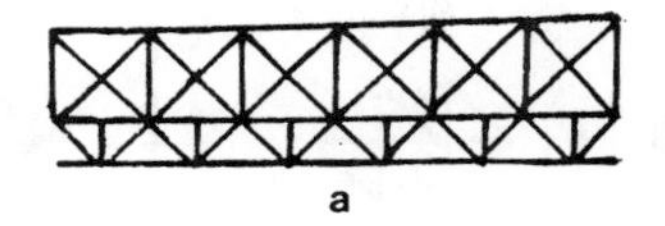

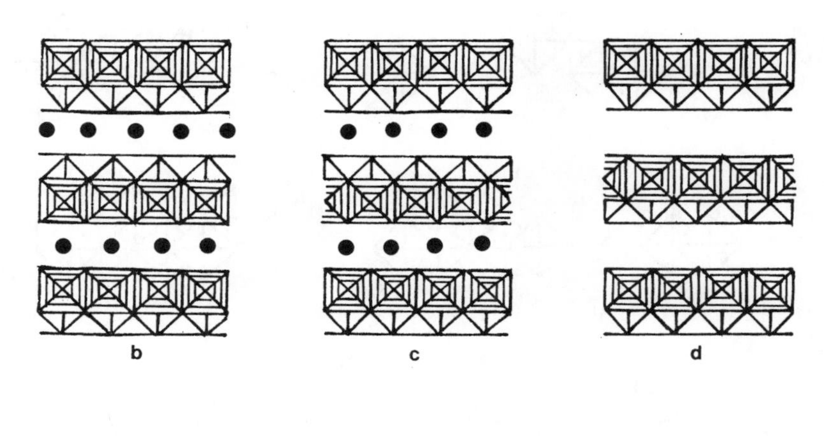

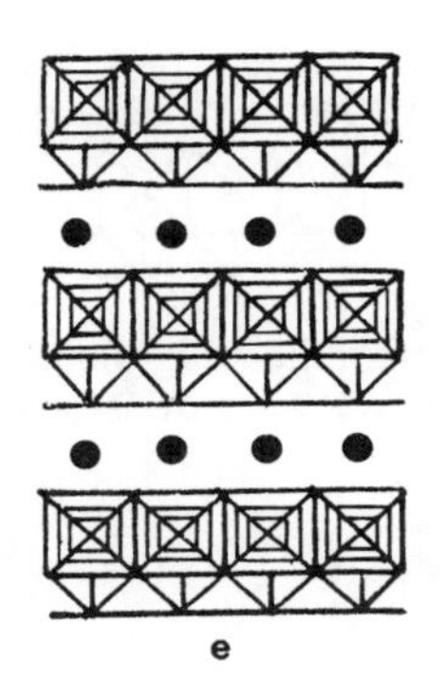

Figure 85. $A_xB_2O_4$: (a) infinite layer of octahedra built up from edge-sharing DRC's according to mode II, (b) $Rb_xB_2O_4$ structure, (c) $K_xB_2O_4$ structure, (d) γ-FeOOH lepidocrocite structure, (e) β-$NaMnO_2$ structure forming a close-packed array with sodium (dots) in octahedral coordination, and (f) β-MnO_2 layers forming octahedral cavities, where Na^+ or Li^+ ions can be located. (After Raveau.[89])

of the DRC's. These structures can also be described as built up of structural units of $2 \times n$ edge-sharing octahedra forming infinite ribbons that are n octahedra wide and connected to each other through the corners of the octahedra. Three oxides correspond to the $n = 3$ members of this series: $Na_2Ti_3O_7$, $CsTi_2NbO_7$, and $A_3Ti_5BO_{14}$ (A = K, Rb, Tl; B = Ta, Nb). These three structures

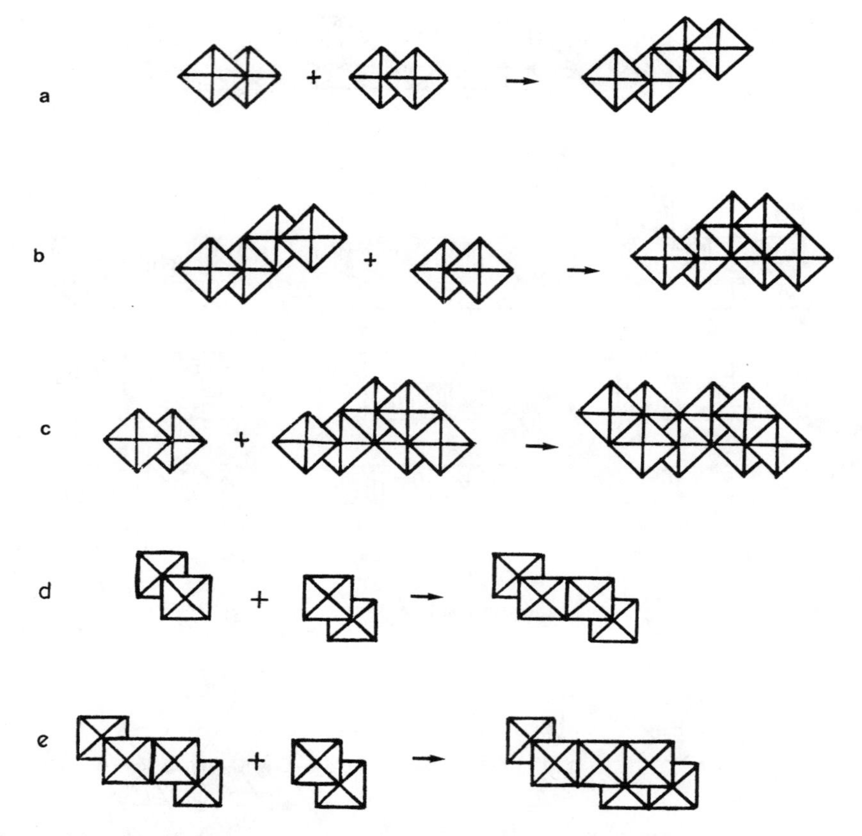

Figure 86. (a)–(c) Association of the DRC's through their edges according to modes I and II, leading to δ-$Ag_xV_2O_5$. (d) and (e) Alternative ways of associating DRC's by modes I and II. (After Raveau.[89])

are characterized by identical layers $[B_3O_7]_\infty$, which result from the association of three DRCs through the edges according to mode II and two DRCs through the corners alternately (Fig. 88a); that is, two edge-sharing mode II operations and one corner-sharing operation occur alternately. Structures of these three compounds differ in the relative positions of the layers. The oxides $Na_2Ti_3O_7$ (Fig. 88b) and $A_3Ti_5NbO_{14}$ (Fig. 88c) have parallel octahedral ribbons of 2 × 3 octahedra parallel and can be related to each other by means of translation by half the height of an octahedron in the direction perpendicular to the projection. These oxides are different from $CsTi_2NbO_7$ because in the latter, two adjacent layers cannot be derived from each other by a simple translation; instead, a glide plane is required. It is noteworthy that the different positions of the layers change

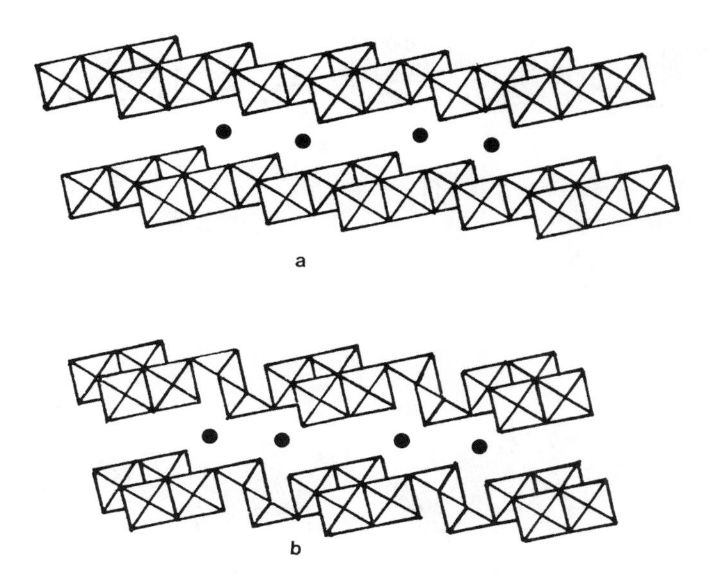

Figure 87. Idealized structure of the vanadium bronze $A_{1+x}V_3O_8$ described in terms of (a) edge-sharing VO_6 octahedra (b) VO_6 octahedra and VO_5 pyramids. (After Raveau.[89])

the coordination of the A ions and also the extent of insertion of the cations. The structure of the $n = 4$ member, $Tl_2Ti_4O_9$, is deduced from that of $A_3Ti_5NbO_{14}$ (Fig. 88c) by simply replacing units of 2×3 octahedra by units of 2×4 edge-sharing octahedra. Thus, the layers of $Tl_2Ti_4O_9$ result from three edge-sharing associations of DRC's (by mode II), combined with one corner-sharing operation.

The $n = 2$ member of this series corresponds to $ATiBO_5$ (A = K, Rb, Tl; B = Ta, Nb) whose $[TiNbO_5]_\infty$ layers are built up of units of 2×2 edge-sharing octahedra (Fig. 89) and result from the combination of one edge-sharing association of the DRC's by mode II with one corner-sharing operation. These oxides are closely related to $CsTi_2NbO_7$.

All the oxides discussed in this subsection show intercalation properties. They also exhibit ion exchange properties in aqueous medium as well as in solid state reactions and molten salt media. Numerous hydrated oxides and protonic oxides have been synthesized by ion exchange reactions: $A_2Ti_4O_9$ (A = Li, Na, K, Rb, Cs, Ag), $H_3OTi_2NbO_7{\cdot}H_2O$, $H_2Ti_4O_9{\cdot}nH_2O$, $HTiNbO_5$, $H_3Ti_5NbO_{14}{\cdot}H_2O$, $A_{1-x}(H_3O)_xTiNbO_5{\cdot}nH_2O$ (A = Li, Na, Ag), $A_{1-x}(H_3O)_xTi_2NbO_7{\cdot}H_2O$ (A = K, Rb, Tl, Ag), $HA_2Ti_5NbO_{14}{\cdot}H_2O$ (A = Li, Na, K, Rb, Tl, Cs), and $A_{0.5}H(TiNbO_5)_2{\cdot}nH_2O$. Acidic properties of the protonic oxides enable them to intercalate alkylamines and diamines. Strong hydrogen bonds ($H \cdots O = 1.5$

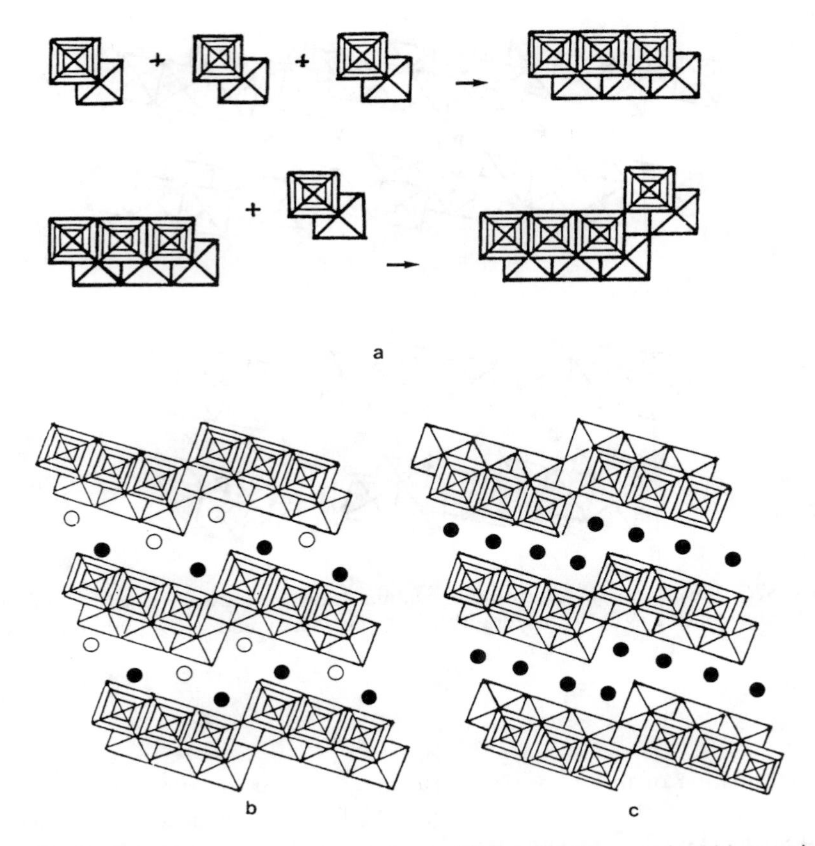

Figure 88. Structure of the $n = 3$ members of the $A_nB_{2n}O_{4n+2}$ series: (a) association of DRC's through the corners and edges alternately, (b) $Na_2Ti_3O_7$, and (c) $A_3Ti_5NbO_{14}$. (After Raveau.[89])

Å) responsible for the cohesion between the $[TiNbO_5]_\infty$ layers notwithstanding, $HTiNbO_5$ appears to have the greatest ability for intercalation. Quantitative intercalation of monoamines and diamines occurs in the form of ammonium ions leading to $[RNH_3]TiNbO_5$ and $[NH_3(CH_2)_nNH_3]_{0.5}TiNbO_5$. The intercalation is accompanied by a drastic expansion of the layer spacing, which results from the orientation of the amine molecules in the direction perpendicular to the layers (Fig. 90). This oxide can intercalate organic nitrogen compounds having pK_A values down to 6, whereas a minimum pK_A of 9 is generally required in the other protonic oxides.

Finally, it must be pointed out that layered oxides of the type $A_xB_2O_4$ resulting from the association of the DRC's through their edges correspond to the $n = \infty$ member of the structural series $A_{n'}B_nO_{4n+2}$.

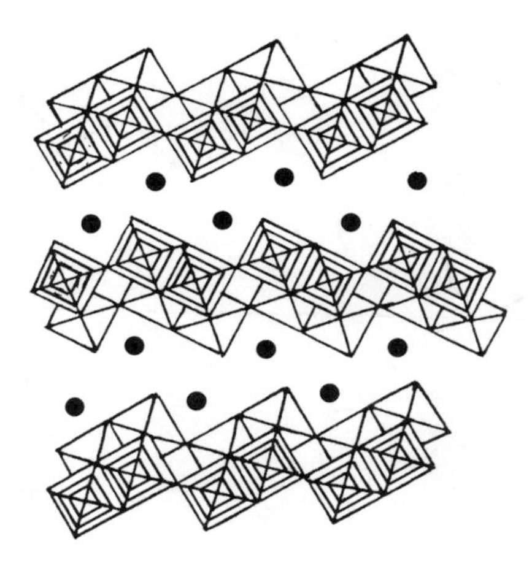

Figure 89. Structure of $KTiNbO_5$, the $n = 2$ member of the $A_nB_{2n}O_{4n+2}$ series. (After Wadsley.[77])

7.2.2 *Layered Oxides Involving Deficiency of Octahedra and Condensed DRCs*

Several families of oxides have their structure built up of double ReO_3-type chains in which some octahedra are missing periodically. The niobates $A_4Nb_6O_{17}{\cdot}xH_2O$ (A = K, Rb, Cs; x 0–5) and the molybdenum bronzes $K_{0.33}MoO_3$ and $Cs_{0.25}MoO_3$ are systems of this description. The layers in these three oxide families are built up of double ReO_3-type chains in which one octahedral chain out of two remains untouched with respect to the ReO_3 structure, whereas in the second, one octahedron out of two is missing (Fig. 91). In $Rb_4Nb_6O_{17}{\cdot}4H_2O$, such deficient DRC's share their corners and edges alternately, exactly as described for $A_3Ti_5NbO_{14}$ and $KTiNbO_5$, respectively (see Figs. 88 and 89). In $K_{0.33}MoO_3$ and $Cs_{0.25}MoO_3$, the deficient DRC's share their corners and edges alternately (Fig. 92).[91] However, the two deficient DRC's share their edges by forming double edge-sharing ReO_3-type chains, unlike the case of $KTiNbO_5$. There are two ways of connecting two deficient DRC's, corresponding to the configurations observed in $K_{0.33}MoO_3$ (Fig. 93a). and $Cs_{0.25}MoO_3$ (Fig. 93b); these two configurations differ from each other in the relative translation of one of the two DRC's by the height of an octahedron along the chain. This mode of association of two DRC's through the edges causes their alternative connection through the corners to be different: only one ReO_3-type chain out of two (the full one) shares corners with its neighbor (Fig. 92a,b).

Double ReO_3-type chains are also observed in ANb_3O_8 (A = K, H_3O) and $K_{0.30}MoO_3$. The layers that form these structures cannot however be completely described by a simple association of DRC's. Like $KTiNbO_5$ (Fig. 89), the structure of KNb_3O_8 (Fig. 94) exhibits two edge-sharing octahedra.[92] It is better

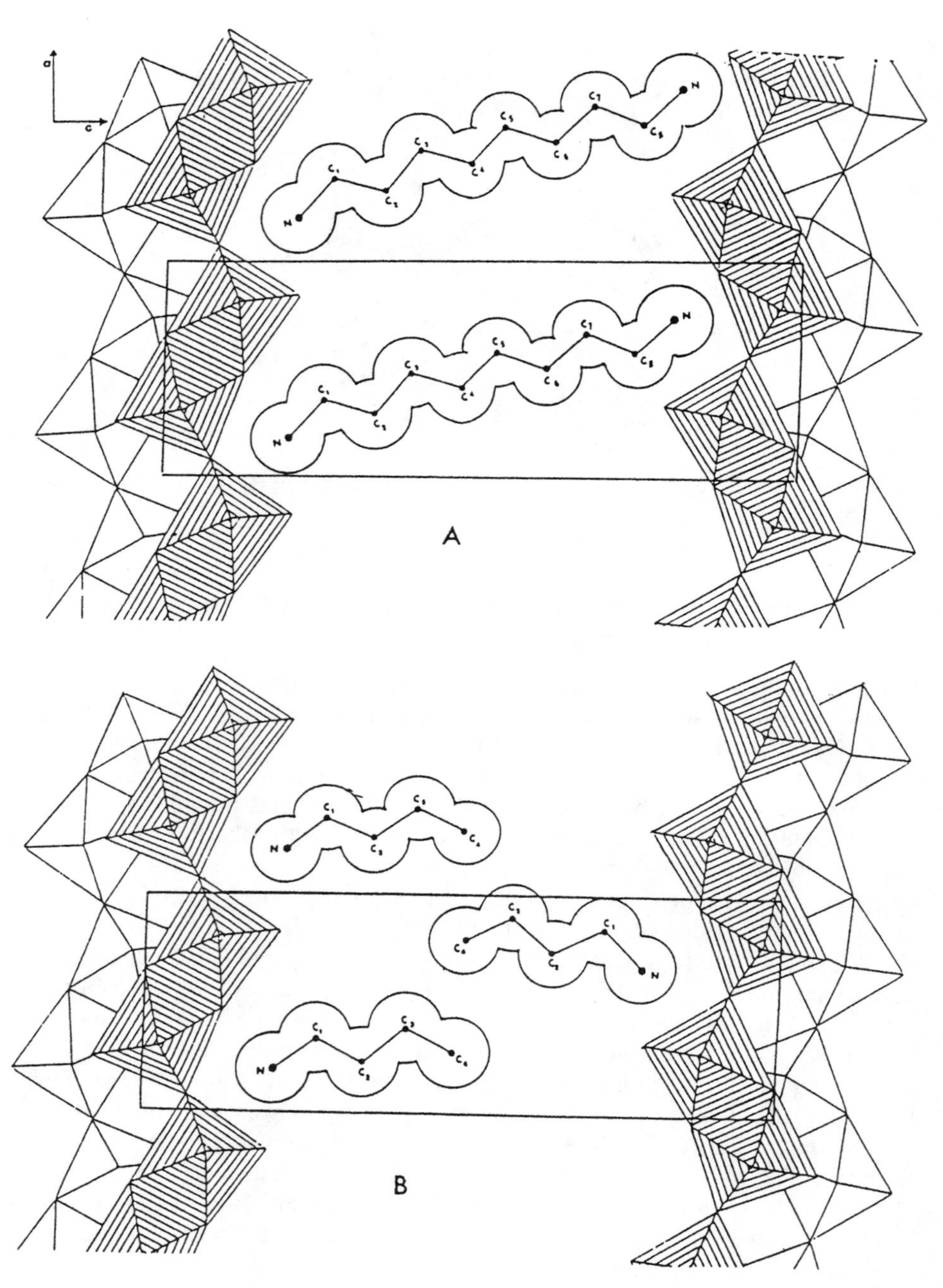

Figure 90. Intercalation of butylamine and octyldiamine in $HTiNbO_5$: (a) $NH_3(CH_2)_8NH_3(TiNbO_5)_2$ and (b) $NH_3(CH_2)_4TiNbO_5$. (After Raveau.[89])

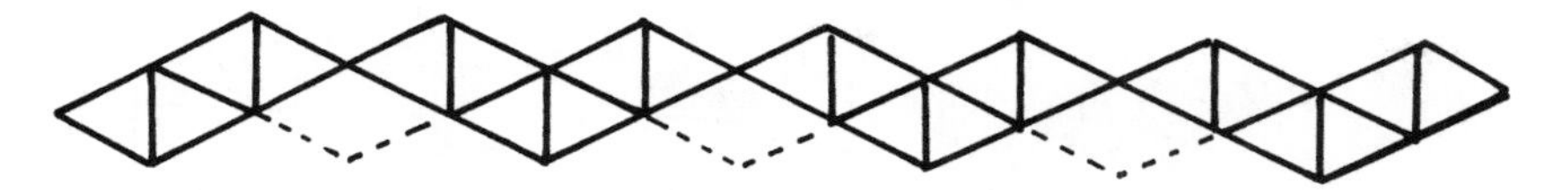

Figure 91. Deficient DRC's: one ReO_3-type chain is untouched, whereas one octahedron out of two is missing in the second chain. (After Raveau.[89])

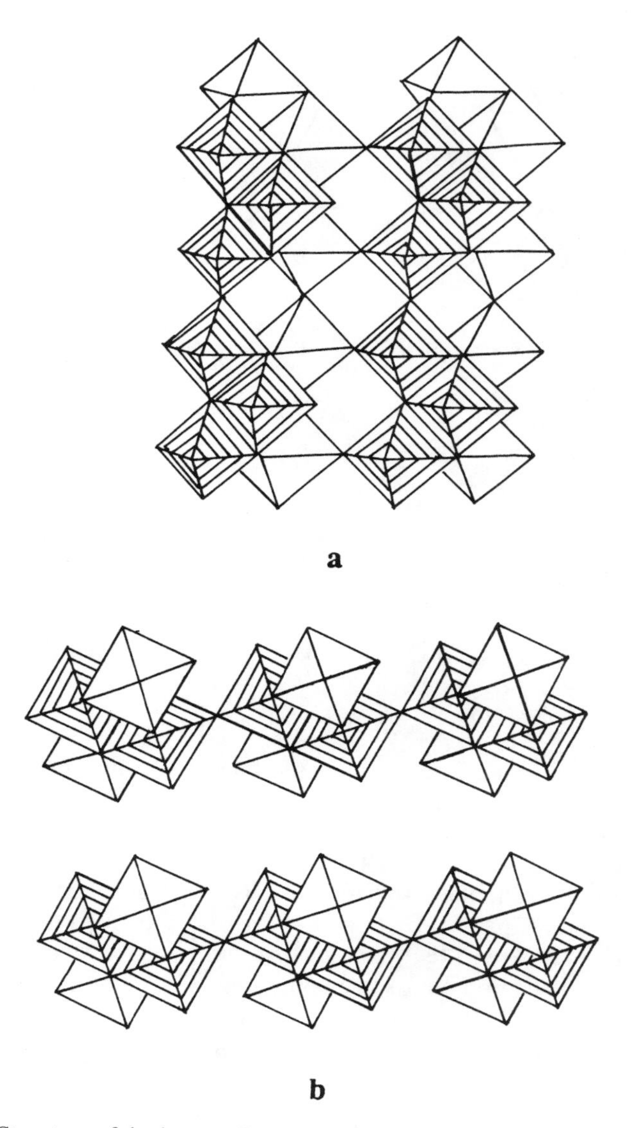

Figure 92. Structure of the bronze $Cs_{0.25}MoO_3$: (a) perspective view of a layer and (b) relative orientations of the layers. (After Mumme and Watts.[91])

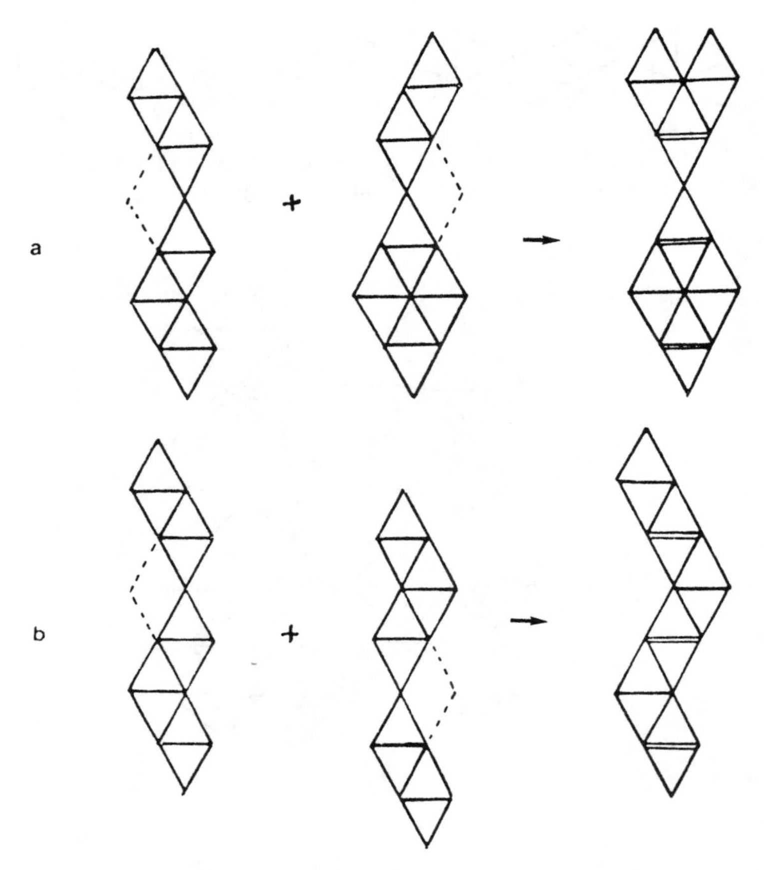

Figure 93. Association of DRC's through their edges in two molybdenum trioxides: (a) $K_{0.33}MoO_3$ and (b) $Cs_{0.25}MoO_3$. (After Raveau.[89])

described by triple ReO_3-type chains of edge-sharing octahedra (TRC), $[Nb_3O_{13}]_\infty$ (Fig. 94a), sharing corners (Fig. 94b) exactly as in the oxides of the $A_nB_nO_{4n+2}$ series. Such $[B_3O_{13}]_\infty$ TRC's can be considered as resulting from the condensation of two DRC's. The structure of the bronze $K_{0.30}MoO_3$ (Fig. 95a) can be attributed to the association of TRC's through the edges and corners of their octahedra (Fig. 95b).[93] In this bronze, the TRC's are deficient of octahedra; in each TRC, two ReO_3-type chains remain untouched, whereas in the third one, one octahedron out of two is missing (Fig. 95a).

7.2.3 Layered Oxides Resulting from the Association of Rutile Chains

Association of rutile chains through the edges of octahedra allows the building of thin layers characterized by great rigidity. The first family belonging to this

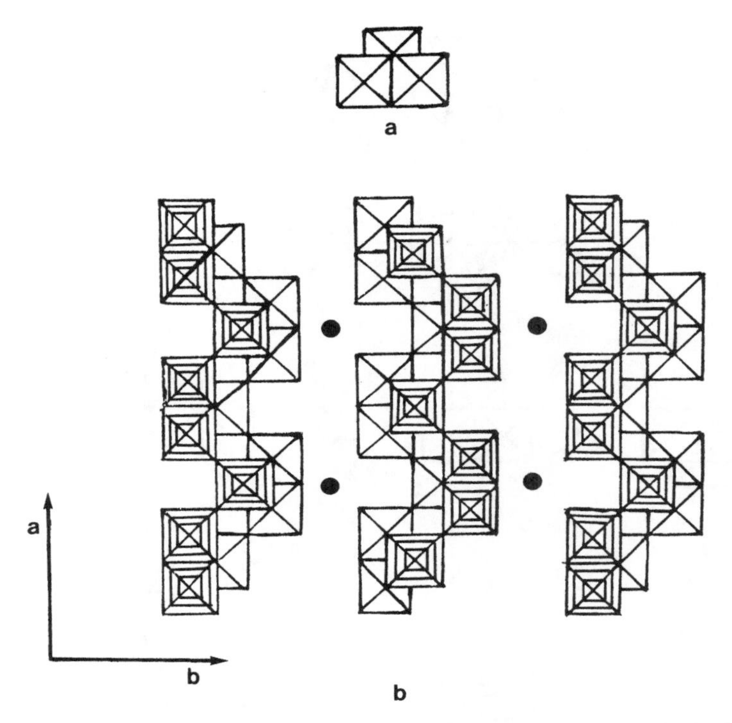

Figure 94. KNb_3O_8: (a) triple ReO_3-type chains of edge-sharing octahedra $|B_3O_{13}|_\infty$ (TRC) and (b) view of the structure along the direction of the TRC's. (After Gasperin.[92])

type of layered oxides corresponds to that of $A_xB_2O_4$ described earlier. In $A_xB_2O_4$ the rutile chains (Fig. 85) share their edges in such a way as to form ReO_3-type chains running along the perpendicular direction. These oxides can be considered to result from the association of DRC's or of rutile chains.

Another large family of oxides, corresponding to the same formulation, is exemplified by A_xBO_2 (A = Na, K; B = Cr, Mn, Co; $x < 1$), $K_xB'_xB''_{1-x}O_2$ (B′ = Sc, In; B″ = Zr, Hf, Sn, Pb; $x \approx 0.7$), and $K_xB_{x/2}Sn_{1-x/2}O_2$ (B = Mg, Ca, Zn; $x < 1$). These oxides exhibit identical layers, one octahedron thick; the rutile chains share edges in such a way that they form a double close-packed oxygen layer. The B ions are thus located in the octahedral cavities of anionic *ab*-type stacking. Such stackings of the BO_2 layers are shown in Figure 96,[94] two of which (*a* and *b*) are similar to the situation in $K_xB_2O_4$. The K^+ ions, which provide the cohesion between the layers, exhibit a prismatic pyramidal coordination. The stacking of the oxygen atoms differs in $K_{0.5}CoO_2$, with the *aabbcc* sequence (Fig. 96a) and $K_{0.67}CoO_2$ with the *aabb* sequence (Fig. 96b). The third structural type observed in α-$NaMnO_2$ (Fig. 96c) can be compared with β-$NaMnO_2$ in that two identical MnO_2 layers are translated in such a way that the Na^+ ions (located between the MnO_2 layers) exhibit a distorted octahedral (tri-

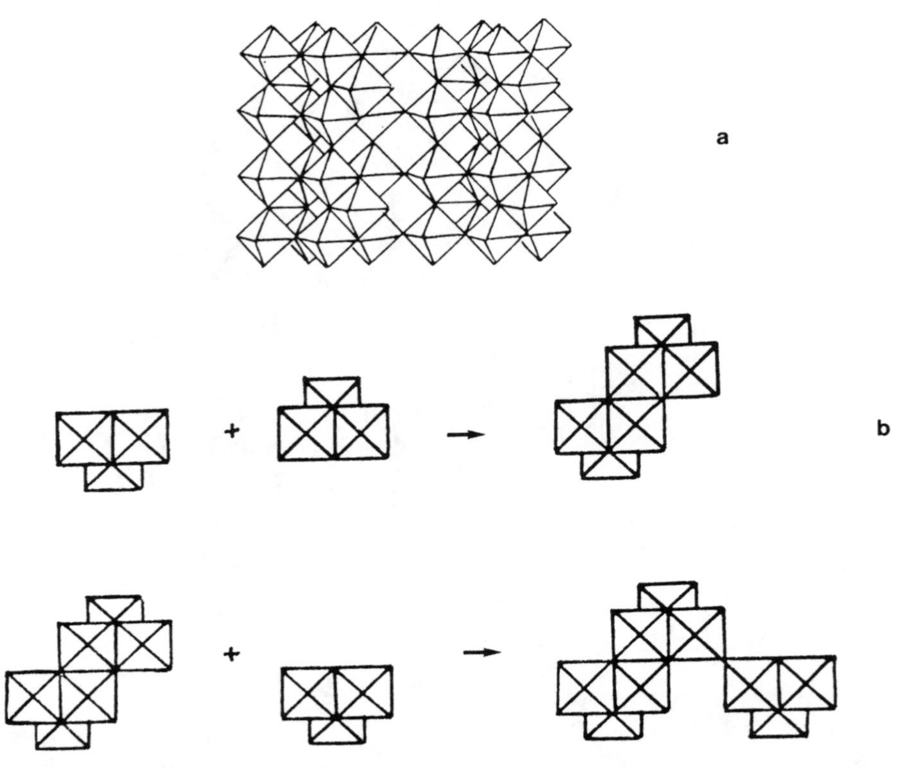

Figure 95. $K_{0.30}MoO_3$: (a) structure of the layers and (b) association of TRC's through the edges and corners of octahedra. (After Graham and Wadsley.[93])

angular antiprismatic) coordination. It is better considered as a close-packed structure derived from α-$NaFeO_2$, characterized by *abcabc* anionic close packing, rather than a layer structure.

Hydrous vanadium oxides, $H_{2n-2}V_nO_{3n-2}$, form an important family of minerals whose layers are built up of rutile chains associated through edges and corners. Cohesion between the layers is through hydrogen bonds. For instance, duttonite, $VO(OH)_2$ ($n = \infty$), has the BO_3 layers formed of rutile chains associated through their corners, so that ReO_3-type chains are formed simultaneously (Fig. 97a). In haggite, $V_4O_6(OH)_6$ ($n = 4$), double edge-sharing rutile chains are linked though octahedral corners (Fig. 97b), whereas in $V_6O_{10}(OH)_6$ ($n = 6$), called "phase B," the layers are formed of single rutile chains and double edge-sharing rutile chains linked through corners (Fig. 97c). The rigidity of the layers in these oxides would be much less than in the oxides described earlier because the corner-sharing rutile chains share only one oxygen per BO_6 octahedron. Since these structures are closely related, longitudinal intergrowths can occur

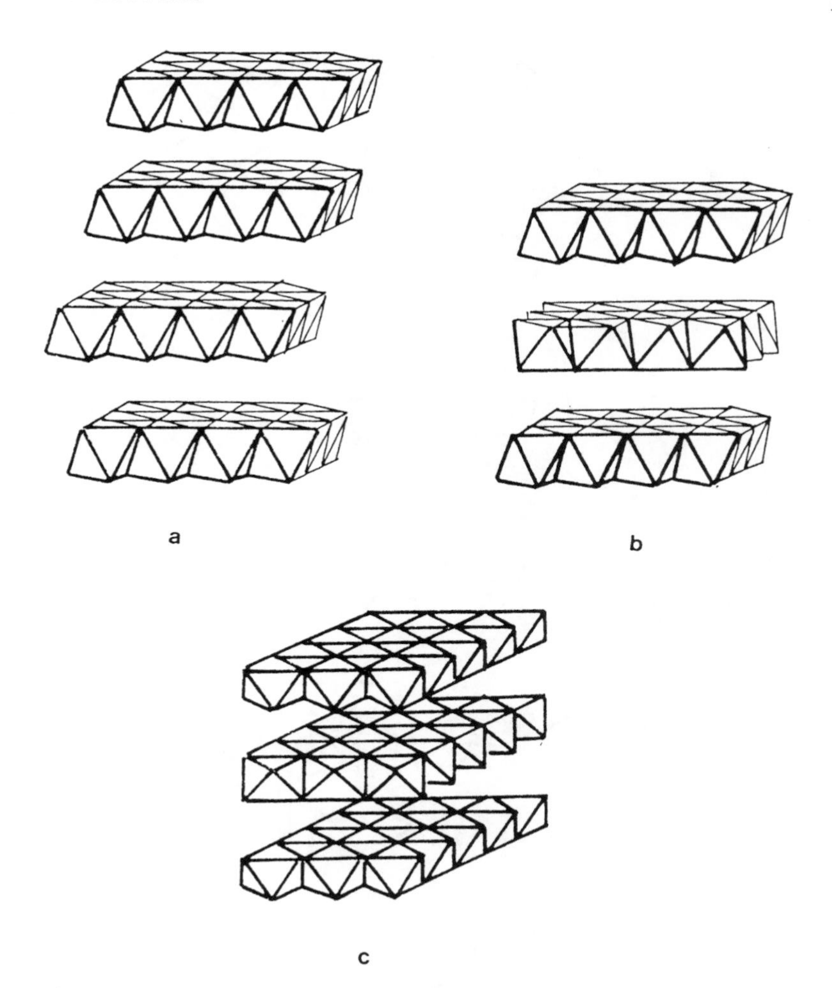

Figure 96. Structure of oxides with close-packed *ab*-type anionic layers: (a) $K_{0.50}CoO_2$-type, (b) $K_{0.67}CoO_2$, and (c) α-$NaMnO_2$-type. (After Fouassier et al.[94])

easily (e.g., intergrowth of phase B and duttonite). The close relationship between these structures and the rutile structure, and especially the diaspore and ramsdellite structures, is remarkable. The distances between the layers permit them to adapt to three-dimensional frameworks. For instance, montroseite (VOOH) and paramontroseite (VO_2), which exhibit tunnel structures (of the ramsdellite type), form intergrowths with the lamellar phase B. All these octahedral layer oxides show interesting magnetic, ion exchange, and intercalation properties.

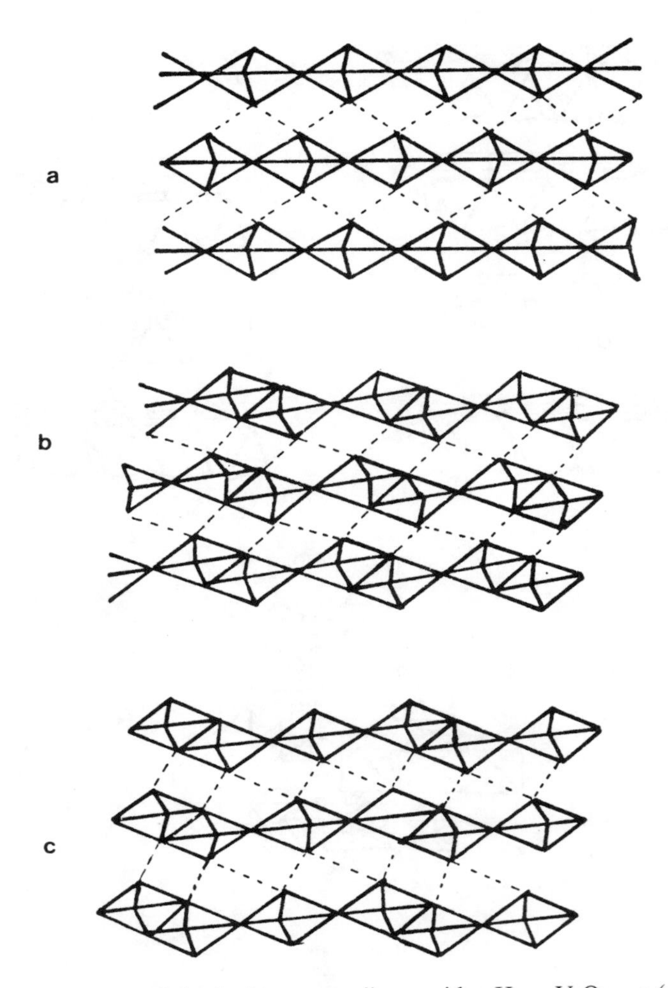

Figure 97. Structure of the hydrous vanadium oxides $H_{2n-2}V_nO_{3n-2}$: (a) duttonite, $VO(OH)_2$ ($n = \infty$), (b) haggite, $V_4O_6(OH_6)_6$ ($n = 4$), and (c) phase B, $V_6O_{10}(OH)_6$ ($n = 6$). (After Wadsley.[37])

8 Close-Packed Oxides: Spinels, Hexagonal Ferrites, and Relatives

Among the oxides containing close-packed arrangements of oxygen atoms, oxides of the formula B_3O_4, where B is a small cation, represent an important group. The most famous family is undoubtedly that of spinels, whose catalytic and magnetic properties have received much attention. Besides the spinels, the well-known olivine and the double-hexagonal (DH) form of $LiFeSnO_4$ have anionic close packing and are closely related to the spinel structure. Existence of nonstoichiometric spinels such as δ-Fe_2O_3 and $Zn_2Mo_3O_8$ shows that defects can be introduced at the metallic sites, maintaining the anionic close packing. In this respect, attention must be drawn to an important family having the stoichiometry B_5O_8, closely related to the DH $LiFeSnO_4$ structure. Hexagonal ferrites possess spinel layers that alternate with close-packed layers where barium partially replaces the oxygen; structures of these ferrites are also related to that of DH $LiFeSnO_4$.

8.1 Close-Packed and Pseudo-Close-Packed B_2O_4 Oxides

The spinel structure has been extensively described by numerous authors. More than 30 ions with radii ranging from 0.5 to 1 Å can be incorporated in spinel-like phases. The cubic structure of these compounds ($a \approx 8$ Å) can be described as a cubic close-packing (*abc*) of the oxide ions with one-eighth of the tetrahedral interstices and half of the octahedral interstices occupied by the metallic ions. If one considers the nature of the B ions, two extreme types of spinel can be distinguished: the normal spinels, $B'B''_2O_4$, in which B′ and B″ are located in the tetrahedral and octahedral sites, respectively (e.g., $MgAl_2O_4$), and the inverse spinels, $B''(B'B'')O_4$, in which the tetrahedral sites are occupied only by the B″ ions and the octahedral sites are occupied half each by the B′ and the B″ ions [e.g., $Fe(MgFe)O_4$, $Zn(SnZn)O_4$]. The intermediate cationic distributions between these two ordered states have been observed and are characterized by the parameter λ (fraction of B atoms in the tetrahedral sites), which describes the degree of inversion, ranging from 0 (normal) to $\frac{1}{2}$ (inverse). The result is that the spinel oxides exhibit numerous order–disorder transitions, often involving the formation of superstructures that influence the physical properties. A helpful way to illustrate the structure as an anionic close packing is to adopt the schematic drawing shown in Figure 98a,[95] which presents the atoms in the mirror plane containing the $\langle 111 \rangle$ spinel axis and shows the anionic layer packing.

To understand the properties of these materials, it is important to visualize the assemblage of polyhedra in the structure: a cation-deficient octahedral layer (Fig. 99a), whose cationic lattice is called kagome, denoted by Oc_3, and a mixed layer built up from corner-sharing octahedra and tetrahedra (Fig. 99b), denoted by Te_2Oc. The junction of two successive layers (Oc_3) and (Te_2Oc) occurs by corner-sharing of the polyhedra. This junction is characterized by the manner of blocking the kagome window (Fig. 99c). An *abc* anionic stacking is obtained by the occlusion of the kagome window by a tetrahedron of the (Te_2Oc) layer; it shares three corners with the octahedra of the kagome window. This type of junction is denoted $J_{K/T}$ (Fig. 99d). All the junctions between two successive layers in the spinel structure are of $J_{K/T}$ type, and the building unit involves the association of two layers: a double layer, denoted by $(Oc_3Te_2Oc)_{DL}$ or S_{DL} as shown in Figure 100a. The spinel structure results from the association of three S_{DL} double-layers through $J_{K/T}$ junctions (Fig. 100b).

A well-known close-packed oxide of B_3O_4 type is olivine, $(Mg,Fe)_2SiO_4$ (Fig. 98b). It may be described as a hexagonal *ab* close packing of oxide layers with

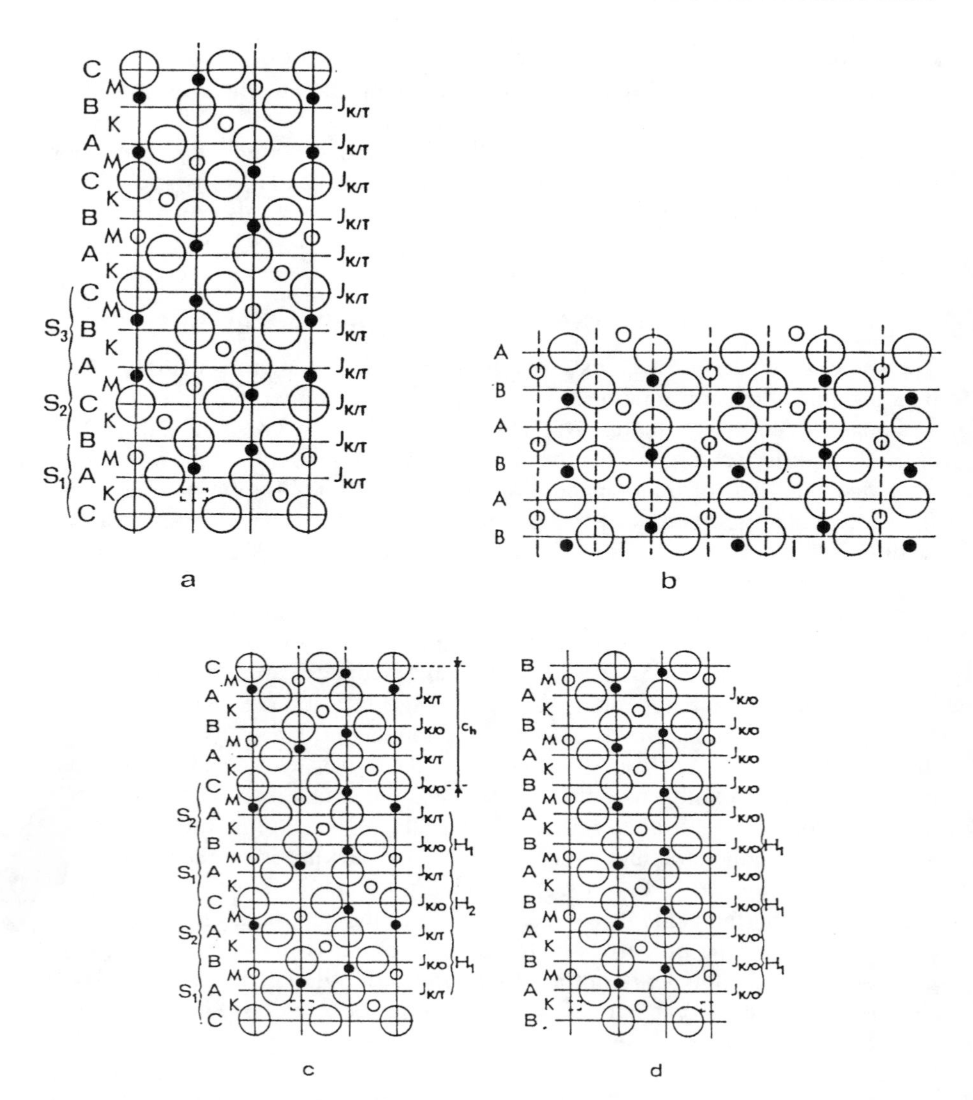

Figure 98. Schematic drawings of the anionic close-packing and cationic distribution in: (a) a spinel, (b) olivine, (c) DH $LiFeSnO_4$, and (d) a hypothetical hexagonal structure. Large circles represent the oxygen atoms and small ones correspond to the metallic atoms. (After Lacorre et al.[95])

Si atoms in the tetrahedral holes and Mg or Fe atoms in the octahedral holes. The fraction of occupied octahedral and tetrahedral holes is identical to that in the spinel structure. When one examines the olivine structure in terms of polyhedra, it becomes apparent that the framework is built up of only one type of polyhedral layer, having the formulation ($TeOc_2$) as shown in Figure 101a. This

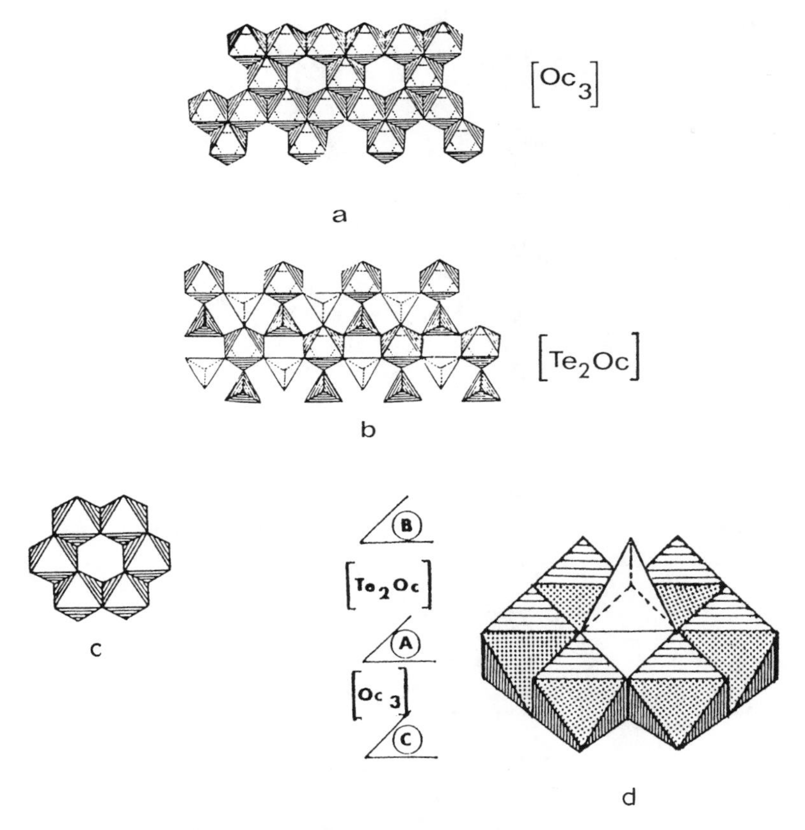

Figure 99. Polyhedral layers: (a) (Oc_3) kagome octahedral layer, (b) (Te_2Oc) mixed layer, and (c) kagome window resulting from the cation deficiency in the octahedral layer and its blocking by a tetrahedron (d) to ensure a $J_{K/T}$ junction. (After Lacorre et al.[95])

layer is characterized by double ribbons of edge-sharing octahedra similar to those in the (Oc_3) octahedral layer and by double ribbons of corner-sharing octahedra and tetrahedra similar to those in the (Te_2Oc) mixed layer. The olivine layer, ($TeOc_2$), can thus be considered as an intergrowth of the (Oc_3) and (Te_2Oc) layers. In such a layer, there is another type of window, built up of four octahedra and one tetrahedron; this layer is denoted K′, by analogy with the kagome window, as illustrated in Figure 101b. The junction of successive layers in the olivine structure is ensured by the occlusion of the K′ window by two octahedra ($J_{K'/O}$-type junction), as shown in Figure 101c.

Close structural relationships can be developed to generate the double hexagonal $LiFeSnO_4$ (DH $LiFeSnO_4$) structure found in $Sn_{2.8}Li_{1.8}M_{1.6}O_8$ and in the low temperature form of $LiFeSnO_4$. The hexagonal structure ($a \approx 6$ Å, $c \approx 9.8$ Å) of DH $LiFeSnO_4$ (Fig. 98c) exhibits a double-hexagonal close packing of the

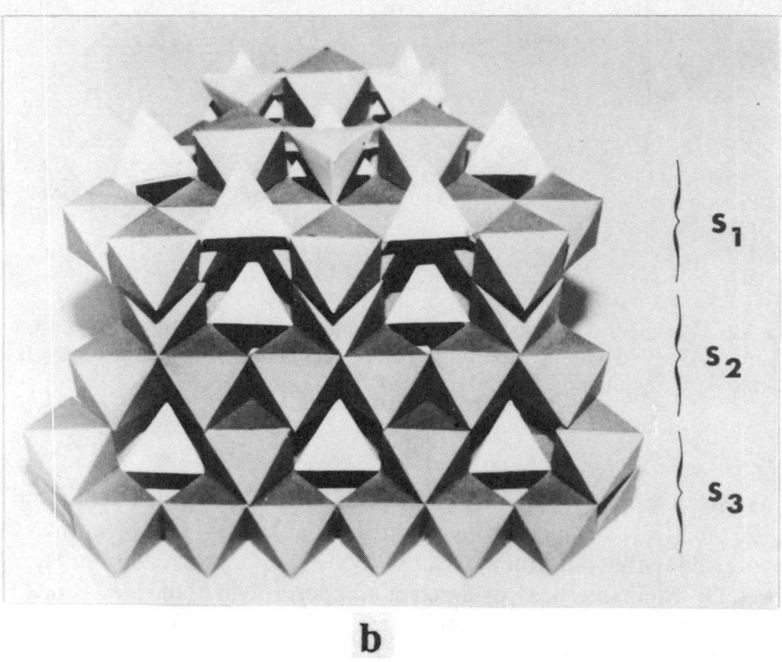

Figure 100. (a) S_{DL} double layer with the composition $(Oc_3Te_2Oc)_{DL}$ and (b) the spinel structure resulting from the association of three S_{DL} double layers through a $J_{K/T}$ junction. (After Lacorre et al.[95])

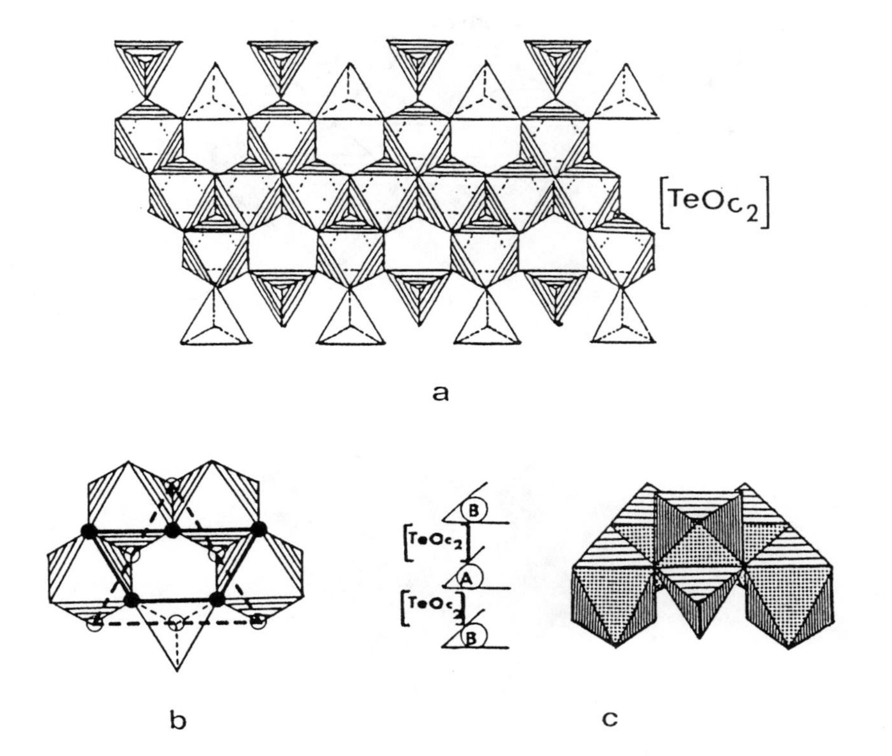

Figure 101. Olivine structure: (a) mixed layer ($TeOc_2$) and (b) type of window appearing in this cation-deficient layer and the occlusion of the window by an octahedron through a $J_{K'/O}$ junction (c). (After Lacorre et al.[95])

oxide ions (*abc*) with an occupancy of the interstitial holes similar to that of the spinel and the olivine structures (one-eighth of the tetrahedral interstices occupied by Li^+ ions). There are two types of tetrahedral site, denoted T_I and T_{II}, and two types of octahedral site, denoted Oc_I and Oc_{II}.

An examination of the coordination polyhedra shows that the DH structure is built up of the same polyhedral layers, (Oc_3) and (Te_2Oc), as the spinel structure (Fig. 99a,b). The *aba* stacking of the three oxygen layers involves the occlusion of the kagome window by one octahedron, leading to a $J_{K/O}$-type junction as shown in Figure 102a. A double-layer H_{DL} is characterized by the association of a kagome layer and a mixed layer by corner-sharing of the polyhedra through a $J_{K/O}$-type junction (Fig. 102b). The *abac* stacking of the oxygen layers involves the existence of junctions of both the $J_{K/O}$ and $J_{K/T}$ types (a kagome window is blocked by an octahedron on one side and a tetrahedron on the other side). Thus, the DH $LiFeSnO_4$ structure can be described as the association of two S_{DL} double layers through $J_{K/O}$ junctions or two H_{DL} double layers through $J_{K/T}$ junctions as shown in Figure 102c and Table 9.

A comparison of the three close-packed structures, spinel, olivine, and DH

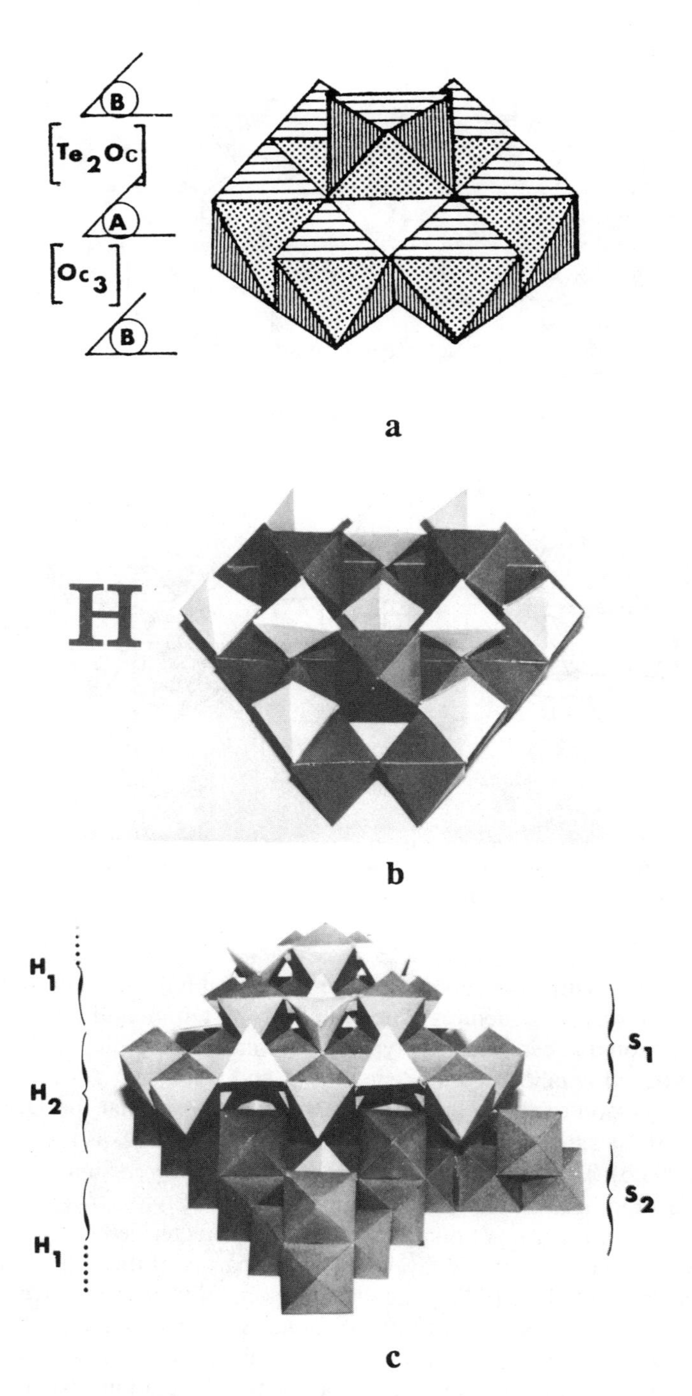

Figure 102. DH $LiFeSnO_4$ structure: (a) $J_{K/O}$ junction, (b) H_{DL} double layer having the composition $(Oc_3Te_2Oc)_{DL}$, and (c) association of two H_{DL} double layers. (After Lacorre et al.[95])

Table 9. Association of Two S or H Double Layers in the DH $LiFeSnO_4$-Type Structure

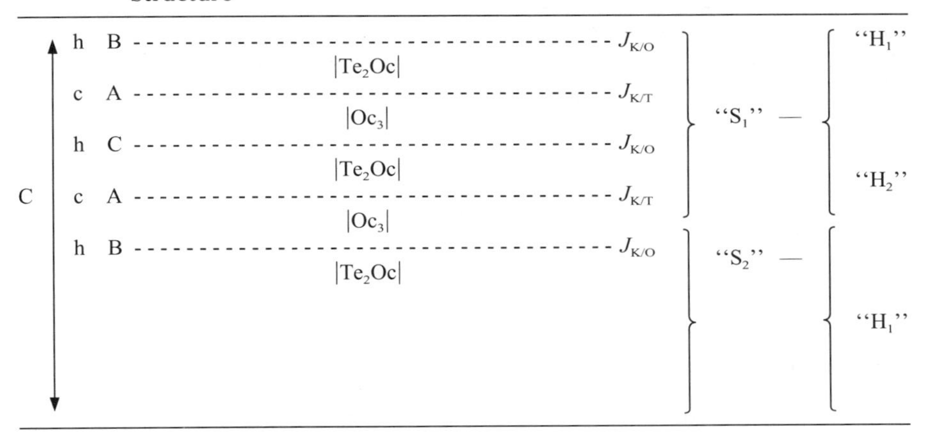

$LiFeSO_4$, shows close relationships but also differences resulting from the different types of junctions of the polyhedral layers. From this analysis, it appears that numerous hypothetical structures can be predicted by combining the layers of polyhedra and the types of junction. We describe here only the simplest hypothetical structure, involving the combination of identical double-layer H_{DL} and/or S_{DL} (Table 10). Besides the previously described spinel and DH $LiFeSnO_4$, we see that a simple hypothetical structure with a hexagonal *ab* close packing of the oxygen atoms can be obtained. This hypothetical hexagonal structure (HH) is built up of double H_{DL} layers through $J_{K/O}$ junctions (Fig. 98d).

At first sight, oxides with the ramsdellite structure do not exhibit any particular relationship with the B_3O_4 oxides because the stoichiometric formulation of the former oxides corresponds to BO_2. However, polymorphism of $LiFeSnO_4$, which gives a high temperature form with the ramsdellite structure, raises the possibility of close relationships between the DH and ramsdellite forms of $LiFeSnO_4$.

The ramsdellite structure (γ-MnO_2) is usually described as formed of multiple octahedral chains and is compared to the structures of rutile and hollandite (see Section 5.1). The high temperature form of $LiFeSnO_4$ exhibits the same $FeSnO_4$ framework, but in addition, the lithium ions occupy the tetrahedral sites in the tunnels of the framework. Analysis of the structure shows that the oxygen atoms form hexagonal pseudo-close-packed layers along two directions: a pseudo-close-packing along $\langle 001 \rangle$ close to that in diaspore, and a pseudo-close-packing *abac* along $\langle 010 \rangle$ similar to that in DH $LiFeSnO_4$. The oxygen framework of the ramsdellite structure differs from that of DH $LiFeSnO_4$ in the slight displacement of the oxygen rows parallel to $\langle 100 \rangle$. The main difference between the two

Table 10 Sequence of Operations Involved in DH $LiFeSnO_4$–Spinel Transformation

Double Layer	Junction	Polyhedral Layer	Oxygen Layer	Operation	Oxygen Layer	Polyhedral Layer	Junction	Double Layer
	$J_{K/O}$	$\vert Te_2Oc\vert$	C		C	$\vert Te_2Oc\vert$	$J_{K/T}$	
"S_2"	$J_{K/T}$	$\vert Oc_3\vert$	A	$R_{T_{II}}$ $\rightarrow$	B	$\vert Oc_3\vert$	$J_{K/T}$	"S_3"
	$J_{K/O}$	$\vert Te_2Oc\vert$	B		A	$\vert Te_2Oc\vert$	$J_{K/T}$	
"S_1"	$J_{K/T}$	$\vert Oc_3\vert$	A	T $\rightarrow$	C	$\vert Oc_3\vert$	$J_{K/T}$	"S_2"
	$J_{K/O}$	$\vert Te_2Oc\vert$	C		B	$\vert Te_2Oc\vert$	$J_{K/T}$	
"S_2"	$J_{K/T}$	$\vert Oc_3\vert$	A	R_{Oc_I} $\rightarrow$	A	$\vert Oc_3\vert$	$J_{K/T}$	"S_1"
	$J_{K/O}$	$\vert Te_2Oc\vert$	B		C	$\vert Te_2Oc\vert$	$J_{K/T}$	
"S_1"	$J_{K/T}$	$\vert Oc_3\vert$	A	T $\rightarrow$	B	$\vert Oc_3\vert$	$J_{K/T}$	"S_3"
	$J_{K/O}$	$\vert Te_2Oc\vert$	C		A	$\vert Te_2Oc\vert$	$J_{K/T}$	
"S_2"	$J_{K/T}$	$\vert Oc_3\vert$	A	R_{T_I} $\rightarrow$	C	$\vert Oc_3\vert$	$J_{K/T}$	"S_2"
	$J_{K/O}$	$\vert Te_2Oc\vert$	B		B	$\vert Te_2Oc\vert$	$J_{K/T}$	
"S_1"	$J_{K/T}$	$\vert Oc_3\vert$	A	I $\rightarrow$	A	$\vert Oc_3\vert$	$J_{K/T}$	"S_1"
	$J_{K/O}$		C		C		$J_{K/T}$	

structures is in the distribution of the metal ions. The passage from one structure to the other can be explained in two steps, using an intermediate hypothetical structure having tunnels with angles of 60° and 120° where Li^+ would be located. A single migration of the metal ions leads to the DH $LiFeSnO_4$ structure.

It appears that the ramsdellite structure should be more stable for B_3O_4-type oxides than for BO_2 oxides, which themselves tend to exhibit the rutile or the columbite structures. Formation at high temperatures of nonstoichiometric ramsdellite, $A_xB_2O_4$ such as $Li_2Ti_3O_7$ ($x = 0.857$) and $Li_{1.42}Fe_{0.75}Sn_{1.08}O_4$ ($x = 1.25$) tends to support this point of view (see Section 8.3).

8.2 Close-Packed B_5O_8 Oxides: Relationship with the DH $LiFeSnO_4$ Structure

The B_5O_8 oxides can be divided into two classes according to their structures: $A_2Mo_3^{IV}O_8$ (A = Mg, Zn, Mn^{II}, Fe^{II}, Co^{II}, Ni^{II}, Cd^{II}) and $A_2Mn_3^{IV}O_8$ (A = Mn^{II}, Ca^{II}, Cd^{II}, Co^{II}, Zn^{II}, Cu^{II}).

8.2.1 $A_2Mo_3O_8$ Oxides

The structural similarity between the nolanite-type $A_2Mo_3O_8$ oxides and DH $LiFeSnO_4$ is remarkable. There is only one difference between the two stackings:

namely, the tetrahedral T_{II} sites in the $A_2Mo_3O_8$ are empty with respect to the DH $LiFeSnO_4$ structure. In $A_2Mo_3O_8$, Mo(IV) ions are located in the octahedra (Oc_{II}) of the kagome layer, whereas the A(II) ions are located in the tetrahedral T_I sites and octahedral (Oc_I) sites of the mixed layer. Analogy between the mixed layers of the two structures is shown in Figure 103. The $A_2Mo_3O_8$ structure can thus be considered as a cation defect DH $LiFeSnO_4$ structure (Fig. 104a). The two compounds exhibit the same space group $P6_3mc$ and have nearly identical parameters ($a \approx 6$ Å; $c \approx 10$ Å).

Substitution of a monovalent/trivalent ion couple for the bivalent A ion is possible without changing the structure. This is the case in $LiScMo_3O_8$ and $LiInMo_3O_8$. On the other hand, $LiLnMo_3O_8$ (Ln = Sm, Gd, Tb, Dy, Ho, Er, Yb) and $LiYMo_3O_8$ are characterized by a different anionic stacking. In these compounds, the stacking of the oxygen atoms is identical to that described for the hypothetical hexagonal structure HH (see Section 8.1, Fig. 98d); that is, it consists of *ab* stacking. The structure of the $LiLnMo_3O_8$ oxides can be described as

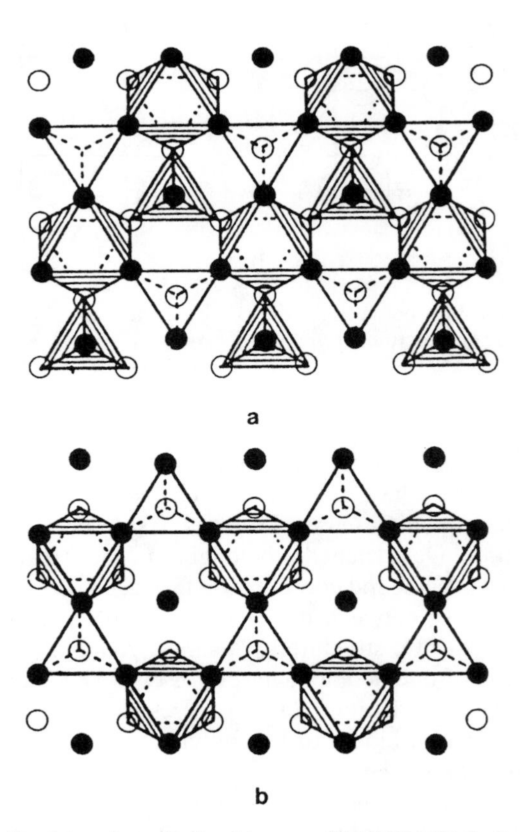

Figure 103. Idealized drawing of mixed layers of (a) DH $LiFeSnO_4$ and (b) $A_2Mo_3O_8$. (After Lacorre et al.[95])

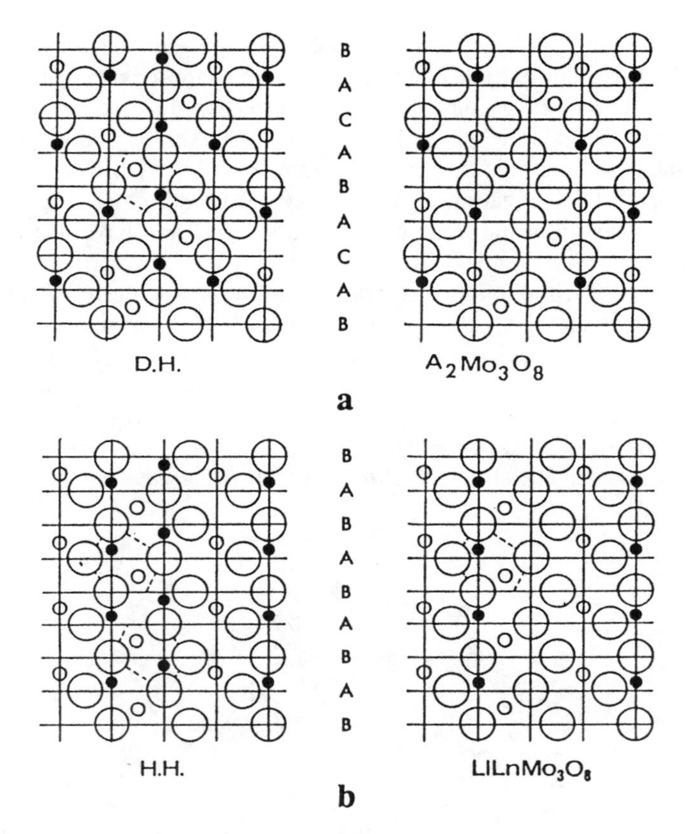

Figure 104. Relationships between (a) the DH $LiFeSnO_4$ and $A_2Mo_3O_8$ structures and (b) HH and $LiLnMo_3O_8$ structures.

an HH structure in which one tetrahedral site out of two in the mixed layer is unoccupied (Fig. 104b). The Ln^{3+} ions are located in the octahedral sites and the Li^{+} ions in the remaining tetrahedral sites. As described earlier, the Mo^{4+} ions are located in the Oc_{II} octahedra belonging to the (Oc_3) layers. It appears that two Ln^{3+} ions in octahedral coordination face each other through a kagome window. This similarity between the HH and $LiLnMo_3O_8$ structures can be deduced from the $A_2Mo_3O_8$ structure by a simple rotation (by $\pi/3$) around the axis of a tetrahedron of one double layer out of two. The $LiLnMo_3O_8$ structure (Fig. 104b), like that of DH $LiFeSnO_4$, possesses short intermetallic distances. Both the structures exhibit blocks of five polyhedra built up of three octahedra and one tetrahedron sharing corners. These blocks are not observed in the other $A_2Mo_3O_8$ oxides in spite of their similarity with DH $LiFeSnO_4$, since the T_{II} sites are empty. Thus, the stability of $LiLnMo_3O_8$ and DH $LiFeSnO_4$ structures with respect to those of $A_2Mo_3O_8$ structures may be due to the small charge of

Li^+ in the tetrahedral sites (due in turn to a decrease in the Coulombic repulsion). It appears that the DH $LiFeSnO_4$ structure is stabilized by the presence of a d^{10} cation as in $Li_2Fe_{3-x}Cr_xSbO_8$.

8.2.2 $A_2Mn_3O_8$ Oxides

The $A_2Mn_3O_8$ oxides are closely related to $A_2Mo_3O_8$ and DH $LiFeSnO_4$. For example, $Co_2Mn_3O_8$ and $Zn_2Mn_3O_8$ have a structure closely related to $A_2Mo_3O_8$ or DH $LiFeSnO_4$ but are orthorhombic ($a \approx 6$ Å, $b \approx 5$ Å, $c \approx 10$ Å). The oxygen atoms form a hexagonal anionic close packing (*abac*) similar to that in DH $LiFeSnO_4$ and $A_2Mo_3O_8$ ($c/a = 1.63$). The structure can be described as arising from the stacking of octahedral layers, which are three-quarters occupied by Mn^{4+}, and of mixed layers of octahedra and tetrahedra containing Co^{2+} and Zn^{2+} ions. Cation distribution within the layers is however different. Figure 105 compares two types of octahedral layers. These arrangements can be described as the intergrowth of rows of edge-sharing octahedra (rutile ribbons) with half-occupied rutile rows. One-quarter of the octahedral sites are empty with respect to a complete octahedral layer of the CdI_2 type. The two structures differ in the relative positions of the octahedral windows, which are distributed in a hexagonal array in DH $LiFeSnO_4$ and $A_2Mo_3O_8$ ($a_H \approx 6$ Å) (Fig. 105a) and in an orthorhombic distribution of the windows in $Co_2Mo_3O_8$ ($a_0 \approx a_H$ and $b_- \approx a_H\sqrt{3}/2$) (Fig. 105b). The outcome is that layers of DH $LiFeSnO_4$ can be described as fully occupied intersecting rutile rows running along the three directions $\langle 100\rangle_H$, $\langle 010\rangle_H$, and $\langle 110\rangle_H$, whereas in $A_2Mo_3O_8$, the rutile rows run only along $\langle 100\rangle_O$ and are isolated. The A^{2+} ions (Co^{2+}, Zn^{2+}) of $A_2Mn_3O_8$ are located (on both sides of the octahedral windows) in octahedral and tetrahedral coordinations exactly as in $A_2Mo_3O_8$. The stacking of the octahedral layers and the relative positions of the A^{2+} ions belonging to the mixed layers with respect to the windows are shown in Figure 106.

The mixed layer of $Co_2Mn_3O_8$ is also related to that in $A_2Mo_3O_8$. The layers in both cases can be described as formed of infinite mixed bands of octahedra and tetrahedra (TeOc) parallel to $\langle 100\rangle_H$ or $\langle 100\rangle_O$. These arrangements can be derived from each other by a translation of an O—O distance. Such an analysis shows that $Co_2Mn_3O_8$ can be considered as a stacking of defect DH $LiFeSnO_4$-type slices, which are two polyhedra thick, whereas $A_2Mo_3O_8$ is formed of identical slices shifted (one out of two) by an O—O distance.

The Mn_5O_8 structure has $(Mn_3O_8)^{4-}$ layers identical to those in $Co_2Mn_3O_8$ (Fig. 106a), with three-quarters of the octahedral sites occupied by manganese, but the stacking of these layers is not dense. Two successive $(Mn_3O_8)^{4-}$ layers have their oxygen planes forming an *aa* stacking. This results in trigonal prismatic cavities where the Mn^{2+} ions are located; two such Mn^{2+} ions face each other through the octahedral window of the $(Mn_3O_8)^{4-}$ layers; stacking of these layers is shown in Figure 106b. $Ca_2Mn_3O_8$ and $Cd_2Mn_3O_8$ exhibit the same stacking of the $(Mn_3O_8)^{4-}$ layers with Ca^{2+} or Cd^{2+} in prismatic coordination, but the unit cell is monoclinic ($a_M \approx 10.8$ Å, $b \approx 5.8$ Å, $c \approx 4.9$ Å, $\beta = 109°$)

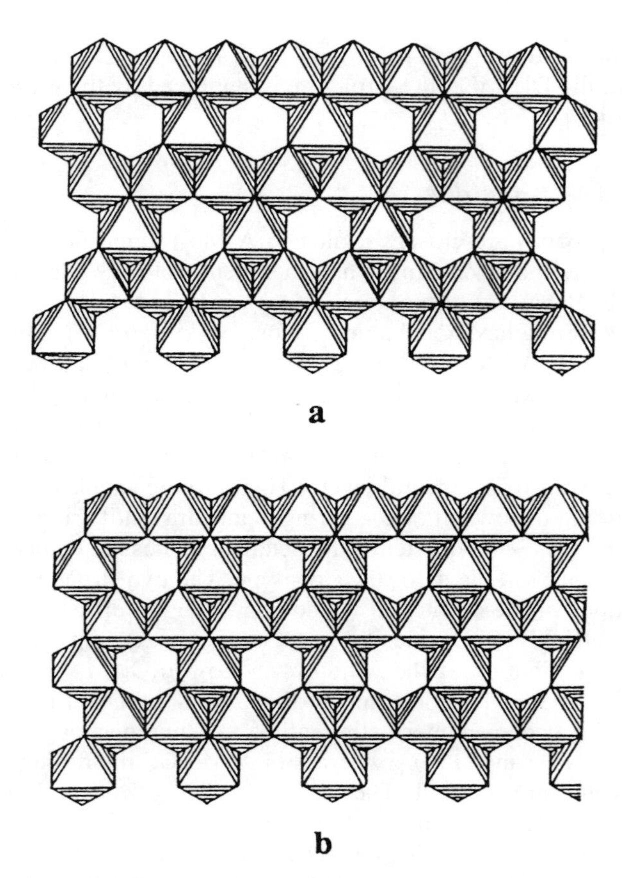

Figure 105. (a) Kagome layer of DH $LiFeSnO_4$ and the $A_2Mo_3O_8$ structure and (b) octahedral layer of $A_2Mo_3O_8$. (After Lacorre et al.[95])

owing to the distortions introduced by the larger size of the Cd^{2+} and Ca^{2+} ions with respect to Mn^{2+}.

8.2.3 $Cu_2Mn_3O_8$

Its characteristic $(Mn_3O_8)^{4-}$ layers notwithstanding, the stacking of the oxygen atoms in $Cu_2Mn_3O_8$ is not dense. Two successive $(Mn_3O_8)^{4-}$ layers form square-pyramidal sites where the Cu^{2+} ions are located. (This arrangement results from the ability of Cu^{2+} to take up fivefold coordination.) This coordination polyhedron of Cu^{2+} is better described as a pentahedron formed of an equilateral triangle (coordination 3 + 2). The three oxygen atoms that form the equilateral triangle belong to the same empty octahedron of the $(Mn_3O_8)^{4-}$ layers. Thus, two Cu^{2+} ions in pentahedral coordination face one another through each octa-

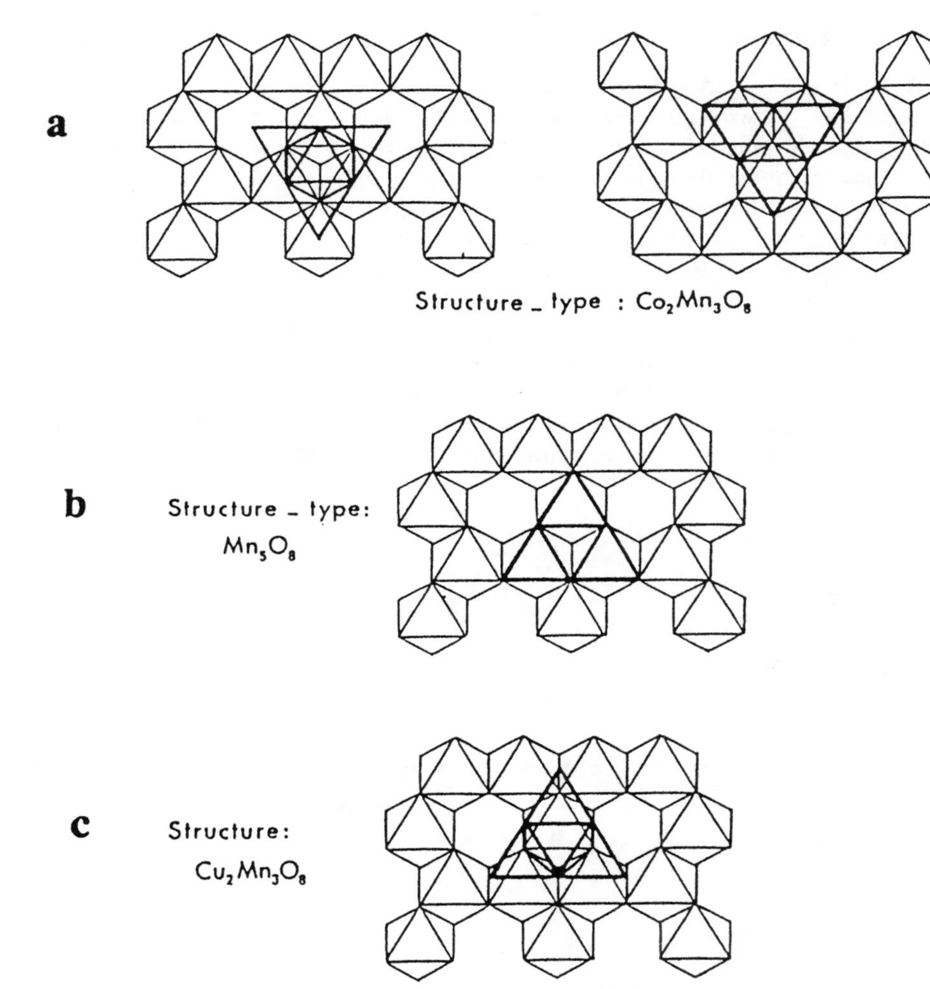

Figure 106. Octahedral layer stacking in $A_2Mn_3O_8$: (a) $Co_2Mn_3O_8$, (b) Mn_5O_8, and (c) $Cu_2Mn_3O_8$ (The largest bold-lined triangle corresponds to the windows of the layer close to the drawn layer.) The A cations are located between the two octahedral layers, inside the triangles; the sites are drawn in bold lines). (After Lacorre et al.[95])

hedral window. This coordination of Cu^{2+} can be considered as intermediate between the tetrahedral (3 + 1) and the octahedral (3 + 3) coordinations. Figure 106c shows the stacking of the octahedral layers and the position of the pentahedral sites with respect to the octahedral windows. This oxide exhibits a monoclinic cell (a = 9.69 Å, b = 5.63 Å, c = 4.91 Å, β = 103°), closely related to the other structures. This is the only example of an oxide with this structure (arising from the specific coordination of Cu^{2+}).

8.2.4 General Remarks

All the $A_2B_3O_8$ oxides describes here possess the cation-deficient octahedral layers $(B_3O_8)^{4-}$, where one octahedral site out of four is unoccupied with respect to the fully occupied CdI_2-type layer. Whatever the distribution of the windows in the layer may be, the windows correspond to a defect of positive charge with respect to the other sites, where the M^{4+} ions are located. Thus, the A^{2+} ions tend to occupy these vacancies, facing one another through the windows. Coordination of the A^{2+} ion determines the stacking of the $(B_3O_8)^{4-}$ layers. Hexagonal windows of the $(B_3O_8)^{4-}$ layers are similar in all these structures. The different structures can be distinguished on the basis of polyhedra facing one another through the windows. These different possibilities for all the structures described earlier (spinel, DH $LiFeSnO_4$, HH, and $A_2Mo_3O_8$) are shown schematically in Figure 107. The various coordinations can adapt to the octahedral windows, intermediate between tetrahedra and the octahedra. It is noteworthy

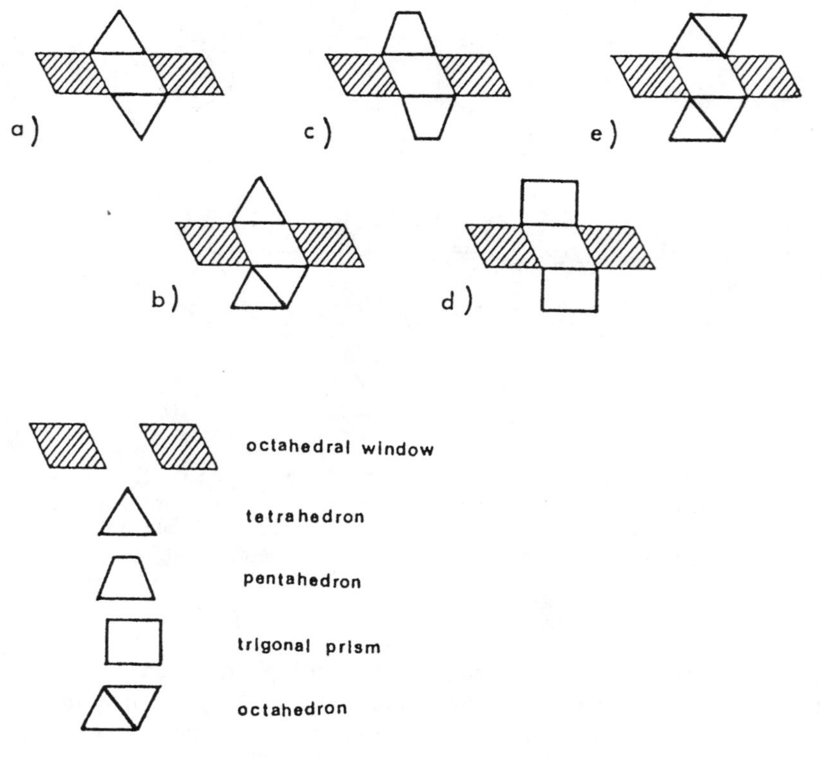

Figure 107. Different possibilities of the polyhedra facing through the octahedral windows: (a) two tetrahedra in spinel, (b) one tetrahedron and one octahedron in DH $LiFeSnO_4$, nolanite, and $Co_2Mn_3O_8$, (c) two pentahedra in $Cu_2Mn_3O_8$, (d) two trigonal prisms in Mn_5O_8, and (e) two octahedra in $LiLnMo_3O_8$ and HH. (After Lacorre et al.[[95]])

that such B_3O_8 layers are also observed in zinc hydroxides such as $Zn_2(OH)_8Cl_2 \cdot H_2O$ and $Zn_5(OH)_8(NO_3)_2 \cdot 2H_2O$, as well as in $U_2Co_3O_8$.

8.3 Hexagonal Ferrites: Relationship with Close-Packed B_5O_8 and B_3O_4 Oxides

Hexagonal ferrites form a large family of oxides belonging to the system AO/Fe_2O_3/MeO (A = Ba, Pb, Sr; Me = Mg, Fe, Co, Ni, Zn). Many coupled substitutions in these oxides, such as the replacement of bivalent Me^{II} by a couple, Li^+/Fe^{3+} or Li^+/Sn^{4+}, are possible without changing the structure. Among these oxides, the barium compounds are the most numerous, with more than 60 different oxides known up to the present time. The structure of the hexagonal ferrites ($a \approx 6$ Å) results from the stacking of close-packed oxygen slabs denoted as O_4 and the A-substituted anionic layers, AO_3 (Fig. 108). Close packing of these anionic layers with the B cations located in their cavities gives rise to structural units of four types, called S, R, T, and Q, defined by the cationic planes perpendicular to the c-axis.

The S units characterize a structural element of the spinel structure corresponding to a cubic face-centered stacking of the oxygen ions (*abc*) with B = Me + Fe in the tetrahedral and octahedral sites (Fig. 109).These units are described by the composition B_6O_8. The R unit, also shown in Figure 109, has composition B_6BaO_{11} and is built up of two O_4 oxygen planes, in *aba* stacking, on both sides of one BaO_3 plane. The T units, with the formulation B_8BaO_{14}, contain two successive BaO_3 planes bounded by two O_4 oxygen planes. In these three types of unit, the octahedral and tetrahedral cavities are occupied by smaller cations. The Q and T units bear a close relationship, but differ in the type of stacking (*abc* and *ab*, respectively), as indicated in Figure 109. Hexagonal barium ferrites can be described as built up of stacks of these four different types of unit along $\vec{c}$. The general formula of these ferrites is given by

$$(B_6O_8)^{S}_{n_1}(B_6BaO_{11})^{R}_{n_2}(B_8Ba_2O_{14})^{T}_{n_3}(B_7Ba_2O_{14})^{Q}_{n_4}.$$

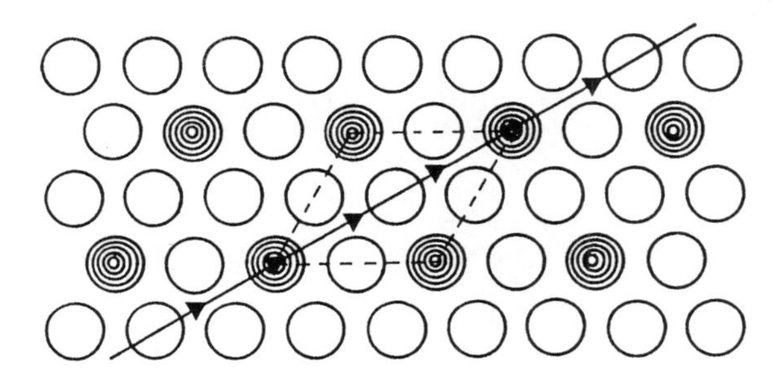

Figure 108. Anionic AO_3 layer of hexagonal ferrites: open circles, oxygen atoms; other circles, A atoms. (After Lacorre et al.[95])

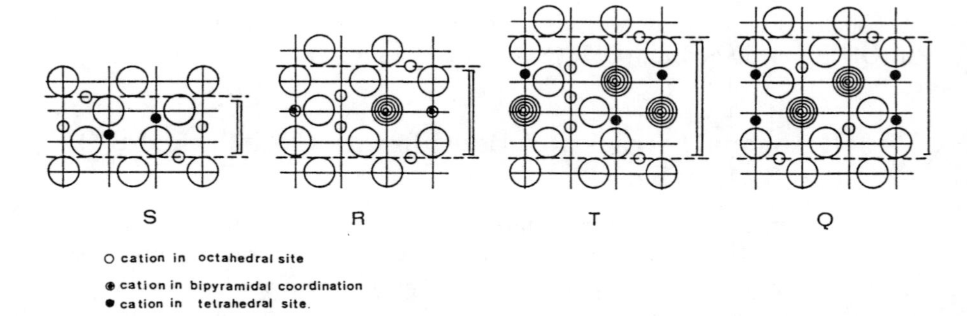

Figure 109. Schematic drawing of the S, R, T, and Q units of hexagonal ferrites. (After Wells.[4])

where n_1, n_2, n_3, and n_4 are integers. The stacking sequences are denoted by the letters S, R, T, and Q and also by the letters with asterisks (S*, R*, T*, and Q*); R* and S* units are obtained from R and S units, respectively, by rotation of 180° around the *c*-axis.

The theoretical number of the possible sequences in these ferrites is infinite, and many of these combinations have been observed. Even in the most complex ferrites, however, certain sequences of stacking are found: $Ba_2Fe_4O_8$(S), $Ba_2Fe_{12}O_{19}$ (M = R, S, R*, S*), and $Ba_2Me_2Fe_{12}O_{22}$ [Y = $(TS)_3$]. These units show preferences in stacking giving rise to two main series, M_nS and M_pY_n. The simplest members of these series are also represented by letters such as W, Z, X, and U. Table 11 presents these sequences. Their diagrammatic representations appear in Figure 110.

In the series M_pY_n (Z = MY), $Ba_3Me_2Fe_{24}O_{41}$ is the first member, with c = 52.3 Å; more complex sequences have been characterized by electron microscopy, examples being $MYMY_2MY_3MY_9$ ($c \approx 793$ Å) and $MY_6MY_{10}MY_7MY_{10}$ ($c = 1577$ Å). Other combinations of units such as QS ($Ba_2Sn_2MeFe_{10}O_{22}$), Q_3 ($Ba_2Cr_{6.5}O_{14}$), and RT* ($BaTi_2Fe_4O_{11}$) have been identified. Hexagonal ferrite structures can be simply considered as resulting from a stacking of polyhedral

Table 11 Different Sequences in Hexagonal Ferrites

Symbol of the Compounds	Formula	Block Sequence	*n* Values		
M	$BaFe_{12}O_{19}$	R S R* S*	$n_1 = 1$	$n_2 = 1$	$n_3 = n_4 = 0$
Y	$Ba_2Me_2Fe_{12}O_{22}$	$(TS)_3$	$n_1 = 1$	$n_2 = 0$	$n_3 = 1 n_4 = 0$
W = MS	$BaMe_2Fe_{16}O_{26}$	R S S R* S* S*	$n_1 = 1$	$n_2 = 0$	$n_3 = n_4 = 0$
X = M_2S	$Ba_2Me_2Fe_{28}O_{46}$	(R S R* S* S*)$_3$	$n_1 = 3$	$n_2 = 2$	$n_3 = n_4 = 0$
Z = MY	$Ba_3Me_2Fe_{24}O_{41}$	R S T S R* S* T* S*	$n_1 = 2$	$n_2 = 1$	$n_3 = 1 n_4 = 0$
U = M_2Y	$Ba_4Me_2Fe_{36}O_{60}$	R S R* S* TS*	$n_1 = 3$	$n_2 = 2$	$n_3 = 1 n_4 = 0$

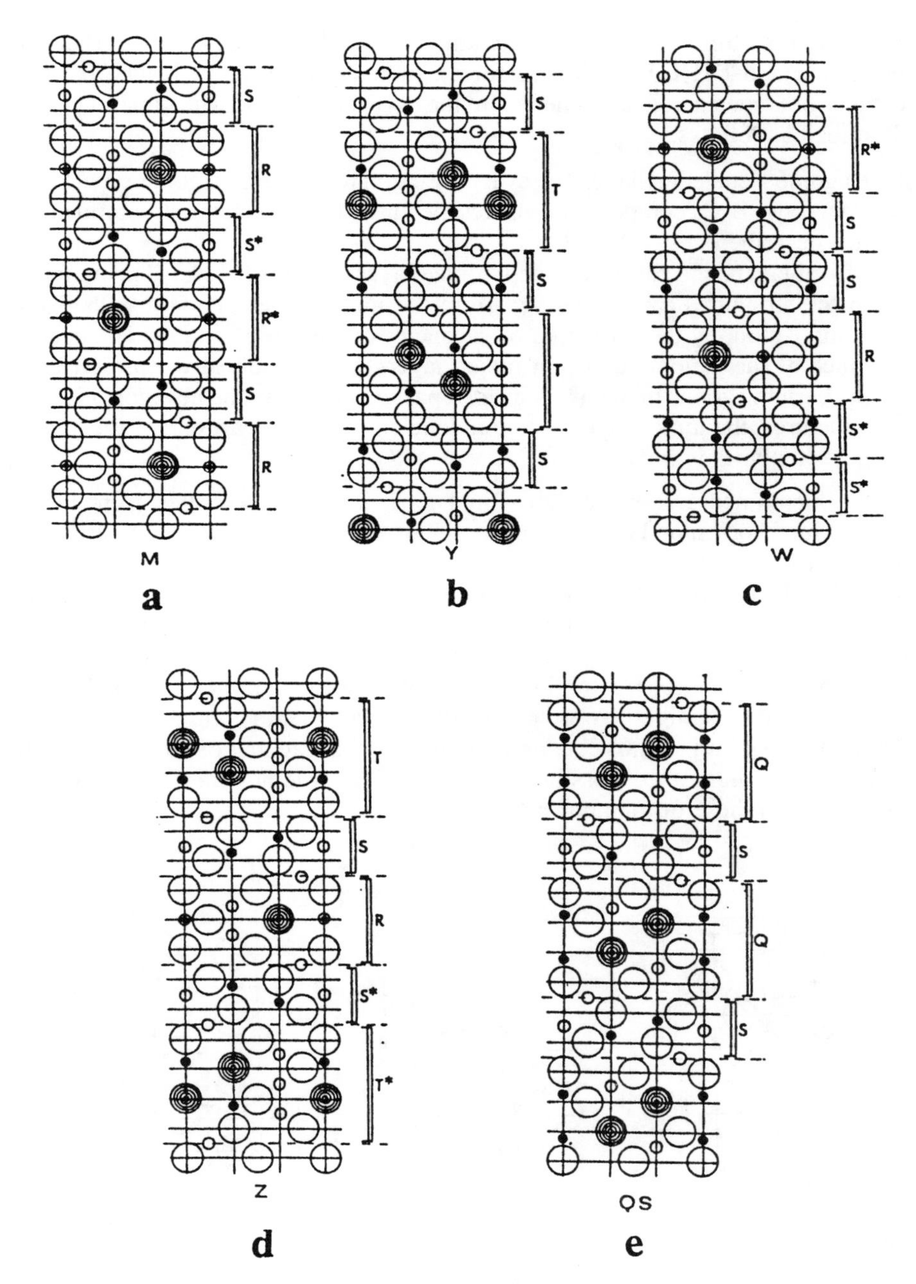

Figure 110. Diagrammatic representation of the hexagonal ferrites (a) M, (b) Y, (c) W, (d) Z, and (e) $Ba_2Sn_2MeFe_{10}O_{22}$. (After Wells.[4])

layers derived from the kagome layer and the mixed layer (described earlier for the spinel and DH $LiFeSnO_4$ structures). In the latter type of layers, barium cations are considered as spheres in the oxygen planes. The latter can be described as follows:

1. Kagome octahedral layer: in every hexagonal ferrite a kagome-type layer (Oc_3) with B_3O_4 composition ensures the junction between the structural units just described.
2. Mixed layers and cation-deficient mixed layers: the other layers encountered in the structures can be derived from the mixed layer (Te_2O_c), with B_3O_4 composition, by the formation of vacancies on the cationic sites. The occurrence of the cationic-deficient mixed layers is the result of the presence of the barium cations in the oxygen close-packed planes. In the hexagonal ferrite structures, the mixed layer is cation deficient when at least one of the close-packed planes is BaO_3.

We mentioned earlier that hexagonal ferrites can be described from the S units, which correspond to a part of the spinel structure. DH $LiFeSnO_4$ structure is similarly built up of units denoted H, bounded by two kagome cationic planes as shown in Figure 111 and can be described as HH*. In the same way, the hypothetical hexagonal stacking is built up from Hy units. The cationic layers of the H and Hy units are similar to those of the S units but differ in their relative positions. These close structural relationships between the three blocks, H, Hy, and S, provide a straightforward connection of the H and Hy units to the ferrite structures. The various possible permutations in the stacking of the S, M, Y, H, and Hy units suggest the possible occurrence of a large new family with structures close to those of the known ferrites.

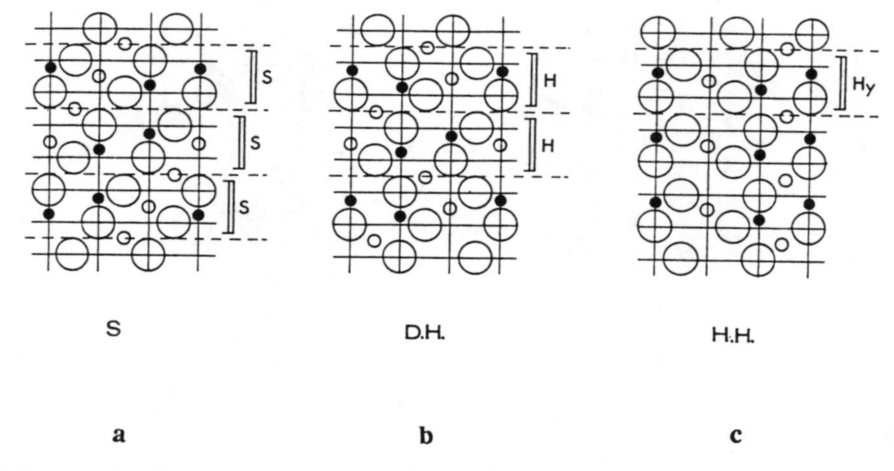

Figure 111. Structures of (a) spinel, (b) DH $LiFeSnO_4$, and (c) HH built up of S, H, and Hy units. (After Lacorre et al.[95])

9 Three-Dimensional Mixed Frameworks Involving Tetrahedra and Octahedra

Oxides formed by a mixed framework involving two kinds of polyhedra, especially BO_6 octahedra and BO_4 tetrahedra, constitute a large class. Silicates, germanates, and the phosphates are the most common among the mixed framework oxides. In Sections 9.1–9.3, we shall examine typical examples from these systems.

9.1 Silicates and Germanates

Among the silicates and germanates of transition elements, six families are of interest not only for their physical properties, but also with respect to the host lattices forming cages and tunnels. These are garnets, nasicons, benitoïtes, wadeites, milarites, and the oxides $Na_5BSi_4O_{12}$ and $A_{6-x}B_6Si_4O_{26}$ (or $A_{6-x}B_6Ge_4O_{26}$).

The garnets form a large family of minerals $A_3B_2Si_3O_{12}$ [A = Ca, Mg, Fe(II); B = Al, Cr, or Fe(III)]. As Table 12 indicates, numerous synthetic oxides of the formula $A_3B'_2B''_3O_{12}$ exhibit this structure. These materials have been extensively studied for nonlinear optical properties. Some of them (e.g., $Y_3Fe_5O_{12}$), exhibit ferromagnetism. This structural type is also found in simple oxides such as $CaGeO_3$ and $CdGeO_3$ prepared under high pressures. In the cubic oxides ($a \approx$ 12 Å), the three-dimensional host lattice $[B'_2B''_3O_{12}]_\infty$ consists of corner-sharing $B''O_6$ octahedra and $B'O_4$ tetrahedra (Fig. 112a). Each $B'O_4$ tetrahedron is linked to four $B''O_6$ octahedra, and reciprocally, each $B''O_6$ octahedron shares its six corners with $B'O_4$ tetrahedra. As a result, rows of polyhedra in which one $B'O_4$ tetrahedron alternates with one $B''O_6$ octahedron run along the $\langle 110 \rangle$ directions (Fig. 112b). The assemblage of these $[B'B''O_4]_\infty$ chains through the corners of the polyhedra forms the $[B'_2B''_3O_{12}]_\infty$ host lattice. This framework has cavities where the A cations are located in eightfold coordination. The presence of A cations is not absolutely essential for the stability of the garnet host lattice. Oxides such as $Al_2W_3O_{12}$ exhibit this structure with a significant distortion of the lattice. The octahedral B″ sites are sometimes partially occupied by the A cations as in $Y_4Al_4O_{12}$. On the other hand, in $Y_3Al_5O_{12}$, which also has the garnet structure, both the B′ and B″ sites are occupied by Al. The same framework is encountered in hydrogarnets such as hydrogrossularite, $Ca_3Al_2(OH)_{12}$, but distortion results in a cation coordination between that of an antiprism and a cube.

Nasicons are silicophosphates and can be classified either as silicates or as phosphates. They have the formula $Na_{1+x}Zr_2P_{3-x}Si_xO_{12}$, with x ranging from 0 to 3. In this oxide, zirconium can be replaced by titanium and germanium or even molybdenum and niobium. These materials are famous as superionic conductors. In this structure (Fig. 113a), each ZrO_6 octahedron shares its six corners with SiO_4 (or PO_4) tetrahedra.[96] This mixed framework does not form tunnels

Table 12 Oxides with the Garnet Structure

Substance	a_0 (Å)	Substance	a_0 (Å)
$Dy_2Fe_5O_{12}$	12.414	$Ca_2ZrNi(GeO_4)_3$	12.50
$Er_3Fe_5O_{12}$	12.349	$CdGd_2Mn_2(GeO_4)_3$	12.473
$Eu_3Fe_5O_{12}$	12.498	$Cd_3Al_2(GeO_4)_3$	12.08
$Gd_3Fe_5O_{12}$	12.44	$Cd_3Cr_2(GeO_4)_3$	12.20
$Ho_3Fe_5O_{12}$	12.380	$Cd_3Fe_2(GeO_4)_3$	12.26
$Lu_3Fe_5O_{12}$	12.277	$Cd_3Ga_2(GeO_4)_3$	12.19
$Nd_3Fe_5O_{12}$	12.60	$CoGd_2Co_2(GeO_4)_3$	12.402
$Sm_3Fe_5O_{12}$	12.530	$CoY_2Co_2(GeO_4)_3$	12.300
$Tb_3Fe_5O_{12}$	12.447	$CuGd_2Mn_2(GeO_4)_3$	12.475
$Tm_3Fe_5O_{12}$	12.325	$MgGd_2Mg_2(GeO_4)_3$	12.31
$Y_3Fe_5O_{12}$	12.376	$MgGd_2Mn_2(GeO_4)_3$	12.395
$Y_2NdFe_5O_{12}$	12.454	$MnY_2Mn_2(GeO_4)_3$	12.392
		$Mn_3Al_2(GeO_4)_3$	11.902
$Yb_3Fe_5O_{12}$	12.291	$Mn_3Cr_2(GeO_4)_3$	12.027
$Dy_3Ga_2(GaO_4)_3$	12.307 (25°C)	$Mn_3Fe_2(GeO_4)_3$	12.087
$Gd_3Ga_2(GaO_4)_3$	12.376 (25°C)	$ZnGd_2Mn_2(GeO_4)_3$	12.427
		$Gd_3Co_2GaGe_2O_{12}$	12.446–12.460
$Nd_3Ga_2(GaO_4)_3$	12.506 (25°C)	$Gd_3Mg_2GaGe_2O_{12}$	12.425
$Y_3Ga_2(GaO_4)_3$	12.277 (25°C)	$Gd_3Mn_2GaGe_2O_{12}$	12.550
$Yb_3Ga_2(GaO_4)_3$	12.200 (25°C)	$Gd_3Ni_2GaGe_2O_{12}$	12.401
$Nd_4Ga_4O_{12}$	12.54	$Gd_3Zn_2GaGe_2O_{12}$	12.464
$Sm_4Ga_4O_{12}$	12.465	$Mn_3NbZnFeGe_2O_{12}$	12.49
$Y_4Ga_4O_{12}$	12.30	$NaCa_2Co_2(VO_4)_3$	12.431
$Sm_3Ga_5O_{12}$	12.432 (25°C)	$NaCa_2Cu_2(VO_4)_3$	12.423
$Bi_4(GeO_4)_3$	10.520	$NaCa_2Mg_2(VO_4)_3$	12.446
$CaGd_2Mn_2(GeO_4)_3$	12.555	$NaCa_2Ni_2(VO_4)_3$	12.373
$CaNa_2Sn_2(GeO_4)_3$	12.430	$NaCa_2Zn_2(VO_4)_3$	12.439
$CaNa_2Ti_2(GeO_4)_3$	12.359	$Ca_3Cr_2(SiO_4)_3$ (uvarovite)	11.999 (25°C)
$Ca_3Cr_2(GeO_4)_3$	12.262 (25°C)	$Gd_3Fe_2Al_3O_{12}$	12.267
$Ca_3Fe_2(GeO_4)_3$	12.325 (25°C)	$Gd_3Al_5O_{12}$	12.12
$Ca_3Ga_2(GeO_4)_3$	12.251 (26°C)	$Mn_3Al_2(SiO_4)_3$	11.64
		$YCa_2Zr_2Fe_3O_{12}$	12.684
$Ca_3TiNi(GeO_4)_3$	12.341		
$Ca_3V_2(GeO_4)_3$	12.324	$Y_3Fe_2Al_3O_{12}$	12.161
$Ca_3ZrCo(GeO_4)_3$	12.54	$Y_3Al_5O_{12}$	12.01

but large cavities (Fig. 113b), where the Na^+ ions are located. Because the cavities communicate with one another, the Na^+ ions can move easily through the large bottlenecks, leading to excellent ionic conducting properties.

The silicates and germanates $ABSi_3O_9$ and $ABGe_3O_9$ (A = K, Rb, Tl, Ba; B = Ta, Nb, Ti, Sn) crystallize in the benitoïte structure. These hexagonal oxides ($a \approx 7$ Å, $c \approx 10$ Å) contain tetrahedral rings, Si_3O_9 or Ge_3O_9, forming layers parallel to the (001) plane (Fig. 114). The tetrahedral rings are linked through BO_6 octahedra, which share their corners with the PO_4 or SiO_4 tetrahedra. This framework gives rise to large tunnels where the A cations are located. The coordination of the A cation corresponds to a distorted octahedron. Although

Figure 112. The garnets $A_3B''_2B''_3O_{12}$: (a) projection of the structure along ⟨110⟩ and (b) projection along ⟨100⟩. (After Wyckoff.[36])

this structure type appears in the form of a tunnel, no deviation from stoichiometry of the A cations has been detected in it.

The silicates and germanates of the wadeite type correspond to the general formula $A_2BSi_3O_9$ or $A_2BGe_3O_9$ (A = K, Rb, Tl, Cs; B = Ti, Sn, Zr). Like the benitoïte, they are hexagonal ($a \approx 7$ Å, $c \approx 10$ Å), and exhibit planar Si_3O_9 (or Ge_3O_9) tetrahedral rings, forming layers parallel to the (001) plane. They differ from the benitoïtes in the relative positions of the Si_3O_9 (or Ge_3O_9) rings. As in benitoïte, the rings share the corners of the tetrahedra with the BO_6 octahedra to form a three-dimensional framework. As a result, there are large cavities with a tetrahedral shape where the A cations are located. The tetragermanates $A_2Ge_4O_9$ (A = K, Rb, Tl) have a structure closely related to that of wadeite,

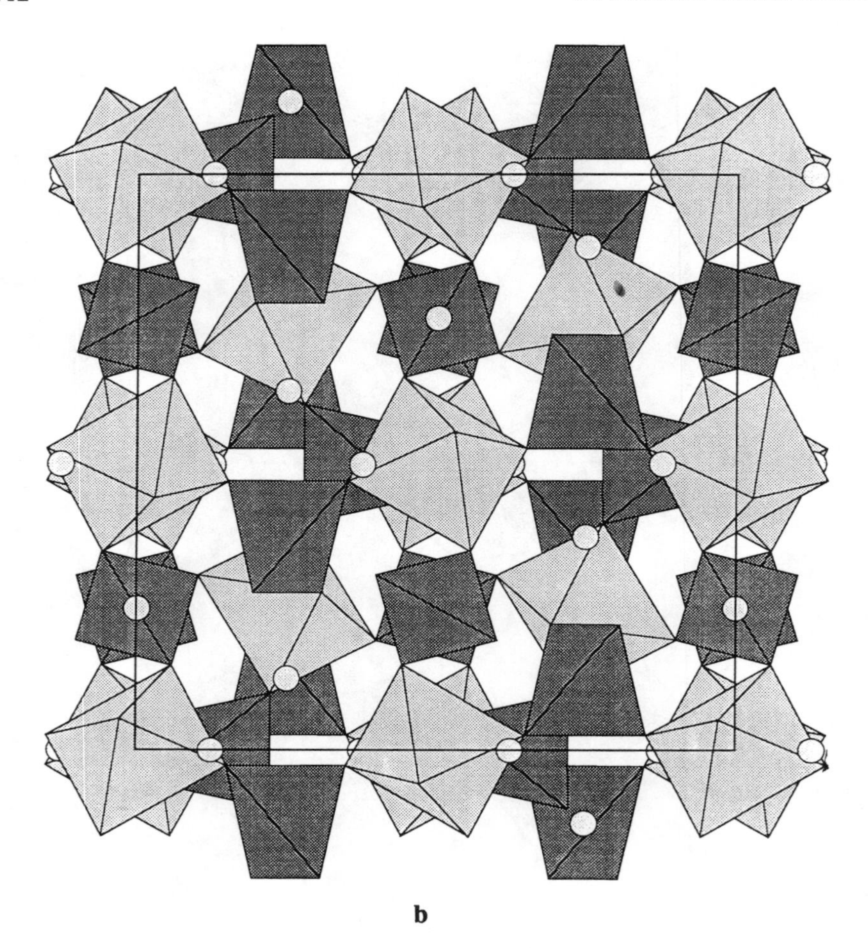

b

Figure 112. *(continued)*

leading to a hexagonal cell ($a \approx a_{\text{wadeite}}\sqrt{3}$ and $c \approx c_{\text{wadeite}}$). The Ge_3O_9 rings however are not planar and are tilted with respect to the (001) plane. This deformation of the Ge_3O_9 ring leads to a distortion of the GeO_6 octahedra; the germanates $A_2BGe_3O_9$ (A = K, Rb; B = Ti, Sn) and $Tl_2SnGe_3O_9$ exhibit the tetragermanate structure.

Milarite type silicates, $A_xB_3B'_2Si_{12}O_{30}$, displaying different extents of intercalation of the A cation have been prepared. Thus we have $AB_3B'_2(Si,Al)_{12}O_{30}$ (A = K, Ba; B = Zn, Mn, Al, Fe; B′ = Mg, Mn, Fe), $A_2B_3B'_2Si_{12}O_{30}$ [A = K, Na, Rb; B = Mg, Zn, Fe(II), Cu; B′ = Mg, Cu, Fe(II)], and $A_3Mg_4LiSi_{12}O_{30}$ (A = Na, K) corresponding to $x = 1$, 2, and 3, respectively. These hexagonal silicates ($a \approx 10$ Å, $c \approx 14$ Å) exhibit an almost tetrahedral host lattice, since only the B ions have an octahedral coordination. The $[B_3B'_2Si_{12}O_{30}]_\infty$ framework

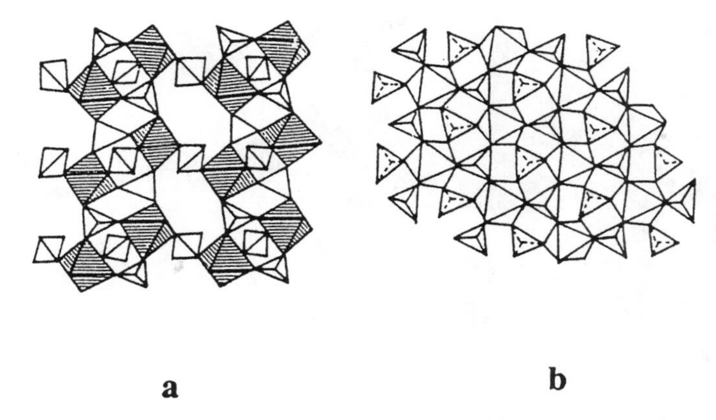

Figure 113. Nasicon, $NaZr_2(P_{3-x}Si_xO_4)_3$: (a) three-dimensional view of the structure along the $\vec{\mathbf{c}}_R$ direction of the rhombohedral cell and (b) projection of half the unit cell along the $\vec{\mathbf{a}}_R$ axis. (After Hagman and Kierkegaard.[96])

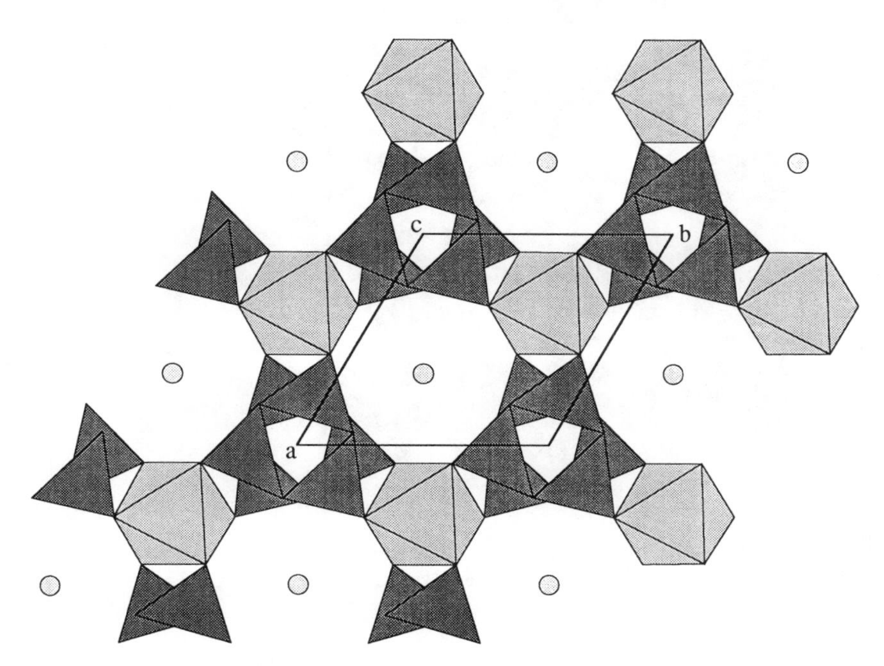

Figure 114. The benitoïte type silicates $ABSi_3O_9$: projection of the structure onto (001).

a

b

Figure 115. The milarites $AB_3B'_2Si_{12}O_{30}$: (a) view of the structure along a direction perpendicular to $\vec{c}$ and (b) view along $\vec{c}$. (After Nguyen et al.[97])

consists of hexagonal $Si_{12}O_{30}$ rings of 2 × 6 SiO_4 tetrahedra (Fig. 115a).[97] These rings are linked to each other through $B'O_4$ tetrahedra and BO_6 octahedra along $\vec{c}$. Within the rings, the $B'O_4$ tetrahedra and BO_6 octahedra share edges. This host lattice gives rise to several sorts of cavities available for the A cations. The stacking of the $Si_{12}O_{30}$ rings along $\vec{c}$ forms large hexagonal tunnels (Fig. 115b), which offer sites of two kinds for the A cations: the A_1 sites located between the tunnels with a 12-fold coordination and the A_2 sites located inside the tunnels with an 18-fold coordination (12 + 6). Outside the $Si_{12}O_{30}$ rings, the assemblage of three $Si_{12}O_{30}$ rings and of two BO_6 octahedra results in A_3 sites with a 9-fold coordination (6 + 3).

The silicates, $Na_5BSi_4O_{12}$ [B = Fe, In, Sc, Y, Ln (= La–Sm)], form an important family of superionic conductors. These hexagonal oxides ($a \approx 22$ Å, $c \approx 13$ Å) have their host lattice built up of $Si_{12}O_{30}$ tetrahedral rings (Fig. 116a).[98] These rings, whose mean plane is parallel to (001) are stacked along $\vec{c}$ forming columns linked through BO_6 octahedra. Each BO_6 octahedron shares

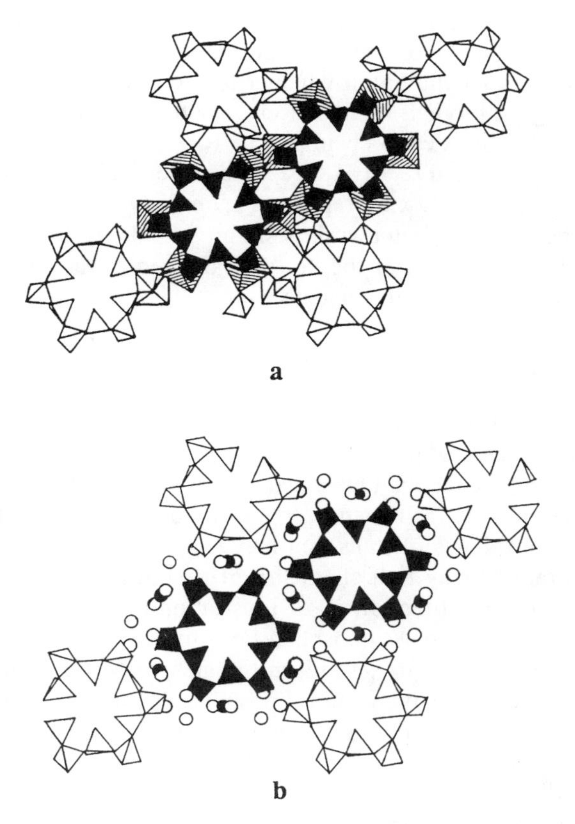

Figure 116. $Na_5BSi_4O_{12}$: (a) the host lattice $[BSiO_4O_{12}]_\infty$ viewed along $\vec{c}$ and (b) two kinds of tunnel where the Na^+ cations are located. (After Shannon.[98])

its six corners with SiO_4 tetrahedra; half the tetrahedra share two corners with BO_6 octahedra, whereas the other half of the tetrahedra share only one corner (the fourth apex being free). This structure exhibits two sorts of tunnel (Fig. 116b)—inside the silicate columns and between them, respectively—all occupied by Na^+ ions. It has been demonstrated that the Na^+ ions located in the tunnels do not move, whereas those located outside are mobile, giving rise to ionic conductivity.

The siliconiobates and -tantalates $A_3B_6Si_4O_{26}$ (A = Ba, Sr; B = Ta, Nb), $K_{6-2x}Ba_xTa_6Si_4O_{26}$, $K_6B_6Si_4O_{26}$ (B = Ta, Nb), and the silicogermanates $K_6B_6Si_{4-x}Ge_xO_{26}$ (B = Ta, Nb) exhibit the same tunnel structure. The $[B_6Si_4O_{26}]_\infty$ host lattice of these hexagonal compounds ($a \approx 9$ Å; $c \approx 8$ Å) consists of B_3O_{15} octahedral units, described earlier, sharing corners with Si_2O_7 disilicate groups. Along $\vec{\mathbf{c}}$, the B_3O_{15} units share their corners, forming triple octahedral rows (Fig. 117a) identical to those of HTB's.[99] Laterally, in the (001) plane, the triple octahedral rows are linked through Si_2O_7 (or Ge_2O_7) groups (Fig. 117b). This framework has pentagonal tunnels running along $\vec{\mathbf{c}}$, similar to those present in the tetragonal tungsten bronzes, (90° and 120° angles). The K^+ and Ba^{2+} ions are located in the tunnels at the level of the apical oxygen of the BO_6 octahedra along $\vec{\mathbf{c}}$. Two kinds of coordination are obtained for the A sites, 10 + 5 and 10 + 3, depending on whether the A cation is located at the same level as the bridging oxygen of the Si_2O_7 group. Note that the possibility of nonstoichiometry of the A cation is important ($0 \leq x < 3$). The boroniobate and -tantalate, $K_3Nb_3B_2O_{12}$ and $K_3Ta_3B_2O_{12}$, have closely related structures that can be derived from $K_6Nb_6Si_4O_{26}$ by replacing each Si_2O_7 group by two triangular BO_3 groups. The resulting framework is slightly distorted ($a \approx 4 \times 9$ Å; $c \approx 4$ Å), but owing to the absence of the bridging oxygen of the Si_2O_7 group, the c parameter is half that of the silicates.

9.2 Phosphates

The ability of phosphate groups to accommodate an octahedral framework was demonstrated in the case of phosphate tungsten bronzes (see Section 4.6). Phosphates of nickel, iron, and cobalt have been extensively studied, leading to novel frameworks such as $FePO_4$, characterized by a pyramidal coordination of Fe(III). The number of phosphates of transition metals synthesized so far is very large indeed. The extraordinarily rich chemistry of the phosphates of molybdenum, niobium, vanadium, and titanium is an illustration of the great potentiality of these systems. We shall discuss phosphates of mixed valent transition elements, analogous to octahedral bronzes, and phosphates of Mo(V) with the unusual character of pentavalent molybdenum. Phosphates of molybdenum give rise to several mixed valent oxides owing to the different possible oxidation states of molybdenum, Mo(VI), Mo(IV), and Mo(III). In the same way, niobium, titanium, and vanadium are potential candidates for the formation of mixed valent phosphates because of the stability of Nb(V)/Nb(IV), Ti(IV)/Ti(III), and V(V)/V(IV)/V(III) species.

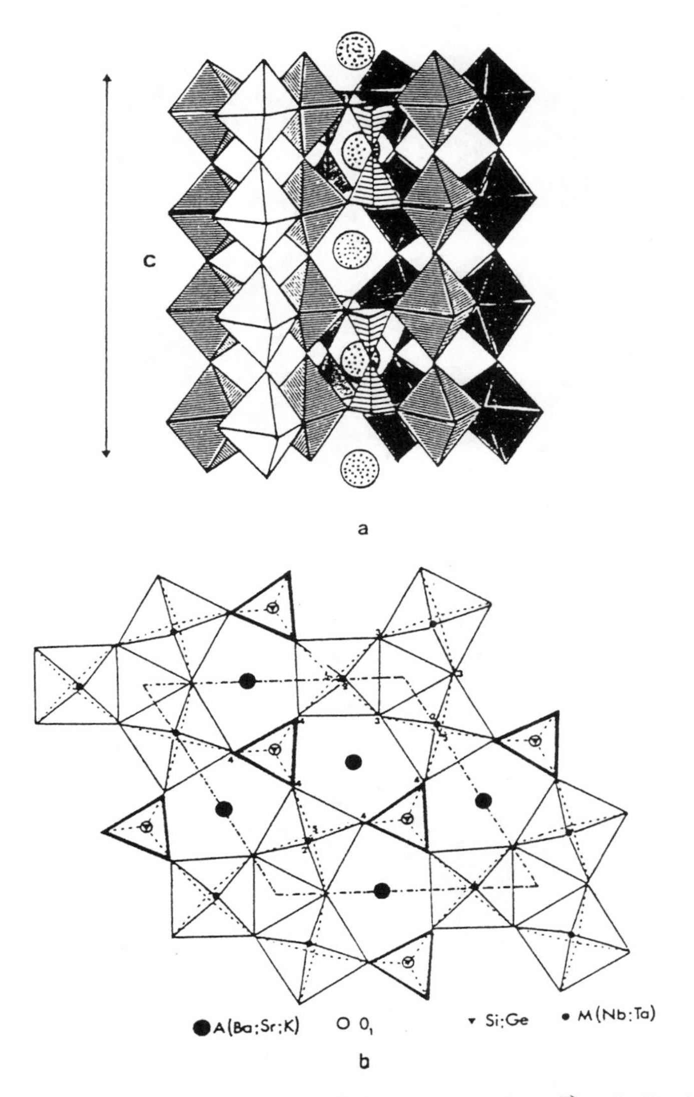

Figure 117. $A_{6-x}Nb_6Si_4O_{26}$: (a) view of the structure along $\vec{a}$ and (b) view of the structure along $\vec{c}$. (After Shannon and Katz[99] and Raveau.[83])

9.2.1 Mixed Valent Molybdenum Phosphates

In the A/Mo/P/O and A/Mo/P/Si/O systems, eight phosphates of molybdenum with the Mo(III)/Mo(IV) mixed valency have been synthesized in the last few years. The first set of compounds is the silicophosphates, $AMo_3P_6Si_2O_{25}$ (A = Rb, Tl, Cs, K). The host lattice in these oxides (Fig. 118a) consists of corner-sharing MoO_6 octahedra, PO_4 tetrahedra, and SiO_4 tetrahedra, which create wide

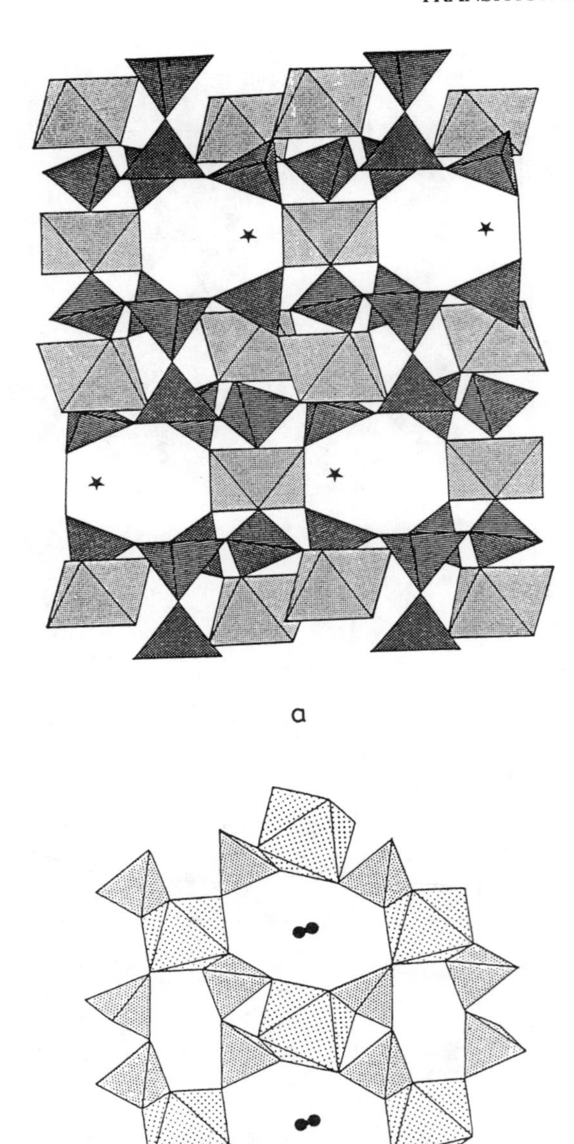

b

Figure 118. (a) Projection of the structure of $AMo_3P_6Si_2O_{25}$ along $\vec{\mathbf{a}}$ showing the tunnels and tetrahedral $P_6Si_2O_{25}$ units linked to isolated MoO_6 octahedra. (After Leclaire et al.[100]) (b) Projection of the $Na_xMoP_2O_7$ structure along $\langle 100 \rangle$ showing tunnels and P_2O_7 groups linked to isolated octahedra. (After Leclaire et al.[101])

tunnels running along the ⟨100⟩, ⟨110⟩, and ⟨010⟩ directions of the hexagonal cell ($a \approx 8.3$ Å, $c \approx 17.4$ Å).[100] The SiO_4 tetrahedra appear in the form of disilicate groups sharing their corners with six PO_4 tetrahedra, resulting in tetrahedral $[P_6Si_2O_{25}]$ units linked together through the corners of MoO_6 octahedra. In such a structure, the MoO_6 octahedra are completely isolated (i.e., they share their six vertices with the PO_4 tetrahedra). Accordingly, there is no delocalization of the *d* electrons even though the material is black. The polyhedra are all regular in these structures, but two types of MoO_6 octahedron differing mainly in Mo—O distances can be distinguished. The two sites, which exhibit Mo—O distances of 2.04–2.06 and 1.97–1.98 Å, can be occupied by Mo(III) and Mo(IV) ions in an ordered manner; there is however no proof for such occupancy from the X-ray diffraction studies.

The rare molybdenum phosphate $Na_xMoP_2O_7$ exhibits nonstoichiometry with a wide homogeneity range ($0.25 \leq x \leq 0.50$). The structure is triclinic ($a \approx 4.9$ Å, $b \approx 7$ Å, $c \approx 8.3$ Å, $\alpha \approx 91.4°$, $\beta \approx 92.5°$, $\gamma \approx 106.5°$). The host lattice $[MoP_2O_7]_\infty$ consists of corner-sharing MoO_6 octahedra and diphosphate groups (Fig. 118b).[101] This framework, in which the MoO_6 octahedra share the six corners with PO_4 tetrahedra, gives rise to large octagonal tunnels where the Na^+ cations are located. Besides these tunnels, smaller empty tunnels with a distorted hexagonal configuration are also observed. Sodium cations are not located at the center of the tunnels but close to the walls. There are two equivalent sites, which can only be half-occupied simultaneously owing to the short distances, leading to the limiting composition $Na_{0.5}MoP_2O_7$. An interesting feature concerns the close relationship between this structure and that of the Mo(III) phosphate $NaMoP_2O_7$, which has the $NaFeP_2O_7$-type structure. Both structures can be described in terms of the stacking of layers of MoO_6 octahedra with layers of diphosphate groups (Fig. 119).[101,102] Such layers of identical composition in the two compounds can be described as $[MoO_3]_\infty$ and $[P_2O_4]_\infty$. These are parallel to the (100) plane in $Na_xMoP_2O_7$—that is, the (001) plane of $NaMoP_2O_7$. The $[MoO_3]_\infty$ layers are practically identical in the two structures, with similar orientations of the MoO_6 octahedra (Fig. 119). The $[P_2O_4]_\infty$ layers in both the structures consist of columns of P_2O_7 groups, running along [010] in $Na_xMoP_2O_7$ and along [100] in $NaMoP_2O_7$. There is only one kind of disposition of the diphosphate groups in $Na_xMoP_2O_7$ against two types of tetrahedral layers in $NaMoP_2O_7$ (labeled Te and Te^E, respectively):

Oc-Te-Oc-Te-Oc	in	$Na_xMoP_2O_7$ (Fig. 119a)
Oc-Te-Oc-Te^E-Oc-Te-Oc	in	$NaMoP_2O_7$ (Fig. 119b)

Cesium molybdenum phosphates form a novel series of phosphates with mixed valent molybdenum, Mo(III)/(IV). For example, $Cs_3Mo_6P_{10}O_{38}$, with a triclinic structure ($a \approx 9.5$ Å, $b \approx 14.2$ Å, $c \approx 6.4$ Å, $\alpha = 91.1°$, $\beta \approx 105.9°$, $\gamma \approx 90°$), has a host lattice (Fig. 120) built up of corner-sharing single MoO_6 octahedra and PO_4 tetrahedra, Mo_2O_{11} octahedral units, and P_2O_7 diphosphate groups.[103,104] Such a framework gives both pentagonal and hexagonal occupied tunnels running along $\vec{c}$, where the Cs^+ cations are located, and more distorted

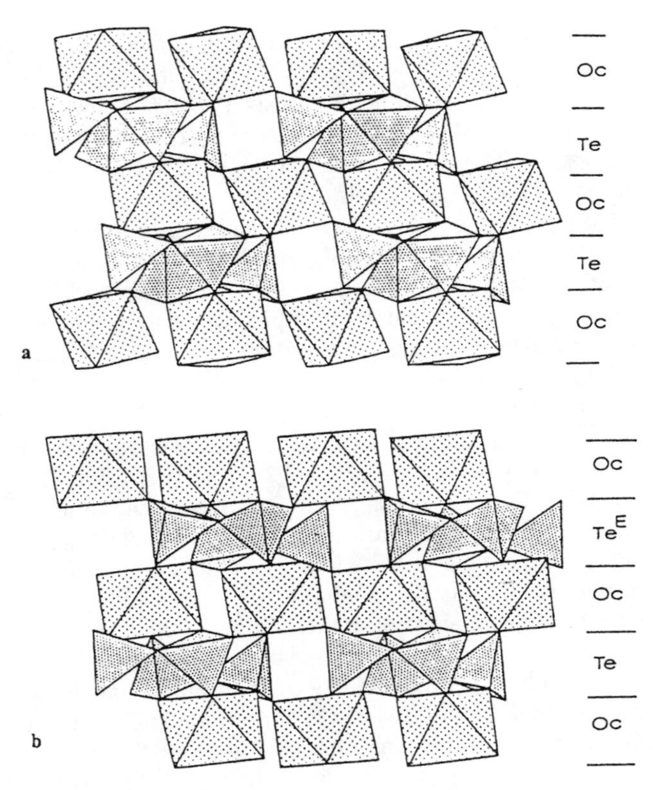

Figure 119. Stacking of layers of MoO_6 octahedra and diphosphate groups in (a) $Na_{0.3}MoP_2O_7$ (after Leclaire et al.[101]) and (b) $NaMoP_2O_7$ (after Leclaire et al.[102])

hexagonal tunnels that are empty. One recognizes Mo_2O_{11} units formed of two corner-sharing octahedra in this structure. The polyhedra are regular and have three types of MoO_6 octahedron. The first two belong to the Mo_2O_{11} units and are characterized by shorter Mo—O distances (1.82–2.06 Å), close to those observed in $AMo_2P_3O_{12}$, suggesting that these sites are occupied by Mo(IV). The third isolated octahedron, on the other hand, is associated with longer Mo—O distances (1.99–2.08 Å) being occupied by Mo(III) and Mo(IV).

Another phosphate, $Cs_4Mo_{10}P_{18}O_{66}$, with a triclinic structure ($a \approx 14.4$ Å, $b \approx 15.6$ Å, $c \approx 6.4$ Å, $\alpha \approx 98.3°$, $\beta \approx 90.8°$, $\gamma \approx 90°$), also has Mo_2O_{11} octahedral units and isolated MoO_6 octahedra, sharing their corners with PO_4 tetrahedra (Fig. 120b,c). Unlike $Cs_3Mo_6P_{10}O_{38}$, this molybdenum phosphate framework does not have single isolated PO_4 tetrahedra; the Mo_2O_{11} units and MoO_6 octahedra are bridged together by diphosphate P_2O_7 and triphosphate P_3O_{10}, groups. This results in two types of pentagonal tunnel running along $\vec{\mathbf{c}}$, of which only one is occupied by Cs^+ ions (Fig. 120b), and one heptagonal tunnel (Fig. 120c), also occupied by Cs^+ ions. The different tunnels do not intersect in this structure.

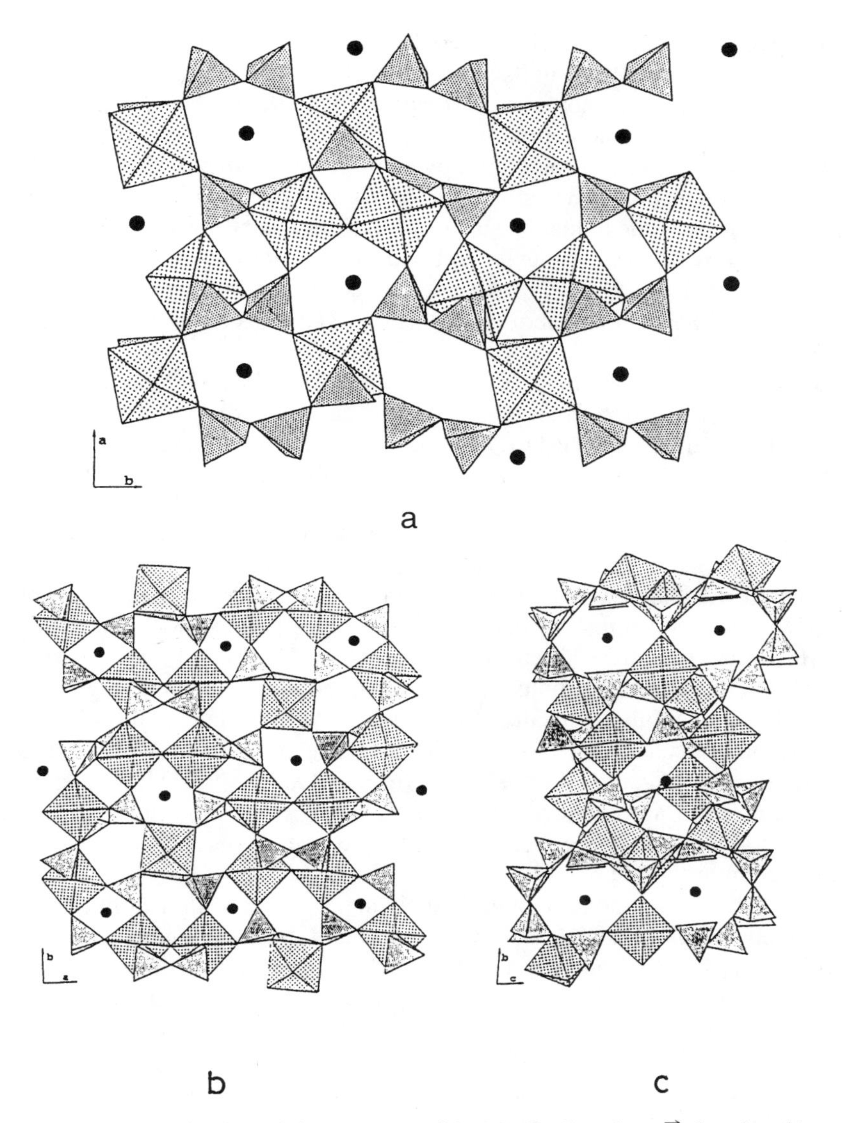

Figure 120. (a) Projection of the structure of $Cs_3Mo_6P_{10}O_{38}$ along $\vec{c}$ showing the pentagonal and hexagonal tunnels. (After Lii and Wang.[103]) Projections of the structure of $Cs_4Mo_{10}P_{18}O_{66}$ (b) along $\vec{c}$ and (c) along $\vec{a}$, showing the diphosphate P_2O_7 and triphosphate P_3O_{10} groups and three sorts of tunnel. (after Haushalter.[104])

The most novel feature of the structure shown in Figure 120a is that three oxygen atoms of the P_3O_{10} units provide the coordination for the isolated MoO_6 octahedra. Thus, in this complex host lattice, there are two sorts of Mo_2O_{11} unit whose Mo—O bonds lie approximately parallel to $\vec{\mathbf{a}}$. These two different octahedral units are linked to each other by P_2O_7 groups, whereas the P_3O_{10} groups ensure the connection between isolated MoO_6 octahedra and the Mo_2O_{11} units. An examination of the oxidation states of Mo shows that the Mo_2O_{11} units should be occupied by Mo(IV), in agreement with the shorter Mo—O bonds observed in these units (1.80–2.06 Å); the isolated octahedra with longer Mo—O bonds (2.03–2.11 Å) are occupied by Mo(III). This leads to the formulation $Cs_4Mo_8^{IV}Mo_2^{III}P_{18}O_{66}$.

Besides the two cesium oxides with bioctahedral molybdenum units just discussed (Fig. 120), two other mixed valent cesium molybdenum phosphates, $Cs_3Mo_4P_3O_{16}$ and $Cs_3Mo_5P_6O_{25}$, have been isolated. Both these phosphates contain cubanelike Mo_4O_4 clusters and exhibit tetrahedral Mo_4 clusters with Mo—Mo distances of 2.56–2.69 Å, characteristic of metal–metal bonds. Each triangular face of the Mo_4 tetrahedron is capped with an oxygen atom forming an Mo_4O_4 cubic unit. The rest of the oxygen atoms linked to molybdenum are distributed in such a way that Mo has an octahedral coordination, leading to Mo_4O_{16} octahedral units consisting of four edge-sharing MoO_6 octahedra. In $Cs_3Mo_4P_3O_{16}$, the three-dimensional framework $[Mo_4P_3O_{16}]_\infty$ results from the connection of the Mo_4O_{16} units by single PO_4 tetrahedra forming large cavities, where the Cs^+ cations are located (Fig. 121a).[105] The host lattice of $Cs_3Mo_5P_6O_{25}$ is more complicated as a result of the presence of isolated MoO_6 octahedra besides the Mo_4O_{16} units. In this structure (Fig. 121b),[106] the Mo_4O_{16} units and MoO_6 octahedra are connected by diphosphate groups forming large tunnels, where the Cs^+ cations are located. Unlike the isolated MoO_6 octahedra, octahedra belonging to the Mo_4O_{16} units are distorted.

Thus far, only one potassium molybdenum phosphate with the mixed valency of molybdenum, Mo(III)/Mo(IV), has been found. This compound has the composition $K_{0.17}MoP_2O_7$ and crystallizes in the tetragonal system ($a \approx 21.2$ Å, $c = 4.9$ Å). Its host lattice consists of corner-sharing MoO_6 octahedra and P_2O_7 groups, forming MoP_2O_{11} units similar to those encountered in $NaMoP_2O_7$, $KMoP_2O_7$, and $Na_{0.30}MoP_2O_7$. This mixed framework has large octagonal tunnels (running along $\vec{\mathbf{c}}$), partially occupied by potassium. Four molybdenum phosphates, $AMo_2P_3O_{12}$ (A = Na, Ba, Ca, Sr), with the nasicon structure (see Section 9.1) have been isolated; three of these (A = Ba, Sr, Ca) exhibit mixed valency of Mo, Mo(III)/Mo(IV), and the fourth contains only Mo(IV).

9.2.2 Mixed Valent Niobium Phosphates

Niobium in the oxidation states Nb(V) and Nb(IV) is a potential element for mixed valency, especially in oxide bronzes. Very few niobium oxides exhibit bronzelike properties. Investigations of niobium phosphates, although recent, show that niobium exhibits a behavior intermediate between that of tungsten

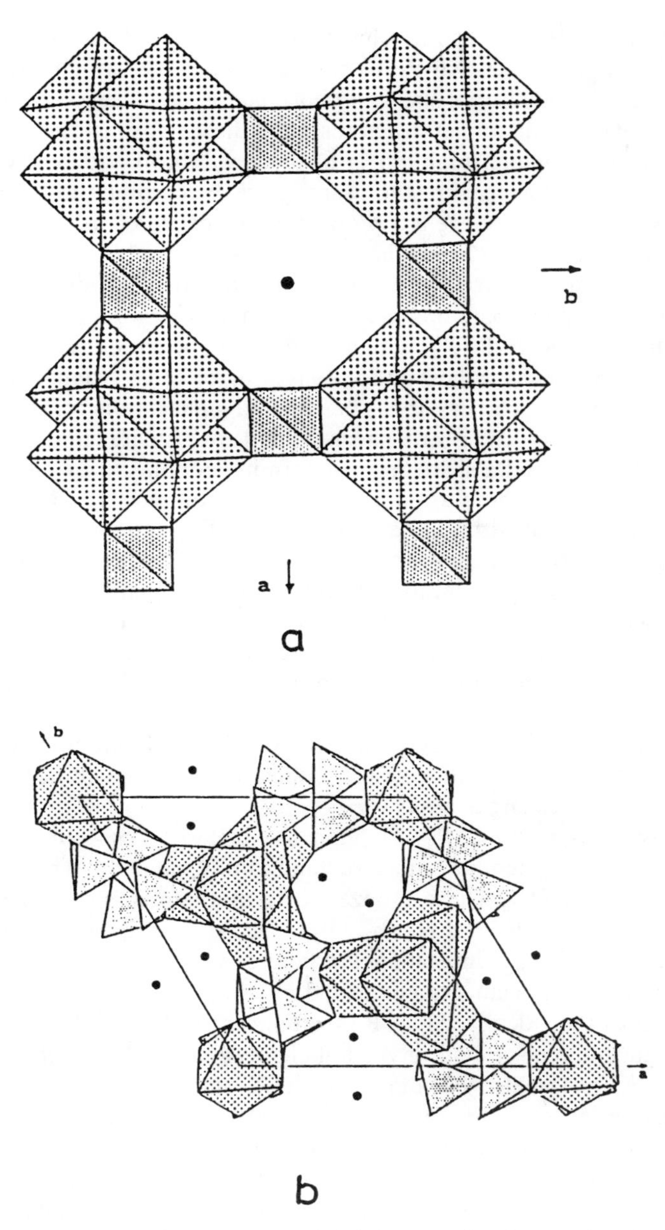

Figure 121. Structure of $Cs_3Mo_4P_3O_{16}$: (a) the coordination environment of Cs in $Cs_3Mo_4P_3O_{16}$ (after Haushalter[105]) and (b) projection along the structure of $Cs_3Mo_5P_6O_{25}$ along $\vec{c}$ (after Lii et al.[106])

and that of molybdenum. The similarity between Nb and Mo is demonstrated by the niobium phosphates $Nb_2P_3O_{12}$ and $Na_{0.5}Nb_2P_3O_{12}$, which have the nasicon structure. The novelty of the first phosphate is the absence of intercalated cations, rare in cage and tunnel structures. Both these phosphates contain regular NbO_6 octahedra in which niobium is off-center; Nb(V) and Nb(IV) ions are distributed at random over the same site according to the formula $Na_xNb^{V}_{1-x}Nb^{IV}_{1+x}P_3O_{12}$. Bronzelike properties are excluded because the NbO_6 octahedra are isolated by PO_4 tetrahedra. The space available for sodium in both the phosphates suggests possible intercalation properties and ionic conductivity.

Three niobium phosphate bronzes with a tunnel structure have been discovered. The first one, $KNb_3P_3O_{15}$, with an orthorhombic structure ($a \approx 13.3$ Å, $b \approx 14.7$ Å, $c \approx 6.4$ Å) has a host lattice consisting of NbO_6 octahedra and single PO_4 tetrahedra sharing corners. It has pentagonal tunnels running along $\vec{\mathbf{c}}$ where the K^+ ions are located and nearly square tunnels, which are empty. In the (001) plane, the NbO_6 octahedra share corners, forming double zigzag chains running along $\vec{\mathbf{b}}$ (Fig. 122a), whereas along $\vec{\mathbf{c}}$, there are infinite $[PNbO_8]_\infty$ chains in which one NbO_6 octahedron alternates with one PO_4 tetrahedron (Fig. 122b), just as in $MoPo_5$, $NbPO_5$, $P_2W_2O_{11}$, $Mo_2P_2O_{11}$, and $NaPWO_6$.[107] Considering the layer of polyhedra parallel to (001), one observes that the double chain of NbO_6 octahedra running along $\vec{\mathbf{b}}$ has exactly the same arrangement as in the TTB structure. This double chain is bordered with PO_4 tetrahedra, which take the place of the octahedra in TTB's. It results in $[Nb_3P_3O_{21}]_\infty$ ribbons in which one recognizes the different rings observed in the TTB structure (i.e., five-sided pentagonal, four-sided square, and triangular windows). The tunnels are not fully occupied and are larger than in the TTB's, opening up the possibility of replacing K^+ by Rb^+ and other larger ions.

The phosphates $K_7Nb_{14+x}P_{9-x}O_{60}$ and $K_3Nb_6P_4O_{26}$, respectively, represent the $n = 2$ and $n = \infty$ members of a structural family having the general formula $(K_3Nb_6P_4O_{26})_nKNb_2PO_8$. These bronzes, of orthorhombic structure ($a \approx 36.9$ Å, $b \approx 10.6$ Å, $c \approx 6.4$ Å for $K_7Nb_{14+x}P_{9-x}O_{60}$; $a \approx 14.7$ Å, $b \approx 31.5$ Å, $c \approx 9.4$ Å for $K_3Nb_6P_4O_{26}$) exhibit tunnel structures with a host lattice built up of corner-sharing NbO_6 octahedra connected by single PO_4 tetrahedra. The view of the structure of the idealized $Nb_{14}P_9O_{60}$ framework along $\vec{\mathbf{c}}$ (Fig. 123a) allows us to distinguish three sorts of polyhedral chain $[PNb_4O_{14}]_\infty$ (labeled a) running along $\vec{\mathbf{a}}$, $[NbO_3]_\infty$ (labeled b1), and $[Nb_2PO_8]_\infty$ (labeled b2) running along $\vec{\mathbf{b}}$. The projection of the structure onto the (100) plane (Fig. 123b) shows $[NbPO_5]_\infty$ chains running along $\vec{\mathbf{c}}$ similar to those observed in several molybdenum and tungsten phosphates.[108] The most striking feature of this latter view is the close similarity with the ITB structure. The $Nb_{14}P_9O_{60}$ framework is indeed formed by the stacking (along $\vec{\mathbf{a}}$) of $[Nb_3P_2O_{13}]_\infty$ layers (Fig. 123b), derived from the $[Mo_5O_{15}]_\infty$ host lattice of $Sb_2Mo_{10}O_{31}$ (Fig. 123c) by replacing two octahedra out of five by PO_4 tetrahedra.[109] This results in two types of hexagonal ring in the $[Nb_3Pb_2O_{13}]_\infty$ layers: HTB rings similar to those in the hexagonal tungsten bronze structure and rings similar to those observed in brownmillerite (BMT)-type oxides such as $Ca_2Fe_2O_5$. The $[Nb_3P_2O_{12}]_\infty$ layers (Fig. 123b) can

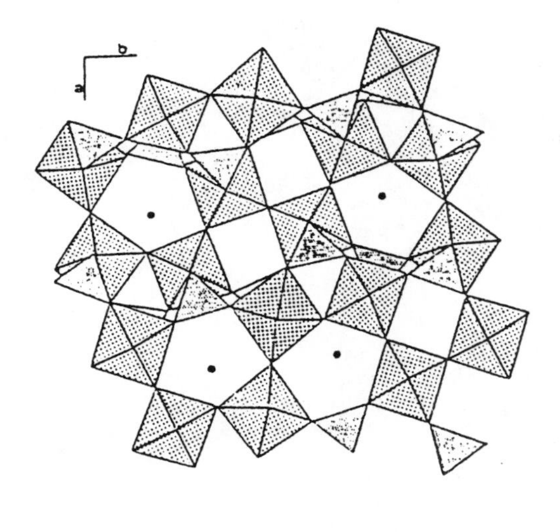

a

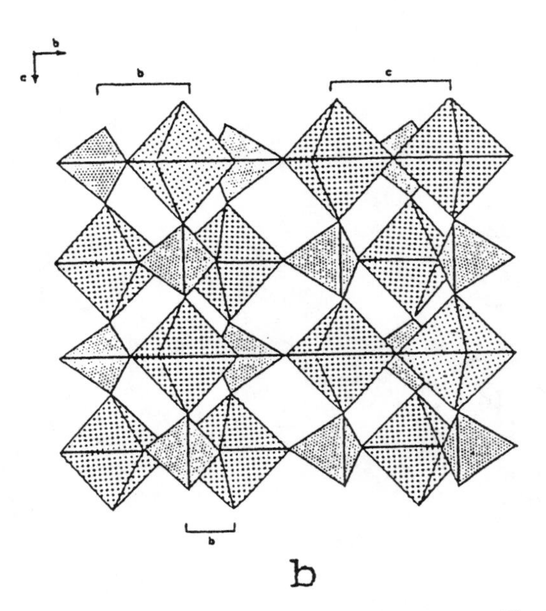

b

Figure 122. Projection of the structure of $KNb_3P_3O_{15}$ (a) along $\vec{c}$, showing the monophosphate PO_4 groups and the pentagonal tunnels, and (b) along $\vec{a}$, showing the $[PNbO_8]_\infty$ chains. (After Leclaire et al.[107])

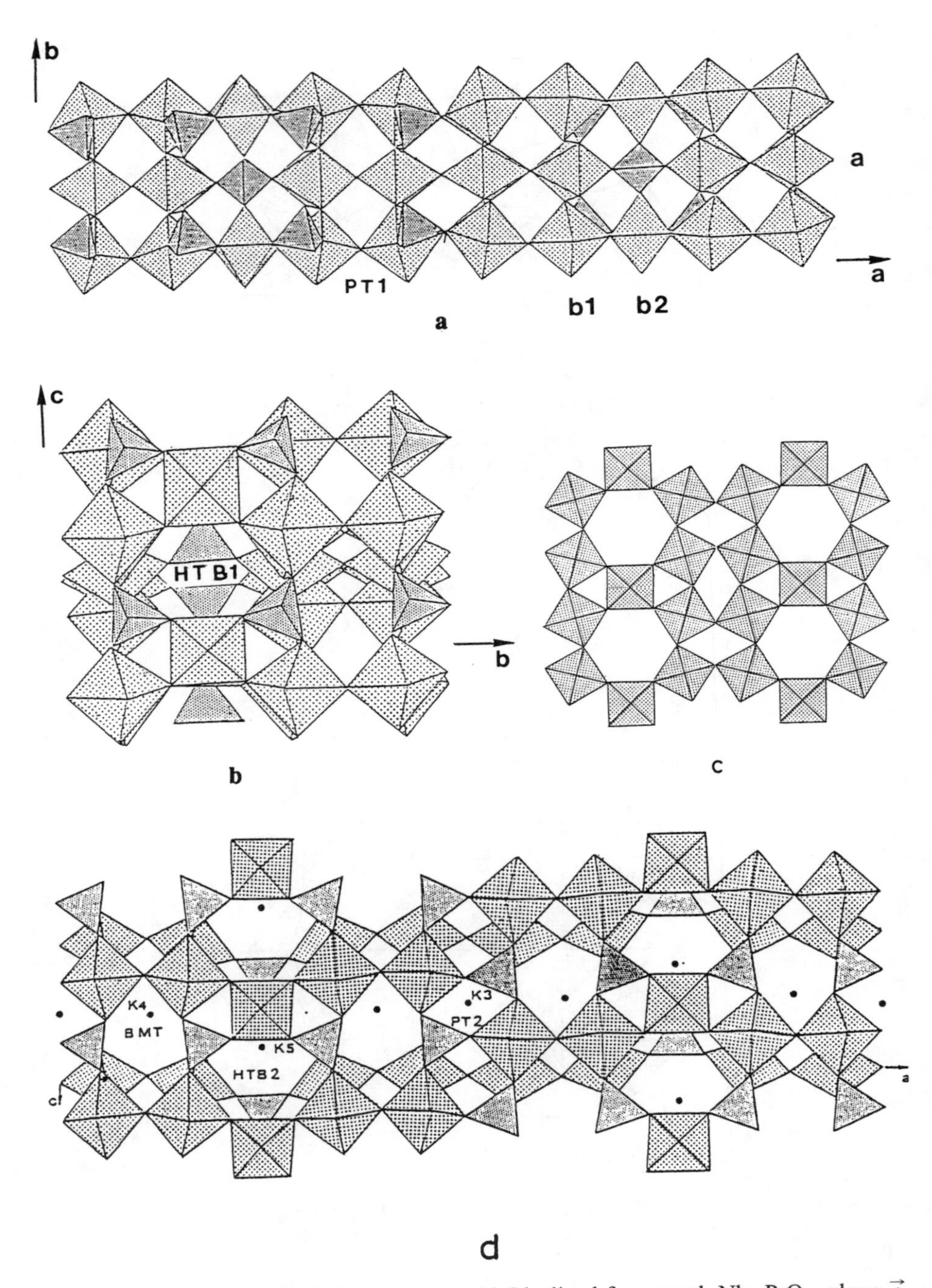

Figure 123. The $K_7Nb_{14}P_9O_{60}$ structure: (a) Idealized framework $Nb_{14}P_9O_{60}$ along $\vec{c}$ showing perovskite-type tunnels (PT1) and chains $[PNb_4O_{14}]_\infty$ (labeled a), $[NbO_3]_\infty$ (labeled b1), and $[Nb_2PO_8]_\infty$ (labeled b2) and (b) projection of the structure of $K_7Nb_{14}P_9O_{60}$ onto the (100) plane. (After Leclaire et al.[108]) (c) The ITB structure of $Sb_2Mo_{10}O_{31}$. (After Parmentier et al.[109]) (d) Projection of the structure of $K_7Nb_{14}P_9O_{60}$ along $\vec{b}$ showing brownmillerite tunnels (BMT), HTB tunnels (HTB2), and perovskite tunnels (PT2). (After Leclaire et al.[108])

thus be described as formed of single ribbons of HTB rings running along $\vec{c}$. Similarity in two dimensions is observed clearly between $K_7Nb_{14}P_9O_{60}$ and $Sb_2Mo_{10}O_{31}$ of the ITB series, along ⟨100⟩. Thus $K_7Nb_{14}P_9O_{60}$ can be described as $[K_3Nb_6P_4O_{26}]_2 \cdot KNb_2PO_8$. The K^+ ions are located in several of the intersecting tunnels in the structure. One observes perovskite-type tunnels (labeled PT1 in Fig. 123a) running along $\vec{c}$ which intersect with brownmillerite tunnels (labeled BMT in Fig. 123d) running along $\vec{b}$. HTB-type tunnels run along $\vec{a}$ (labeled HTB1 in Fig. 123b) and $\vec{b}$ (labeled HTB2 in Fig. 123d), while square (perovskite-type) tunnels resulting from corner-sharing PO_4 tetrahedra and NbO_6 octahedra (labeled PT2 in Fig. 123d) run along $\vec{b}$. It is interesting that some of the P atoms are replaced by niobium, leading to the formula $K_7Nb_{14+x}P_{9-x}O_{60}$ ($x \approx 0.13$); the ability of niobium to be located in the tetrahedral sites is not a recent discovery and has been found in PNb_9O_{25}.

$K_3Nb_6P_4O_{26}$ involves stacking of $[Nb_3P_2O_{13}]_\infty$ layers (along $\vec{b}$) with great similarity to the (100) layers in $K_7Nb_{14}P_9O_{60}$. This bronze has six-sided rings derived from the HTB structure by replacing two octahedra out of six by PO_4 tetrahedra. Since, however, the geometry of the octahedral chains running along $\vec{c}$ is different in the two structures, only one type of hexagonal tunnel is observed in $K_3Nb_6P_4O_{26}$, against two types of tunnels, HTB and BMT, in $K_7Nb_{14}P_9O_{60}$. The $Nb_6P_4O_{26}$ framework results from the stacking of identical $[Nb_3P_2O_{13}]_\infty$ layers by sharing the corners of their polyhedra. The complete stacking of the cell is obtained by the assemblage of (2×4) $[Nb_3P_2O_{13}]_\infty$ layers. Along $\vec{c}$, the framework has hexagonal tunnels similar to those encountered in brownmillerite (or similar to those running along $\vec{b}$ in $K_7Nb_{14}P_9O_{60}$), while four-sided small tunnels are present along $\vec{a}$.

9.2.3 Mixed Valent Titanium and Vanadium Phosphates

The study of phosphates of mixed valent titanium and vanadium is still in its infancy, and only a few phases have been isolated so far. Mixed valency of Ti, Ti(III)/Ti(IV), is also observed in nasicon-type compounds $Na_{1+x}Ti_2P_3O_{12}$ and $ATi_2P_3O_{12}$ (A = Ca, Sr, Ba). Nonstoichiometric $K_{2-x}Ti_2P_3O_{12}$, with the langbeinite structure, suggests the possibility of having a wide homogeneity range ($0 \leq x \leq 1$). A partial projection of this structure (Fig. 124a)shows its similarity to the Nasicon, with wide cages where the K^+ are located.[110] Each TiO_6 octahedron is isolated (i.e., linked only to PO_4 tetrahedra), and electron delocalization cannot occur. Two types of octahedral site differing in size are available for titanium. The one characterized by shorter Ti—O distances would be occupied by Ti(IV), whereas the other (of a larger size) would be simultaneously occupied by Ti(IV) and Ti(III).

Several vanadium phosphates and hydroxyphosphates characterized by mixed valence, V(II)/V(IV), have been prepared: $V_3P_4O_{15}$, $KV_3P_4O_{16}$, $Cd_5V_3P_6O_{25}$, and $V_4P_3H_2O_{15}$. $Na_{2.44}V_4P_4O_{17}OH$, $V_4O(OH)_2(PO_4)_3 \cdot V_3P_4O_{15}$, has an orthorhombic structure ($a \approx 17.5$ Å, $b \approx 12.2$ Å, $c \approx 5.2$ Å) and consists of $[V_3O_{13}]$ units connected by P_2O_7 groups (Fig. 124b).[111] The V_3O_{13} groups are themselves built

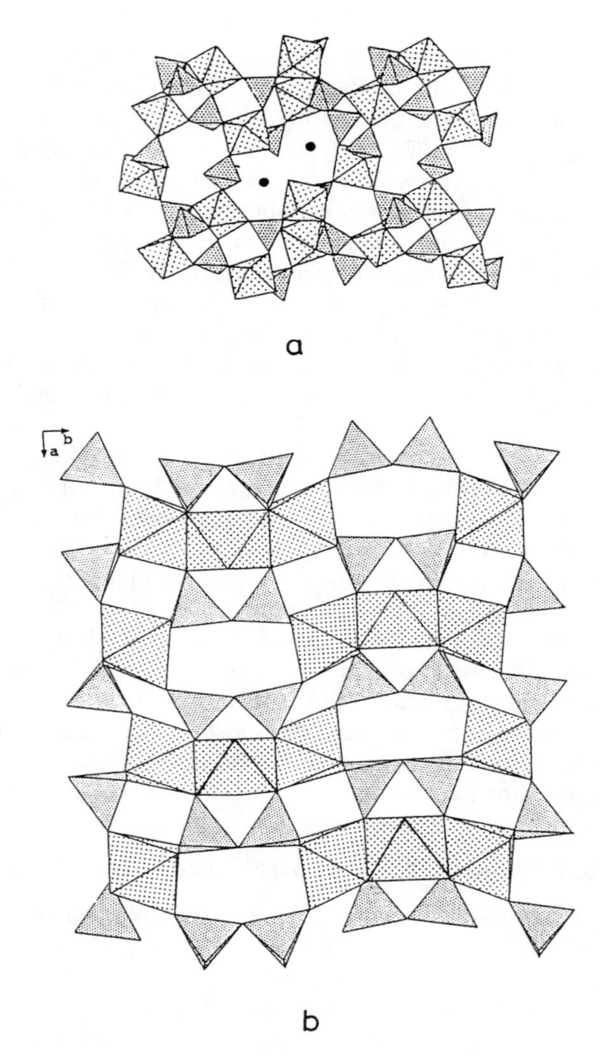

Figure 124. (a) Partial projection of the structure of $K_{2-x}Ti_2(PO_4)_3$ showing the cages where the K atoms lie. (After Leclaire et al.[110]) (b) Structure of $V_2(VO)(P_2O_7)_2$ viewed down the crystallographic *c*-axis showing the $[V_3O_{13}]$ units connected by P_2O_7 groups. (After Johnson et al.[111])

up from one VO_4 pyramid sharing its two opposite edges with two VO_6 octahedra. An ordered distribution of the V(III) and V(IV) ions according to the formula $V_2^{III}V^{IV}P_4O_{15}$ is proposed, consistent with the known V—O distances. While V(IV) would be located in the VO_5 pyamids, the octahedral sites (with larger V—O distances) would be occupied by V(III).

$Cd_5V_3P_6O_{25}$[113] crystallizes in the monoclinic system ($a \approx 15.9$ Å, $b \approx 4.7$ Å, $c \approx 24.2$ Å, $\beta \approx 103°$). Its host lattice consists of VO_5 pyramids and VO_6 octahedra occupied by V(IV) and V(III) respectively, and connected through monophosphate groups. This host lattice forms tunnels running along $\vec{\mathbf{b}}$, where the Cd^{2+} ions are located with a distorted octahedral coordination. The apical V—O distance of the VO_5 pyramid is abnormally short, characteristic of the vanadyl ion. $KV_3P_4O_{16}$, with a monoclinic structure ($a \approx 5.2$ Å, $b \approx 12.7$ Å, $c \approx 9.5$ Å, $\beta \approx 94.1°$), exhibits also a tunnel structure (Fig. 125a), and its host lattice consists of pairs of edge-sharing VO_6 octahedra connected by corners to form strings along the *b*-axis.[112] The latter strings are connected through P_2O_7 groups, forming layers parallel to the (001) plane. The layers are linked through VO_5 pyramids. Again, in this structure one observes an abnormally short apical V—O bond characteristic of the vanadyl ion, in agreement with the occupation of the pyramidal site by V(IV), whereas the octahedral strings are occupied by V(III) and V(IV) ions.

Several phosphates having the V(V)/V(IV) type of mixed valence are also known: $RbV_3P_4O_{17+x}$, $Rb_6V_6P_6O_{31}$, $KV_3P_4O_{17}$, $AgV_2P_2O_{10}$, $Na_{0.5}VOPO_4{\cdot}2H_2O$, and $Tl_3V_2O_3(VO)(PO_4)_2(HPO_4)$. We shall consider here only the anhydrous phosphates.

The diphosphate $RbV_3P_4O_{17+x}$[112] crystallizes in the trigonal system ($a \approx 13.6$ Å, $c \approx 7.3$ Å); its host lattice is built up of infinite ReO_3-type chains running along $\vec{\mathbf{c}}$ and linked by P_2O_7 groups to form a three-dimensional structure. Along ⟨110⟩, there are strings of four VO_6 octahedra, linked transversely by P_2O_7 groups. Two vanadium sites out of three show an abnormally short apical V—O bond, characteristic of pyramidal coordination (i.e., of a vanadyl ion). This oxide cannot be considered as having a tunnel structure, but nevertheless the framework forms cages, larger than the size of rubidium.

The structure of $KV_3P_4O_{17}$ (orthorhombic, $a \approx 7.9$ Å, $b \approx 10.01$ Å, $c \approx 16.3$ Å) is also built up of ReO_3-type infinite chains running along $\vec{\mathbf{a}}$ and linked one to the other by P_2O_7 groups (Fig. 125b),[113] forming six-sided tunnels where potassium is located. This phosphate exhibits structural similarities to many other oxides. Its six-sided tunnels are composed of rings of three octahedra sharing corners with three tetrahedra, leading to an arrangement related to that observed in hexagonal tungsten bronzes and in intergrowth tungsten bronzes. Indeed by replacing three octahedra out of six by PO_4 tetrahedra, one can observe rows of polyhedra $[V_6P_6O_{42}]_\infty$ running along $\vec{\mathbf{b}}$ and $\vec{\mathbf{c}}$, derived from rows of pure octahedral tubes $[M_{12}O_{50}]_\infty$, found in the $Sb_2Mo_{10}O_{31}$ structure (Fig. 123c). There is also great similarity to the diphosphate tungsten bronzes, $K_xP_2O_4(WO_3)_{2m}$, with hexagonal tunnels ($DPTB_H$) (Fig. 46). Four-sided tunnels related to the perovskite and involving two PO_4 tetrahedra and two VO_6 octahedra are also present. The disposition of the hexagonal tunnels along $\vec{\mathbf{b}}$, forming zigzag chains along this direction in $KV_3P_4O_{17}$, is however different. Although they contain only monophosphate groups, the niobium phosphate bronzes $(K_3Nb_6P_4O_{26})_n$, $KNbP_2O_8$ bear some resemblance to $KV_3P_4O_{17}$. For example,

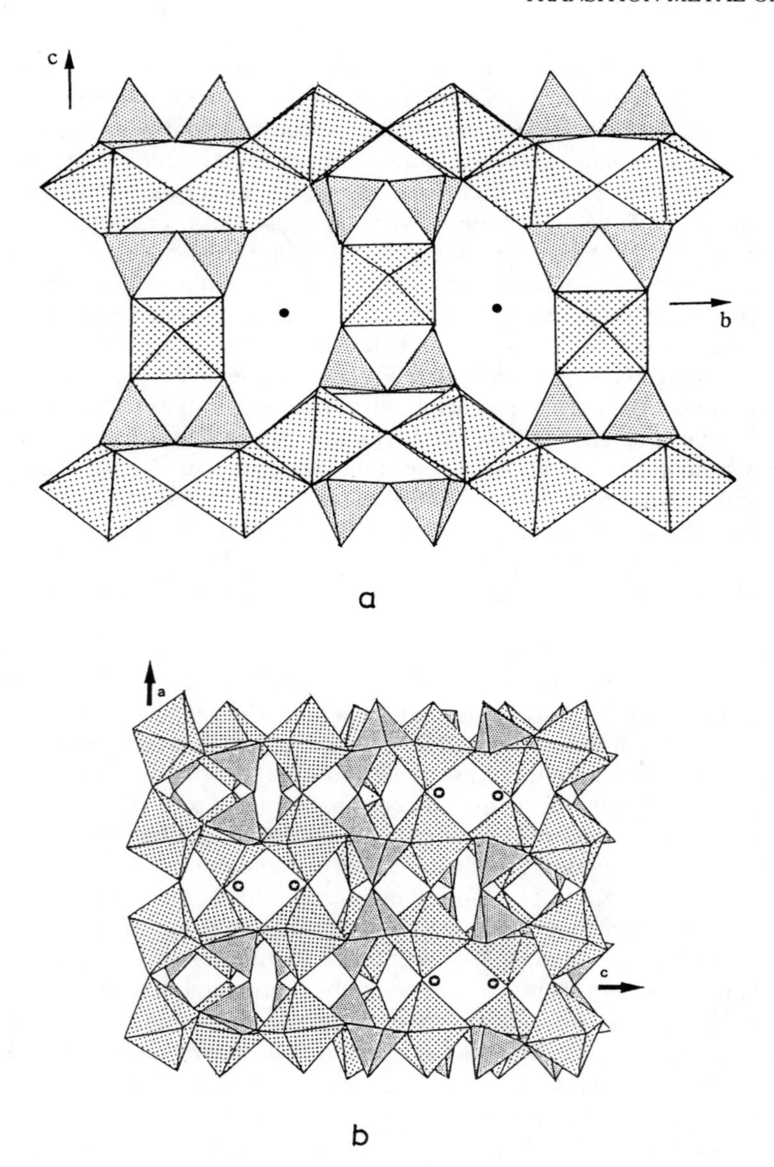

Figure 125. (a) $KV_3P_4O_{16}$: view of the structure along $\vec{a}$ showing three pairs of VO_6 octahedra and the VO_5 pyramids connected through P_2O_7 groups leading to tunnels where the K^+ ions are located. (After Lee and Lii.[112]) (b) $KV_3P_4O_{17}$: projection of the structure along $\vec{b}$ showing infinite ReO_3-type chains (VO_3) linked by P_2O_7 groups. (After Benhamada et al.[113])

they form six-sided tunnels closely related to HTB's, built up of rings of four NbO_6 octahedra and two PO_4 tetrahedra. Brownmillerite tunnels built up of two PO_4 tetrahedra and four NbO_6 octahedra are found in niobium phosphates (Fig. 123), whereas zigzag tunnels involving brownmillerite rings are observed in the $KV_3P_4O_{17}$ structure.

The phosphate $Rb_6V_6P_6O_{31}$ is orthorhombic ($a \approx 7.1$ Å, $b \approx 13.5$ Å, $c \approx 14.4$ Å), and its three-dimensional framework $[V_6P_6O_{13}]_\infty$ can be described by the assemblage of ReO_3-type chains $[VO_3]_\infty$ of corner-sharing octahedra running along $\vec{a}$, and $[V_2P_2O_{10}]_\infty$ chains running along $\vec{b}$, built up of VO_5 pyramids sharing their corners with PO_4 tetrahedra (Fig. 126a). The $[VO_3]_\infty$ chains share the corners of their octahedral with PO_4 tetrahedra. Cohesion of the structure is achieved by the P_2O_7 groups, which share their corners with V_2O_9 units (Fig. 126b).[114] The view of the structure along $\vec{a}$ (Fig. 126b) shows that this framework gives rise to five-sided tunnels running along $\vec{a}$ similar to those observed in the diphosphate tungsten bronzes, $P_2O_4(WO_3)_{2m}$, and in $Na_4Nb_8P_6O_{35}$. Similar rings are also present in $KV_3P_4O_{17}$. The five-sided rings of $Rb_6V_6P_6O_{31}$ however differ from the other phosphates in that two NbO_6 (or WO_6) octahedra are replaced by two VO_5 pyramids. Another novel feature of this structure has to do with the existence of large tunnels (Fig. 126a) running along $\vec{b}$, so that this oxide can also be described as an intersecting tunnel structure. There is one abnormally short V—O bond in both VO_6 octahedra and VO_5 pyramids, characteristic of vanadyl ions.

$AgV_2P_2O_{10}$, which is mixed valent and monoclinic ($a \approx 5.3$ Å, b ≈ 8.1 Å, c ≈ 16.9 Å, $\beta = 91.5°$) is derived from the V(IV) phosphate $BaV_2P_2O_{10}$ by

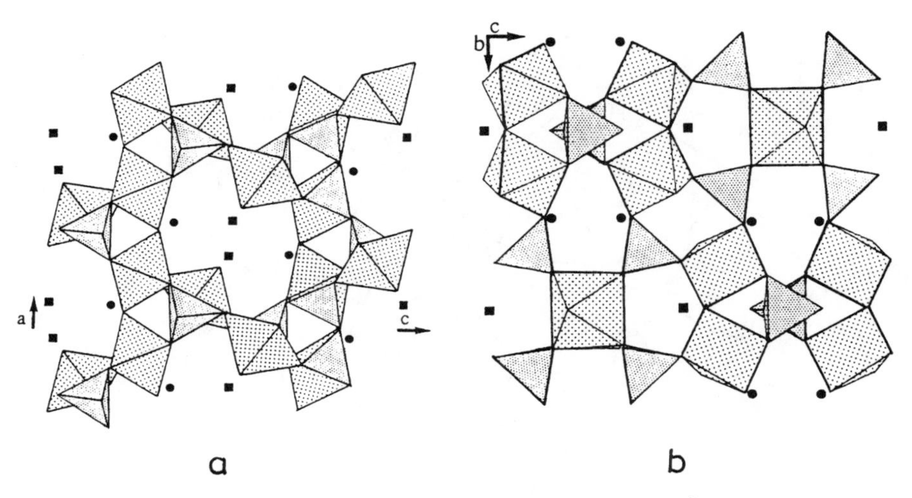

Figure 126. (a) Projection of the structure of $Rb_6V_6P_6O_{31}$ along $\vec{b}$ showing ReO_3-type chains $[VO_3]_\infty$. (b) Projection of the structure of $Rb_6V_6P_6O_{31}$ along $\vec{a}$ showing the V_2O_9 units linked by P_2O_7 groups. (After Benhamada et al.[114])

contraction of the b parameter by about 1 Å and an expansion of the **c** parameter by about 0.6 Å. Projection of the structure of the two phosphates along $\vec{\mathbf{a}}$ (Fig. 127) demonstrates the similarity and difference between the two. In $BaV_2P_2O_{10}$ (Fig. 127a), the $[V_2P_2O_{10}]_\infty$ host lattice is built up of VO_5 pyramids and VO_6 distorted octahedra forming an elongated tunnel running along $\vec{\mathbf{a}}$.[115] The VO_5 pyramid shares the four corners of its basal plane with PO_4 tetrahedra, whereas the fifth one is linked to a VO_6 octahedron; the VO_6 octahedron shares five of

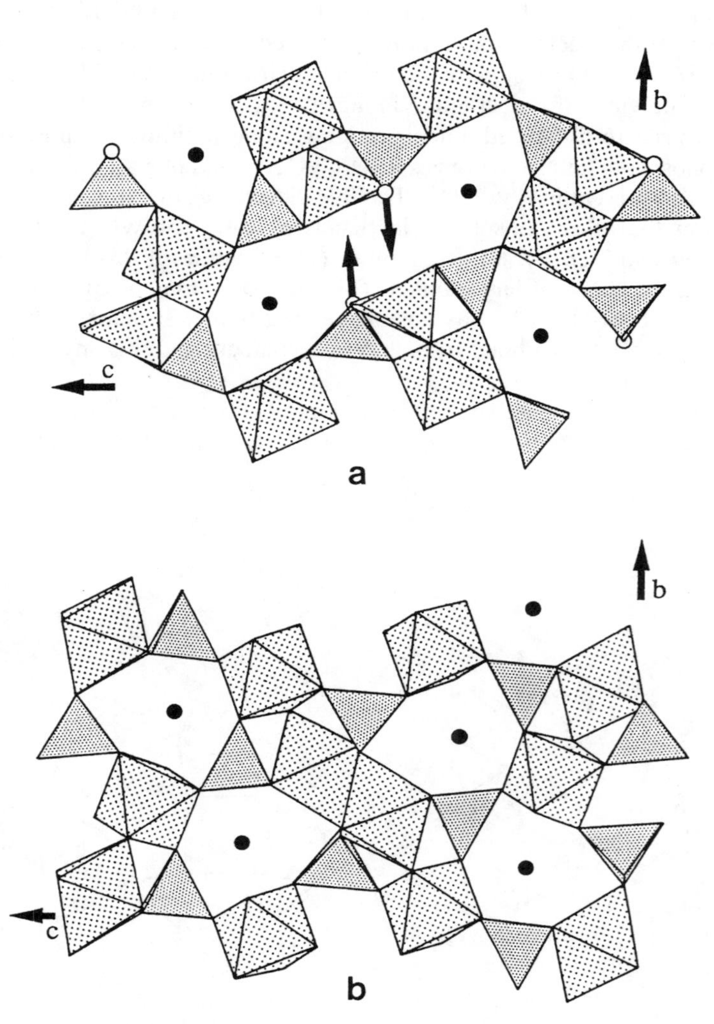

Figure 127. Projection of the structures of (a) $BaV_2P_2O_{10}$ along $\vec{\mathbf{a}}$ (after Grandin et al.[115]) and (b) $AgV_2P_2O_{10}$ (after Grandin et al.[116]). Arrows show the displacement of the basal oxygen atoms of the VO_5 pyramids when Ba is replaced by Ag.

its apices with four PO_4 tetrahedra and one VO_5 pyramid, the sixth corner being free. As a result, the structure of $Ba_2V_2P_2O_{10}$ can be described as $[VP_2O_9]_\infty$ columns of corner-sharing VO_5 pyramids and PO_4 tetrahedra running along $\vec{\mathbf{a}}$ (Fig. 127a) and linked to each other through VO_6 octahedra. In $AgV_2P_2O_{10}$ (Fig. 127b), one observes two kinds of distorted VO_6 octahedron, forming six-sided tunnels with the PO_4 tetrahedra, running along $\vec{\mathbf{a}}$.[116] This structure is obtained from that of $BaV_2P_2O_{10}$ by a distortion of the $[V_2P_2O_{10}]_\infty$ framework in such a way that the VO_5 pyramid is replaced by a VO_6 octahedron. Indeed, the oxygen atoms belonging to the basal planes of the pyramids facing each other in $BaV_2P_2O_{10}$ are brought closer (see arrows in Fig. 127a), with the result that the coordination of vanadium atoms becomes a distorted octahedron or an asymmetrical square bipyramid. Consequently, the elongated tunnels of $BaV_2P_2O_{10}$ are split into two smaller six-sided tunnels in $AgV_2P_2O_{10}$. The structure of the silver phosphate also differs from the barium one by the existence of V_2O_{10} units built up of two edge-sharing VO_6 octahedra. The structure of $AgV_2P_2O_{10}$ can therefore be described as built up of $[V_4O_{20}]_\infty$ strings of edge- and corner-sharing VO_6 octahedra, linked through monophosphate groups. One finds again a short apical V—O bond per octahedron, characteristic of the vanadyl ion. Silver sits in the tunnels with an eightfold coordination. A structure intermediate between $BaV_2P_2O_{10}$ and $AgV_2P_2O_{10}$ is obtained in $PbV_2P_2O_{10}$, suggesting that the size of the intercalated cation is at the origin of the progressive evolution of the coordination of vanadium from a VO_5 pyramid to a VO_6 octahedron.

9.2.4 *Molybdenum(V) Phosphates*

Although pentavalent molybdenum is not commonly observed in octahedral structures, 18 Mo(V) phosphates, possessing 13 different types of mixed framework, have been isolated. Compositions and crystallographic data of typical Mo(V) phosphates are summarized in Table 13. The Mo/O/P framework in these compounds consists of MoO_6 octahedra sharing corners with isolated PO_4 tetrahedra (monophosphates), P_2O_7 groups (diphosphates), or P_4O_{13} groups (tetraphosphates). Taking the presence of such tetrahedral groups into consideration, and making use of the symbol (MoO) to represent MoO_6 octahedra, formulations of these phosphates have been presented. When the phosphate groups exhibit one or several free corners [e.g., $K_2O(MoO)_2(PO_4)_2$], when there is mixed valence of molybdenum [e.g. $NaMoO_2(MoO)_2(PO_4)_3$], or when another element of a size similar to Mo(V) is present in the structure [e.g., $Al(MoO)(PO_4)_2$], such a representation is not very satisfactory. However it has the merit of describing the nature of phosphate groups and shows that there are three pure diphosphates, four pure monophosphates, one pure tetraphosphate of Mo(V), and eleven mixed phosphates (involving both P_2O_7 and PO_4 groups), as can be seen from Table 13. Host lattices of these phosphates contain characteristic $Mo(V)O_6$ octahedra. The O_6 octahedron is regular, but the Mo atom is off-center out of the basal plane, toward an oxygen apex, leading to an abnormally short Mo—O bond (~1.65 Å). This is like a molybdenyl ion, written as $(MoO)^{3+}$. The four equa-

Table 13 Mo(V) Phosphates

Crude Formula	Developed Formula	Cell Parameters: Dimensions, a, b, c (Å)	Cell Parameters: Angles (deg)	Space Group	Structural Family
$\alpha KMo_2P_3O_{13}$ [1]	α-$K(MoO)_2(P_2O_7)(PO_4)$	$a = 10.743(2)$ $b = 14.0839(9)$ $c = 8.8519(7)$	$\beta = 126.42(1)$	$C2/c$	I
$\beta KMo_2P_3O_{13}$ [2]	β-$K(MoO)_2(P_2O_7)(PO_4)$	$a = 9.701(3)$ $b = 18.848(2)$ $c = 6.389(5)$	$\beta = 106.96(3)$	$P2_1/c$	I
$\delta KMo_2P_3O_{13}$ [3]	δ-$K(MoO)_2(P_2O_7)(PO_4)$	$a = 8.846(8)$ $b = 8.846(9)$ $c = 10.01(1)$	$\alpha = 56.488(8)$ $\beta = 55.588(7)$ $\gamma = 68.868(7)$	$P\bar{1}$	I
$\alpha CsMo_2P_3O_{13}$ [4]	α-$Cs(MoO)_2(P_2O_7)(PO_4)$	$a = 7.479(3)$ $b = 8.461(5)$ $c = 9.018(2)$	$\beta = 101.99(3)$	$P2/c$	I
$\beta CsMo_2P_3O_{13}$ [5]	β-$Cs(MoO)_2(P_2O_7)(PO_4)$	$a = 6.398(1)$ $b = 19.497(6)$ $c = 9.835(2)$	$\beta = 107.06(3)$	$P2_1$	I
$\gamma CsMo_2P_3O_{13}$ [6]	γ-$Cs(MoO)_2(P_2O_7)(PO_4)$	$a = 6.342(1)$ $b = 9.676(9)$ $c = 10.0349(8)$	$\alpha = 83.083(7)$ $\beta = 97.42(1)$ $\gamma = 108.30(1)$	$P\bar{1}$	I
$\beta RbMo_2P_3O_{13}$ [7]	β-$Rb(MoO)_2(P_2O_7)(PO_4)$	$a = 6.3847(9)$ $b = 19.088(2)$ $c = 9.7366(9)$	$\beta = 107.05(1)$	$P2_1$	I
$\beta TiMo_2P_3O_{13}$ [8]	β-$Tl(MoO)_2(P_2O_7)(PO_4)$	$a = 9.7536(3)$ $b = 19.064(2)$ $c = 6.3945(7)$	$\beta = 107.099(1)$	$P2_1/c$	I

$\varepsilon NaMo_2P_3O_{13}$ [9]	ε-$Na(MoO)_2(P_2O_7)(PO_4)$	$a = 6.352(3)$ $b = 7.448(5)$ $c = 10.991(6)$	$\alpha = 75.08(5)$ $\beta = 85.33(4)$ $\gamma = 79.10(4)$	$P\bar{1}$	I
$\xi NaMo_2P_3O_{13}$ [10]	ξ-$Na(MoO)_2(P_2O_7)(PO_4)$	$a = 6.3682(5)$ $b = 22.255(1)$ $c = 8.6172(8)$	$\beta = 126.139(7)$	$P2_1/c$	I
$Cs_2K_2Mo_8P_{12}O_{52}$ [11]	$Cs_2K_2(MoO)_8(P_2O_7)_4(PO_4)_4$	$a = 6.388(2)$ $b = 18.901(2)$ $c = 18.805(2)$	$\beta = 92.07(2)$	$P2_1/c$	I
$AgMo_5P_8O_{33}$ [12]	$Ag(MoO)_5(P_2O_7)_4$	$a = 23.050(8)$ $b = 4.831(4)$ $c = 22.935(9)$	$\beta = 90.42(5)$	$I2/a$	II*
$MoPO_5$ [13]	$(MoO)(PO_4)$	$a = 6.1768(3)$ $c = 4.2932(3)$		$P4/n$	II
$AlMoP_2O_9$ [14]	$Al(MoO)(PO_4)_2$	$a = 8.18030(8)$ $c = 8.6970(6)$		$P4/ncc$	II
$KMoP_2O_8$ [15]	$K(MoO)(P_2O_7)$	$a = 5.0862(4)$ $b = 11.720(1)$ $c = 11.486(1)$	$\beta = 90.914(8)$	$P2_1/n$	I
$BaMo_2P_4O_{16}$ [16]	$Ba(MoO)_2(P_2O_7)_2$	$a = 6.4394(4)$ $b = 12.378(1)$ $c = 9.1613(1)$	$\beta = 123.92(1)$	$P2_1/c$	I
$Mo_2P_4O_{15}$ [17]	$(MoO)_2P_4O_{13}$	$a = 8.3068(8)$ $b = 6.5262(6)$ $c = 10.718(2)$	$\beta = 106.70(1)$	$P2_1/c$	I
$K_2Mo_2P_2O_{11}$ [18]	$K_2O(MoO)_2(PO_4)_2$	$a = 9.867(2)$ $b = 10.122(1)$ $c = 9.903(2)$	$\beta = 97.95(1)$	$P2_1/c$	III
$NaMo_3P_3O_{16}$ [19]	$NaMoO_2(MoO)_2(PO_4)_3$	$a = 6.4023(6)$ $b = 7.610(1)$ $c = 12.739(1)$	$\alpha = 80.03(1)$ $\beta = 79.039(9)$ $\gamma = 83.52(1)$	$P\bar{1}$	III*

torial Mo—O distances are generally close to 2.00 Å, but the sixth Mo—O bond opposite to the short molybdenyl bond is considerably longer and varies from one compound to the other. This longer distance (~2.15 Å) corresponds to an Mo—O—P-type bond; it becomes significantly longer in Mo—O—Mo bonds ranging from 2.40 Å in Mo_2O_{10} units to 2.70–2.90 Å in infinite octahedral $[MoO_3]_\infty$ rows. We can distinguish three families according to the way the MoO_6 octahedra are connected to the other polyhedra, as detailed in the subsections that follow.

Family I with Isolated MoO_6 Octahedra

Family I is characterized by the presence of isolated MoO_6 octahedra (i.e., each octahedron shares five corners with PO_4 tetrahedra, the sixth corner being free), giving rise to a short Mo—O bond. There are 14 compounds in this family (Table 13), and they are generally mixed phosphates of the type $A(MoO)_2P_2O_7(PO_4)$. The free apex of the MoO_6 octahedron allows great flexibility to the structures and is certainly at the origin of the polymorphism of the $[Mo_2P_3O_{13}]_\infty$ framework in these phosphates. The MoP_2O_{11} and $Mo_2P_2O_{15}$ units can be delineated in these materials. The MoP_2O_{11} unit is built up of one P_2O_7 group sharing two of its corners with the same MoO_6 octahedron (Fig. 128a).[117] The $Mo_2P_2O_{15}$ unit consists of one MoP_2O_{11} unit sharing two of the remaining apices of the P_2O_7 group with a second isolated octahedron (Fig. 128b). Taking these units into consideration, this family can be subdivided into three groups, IA involving the $Mo_2P_2O_{15}$ unit, IB involving the $Mo_2P_2O_{11}$ unit, and IC, which does not contain these units. The way these units and the remaining polyhedra are associated in the host lattices varies from one framework to another.

In the group IA phosphates, the basic units are $Mo_2P_2O_{15}$ and PO_4 tetrahedra. Their association leads to two kinds of structure:

1. A three-dimensional framework forming tunnels and corresponding to both α- and δ-$K(MoO)_2(PO)_4(P_2O_7)$, represented in Figure 129a.[117] (These two

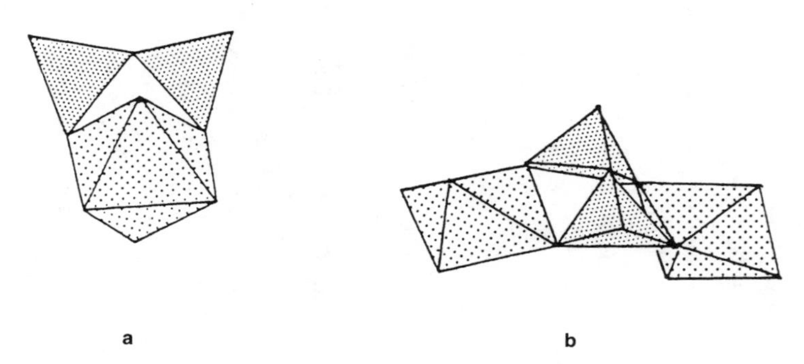

Figure 128. The structural units (a) MoP_2O_{11} and (b) $Mo_2P_2O_{15}$. (After Leclaire et al.[117])

phases correspond to different synthesis temperatures, 1070 and 1270 K, respectively.

2. A layer structure corresponding to α-$Cs(MoO)_2(PO_4)(P_2O_7)$ built up of layers connected through Cs^+ cations (Fig. 129b).[118] The $[Mo_2P_3O_{13}]_\infty$ layers can be generated from α-$K(MoO)_2(PO_4)(P_2O_7)$ by displacing the blocks building the tunnel structure.

In the group IB phosphates, the MoP_2O_{11} units share their corners, forming infinite $[MoP_2O_{10}]_\infty$ chains (Fig. 130).[119] Association of these chains in different ways leads to three-dimensional frameworks. $Ba(MoO)_2(P_2O_7)_2$, ξ-$Na(MoO)_2(P_2O_7)(PO_4)$, β-$A(MoO)_2(P_2O_7)(PO_4)$, γ-$Cs(MoO)_2(P_2O_7)(PO_4)$, and ε-$Na(MoO)_2)(P_2O_7)(PO_4)$ are members of this group, and projections of the structures of the first two appear in Figure 131.[120] Polymorphism of the $[Mo_2P_3O_{13}]_\infty$ framework is evidenced in this group, and different host lattices correspond to the same $A(MoO)_2(P_2O_7)(PO_4)$ formulation. The β-$[Mo_2P_3O_{13}]_\infty$ framework is stabilized by cations such as K^+, Rb^+, Tl^+, and Cs^+ (or Cs^+ and K^+ together).

The group IC phosphates consist of $(MoO)_2P_4O_{13}$ and $KMoO(P_2O_7)$. The $[Mo_2P_4O_{15}]_\infty$ framework in the first phosphate is built up of isolated octahedra and P_4O_{13} tetraphosphate groups (Fig. 132).[121] This is the first molybdenum(V) phosphate involving P_4O_{13} groups. This framework is not stabilized by the insertion of cations. $KMoO(P_2O_7)$ results from the assemblage of isolated octahedra and P_2O_7 groups and, like $Ba(MoO)_2(P_2O_7)_2$ (group IB), it is a pure diphosphate.

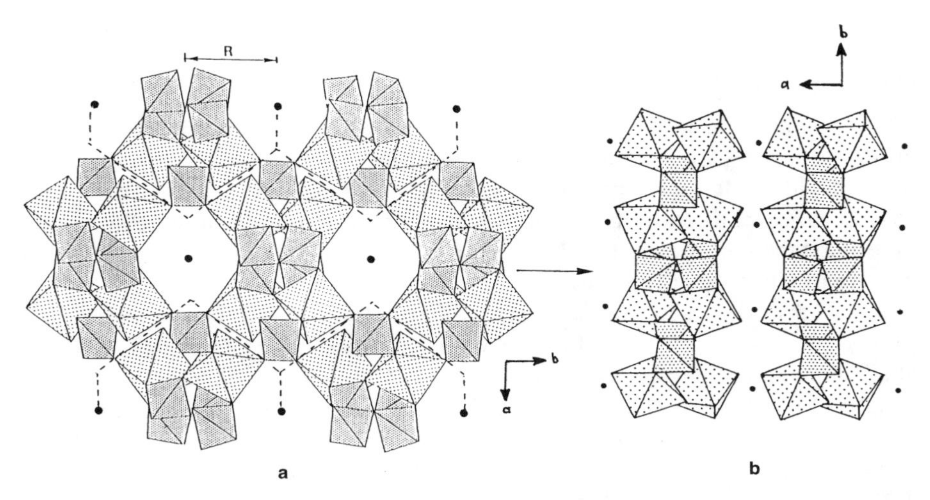

Figure 129. Projections of the structures of (a) δ-$K(MoO)_2(PO_4)(P_2O_7)$ along ⟨101⟩ (after Leclaire et al.[117]) and (b) α-$Cs(MoO)_2(PO_4)(P_2O_7)$ along $\vec{c}$ (after Lii and Haushalter.[118])

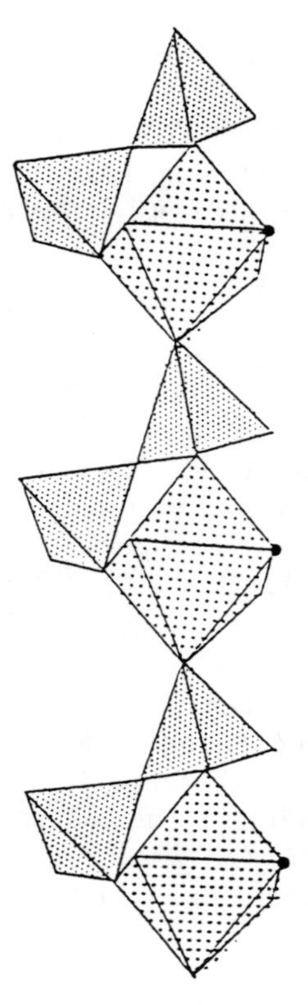

Figure 130. $[MoP_2O_{10}]_\infty$ chains in Mo(V) phosphates. (After Costentin.[119])

Family II with Infinite Octahedral Chains

The phosphates of family II have MoO_6 octahedra sharing two of their corners to form infinite $[MoO_3]_\infty$ chains. One short Mo—O bond (~1.65 Å) alternates with one long Mo—O bond (largest is >2.4 Å) along these chains. As a result, unlike the case of family I, the MoO_6 octahedra of family II do not exhibit any free apex; the other apices are shared with PO_4 tetrahedra. Only three phosphates belong to this series (Table 13).

Two kinds of structure can be distinguished, IIA and IIB. In IIA, part of the MoO_6 octahedra share two of their corners, forming $[MoO_3]_\infty$ chains, whereas the remaining octahedra are isolated. This is the case in $Ag(MoO)_5(P_2O_7)_4$, whose MoP_2O_{11} units (Fig. 133) give rise to tunnels by sharing their corners.[122] The host lattice is formed by the association of columns formed by MoP_2O_{11} units with $[MoO_3]_\infty$ octahedral chains. In group IIB, all the MoO_6 octahedra form $[MoO_3]_\infty$ chains. The two compounds belonging to this group, $(MoO)(PO_4)_2$ and $Al(MoO)PO_4$, can be differentiated by the manner in which the $[MoO_3]_\infty$ chains are connected with the PO_4 tetrahedra; in $(MoO)PO_4$, how-

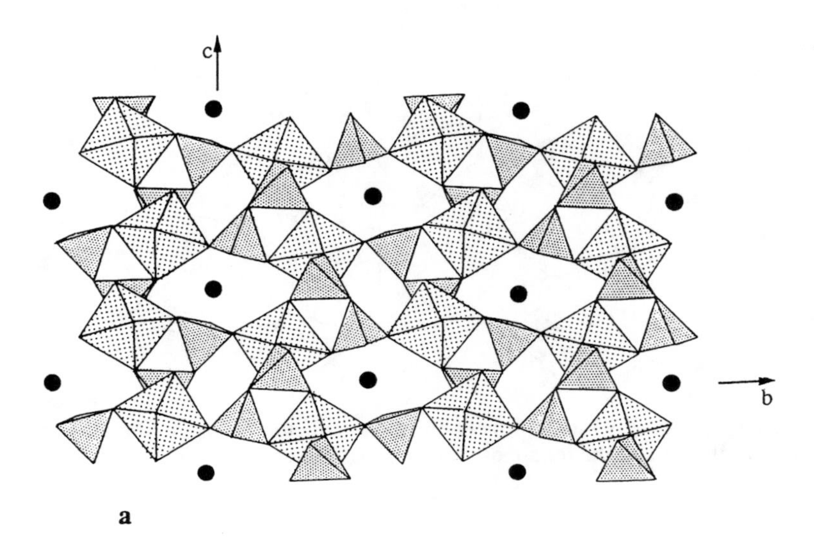

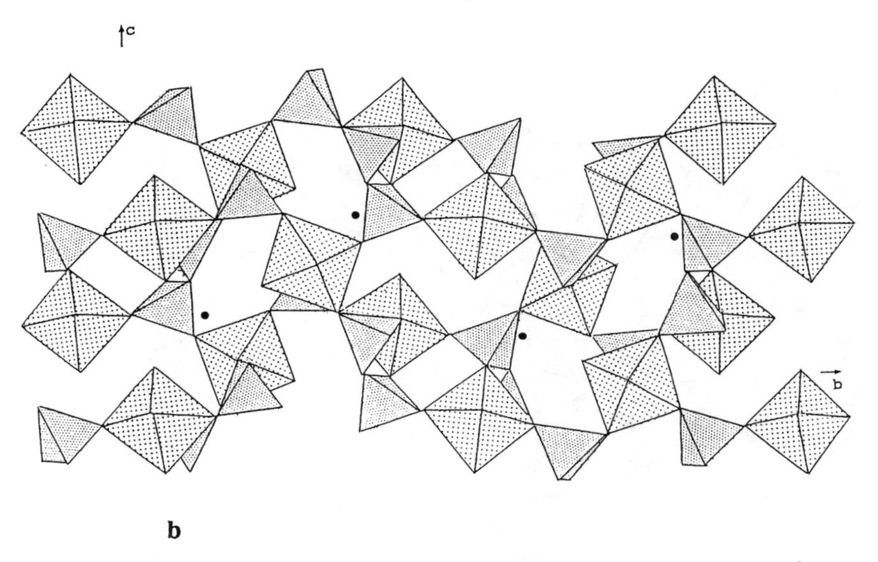

Figure 131. (a) $Ba(MoO)_2(P_2O_7)_2$: view of the $[MoP_2O_9]_\infty$ layers along $\vec{a}$. (b) ξ-$Na(MoO)_2(P_2O_7)(PO_4)$: projection of the structure along $\vec{a}$. (After Costentin et al.[120])

ever, each tetrahedron shares its four corners with four different $[MoO_3]_\infty$ rows (Fig. 134a),[123] whereas in $Al(MoO)(PO_4)_2$ each PO_4 tetrahedron is linked only to two different $[MoO_3]_\infty$ rows, the two remaining corners being shared with two AlO_4 tetrahedra (Fig. 134b).[124] The isolated octahedra in $Ag(MoO)_5(P_2O_7)_4$ exhibit the same geometry as in the compounds of the first family, with a long

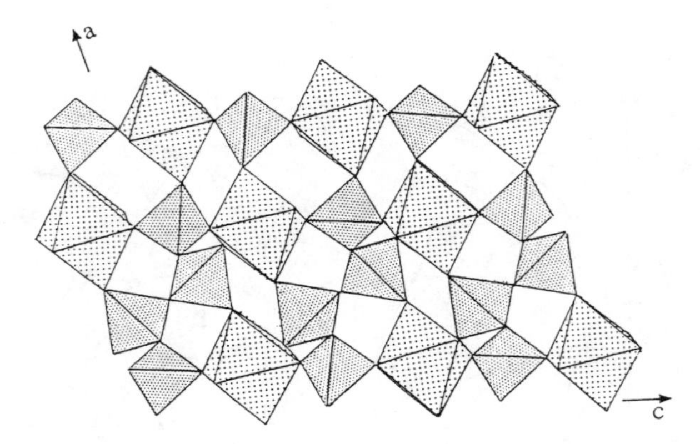

Figure 132. $(MoO)_2P_4O_{13}$: projection of the structure along $\vec{b}$ showing the isolated MoO_6 octahedra and the tetraphosphate groups. (After Costentin et al.[121])

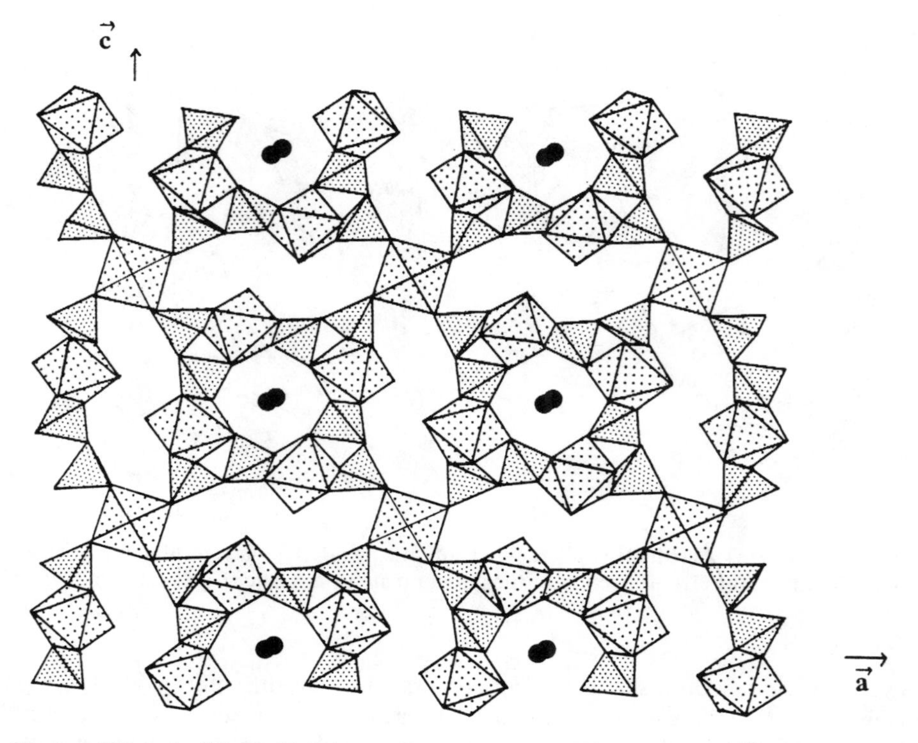

Figure 133. $Ag(MoO)_5(P_2O_7)_4$: projection of the structure along $\vec{b}$ showing the MoP_2O_{11} groups delimiting tunnels and their association with $[MoO_3]_\infty$ chains to form the host lattice. (After Lii et al.[122])

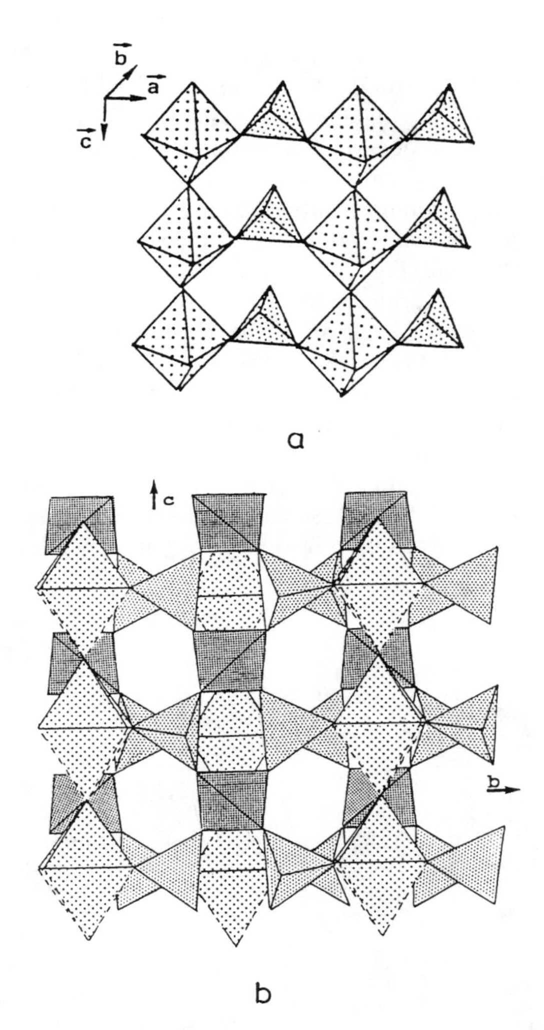

Figure 134. (a) (MoO)PO_4: files $[MoO_3]_\infty$ linked by PO_4 tetrahedra. (After Kierkegaard and Westerlund.[123]) (b) $AlMoO(PO_4)_2$: assemblage of $[MoO_3]_\infty$ chains with PO_4 tetrahedra. (After Leclaire et al.[124])

Mo—O bond of 2.22 Å. In the $[MoO_3]_\infty$ chains, the molybdenum atom is displaced markedly from the basal plane of the octahedra, leading to a very long Mo—O bond (2.66–2.94 Å).

Family III Involving the Association of a Finite Number of Molybdenum Polyhedra

Junction between the polyhedra in family III is ensured by an oxygen belonging either to the longest apical Mo—O bond or to the basal plane of the MoO_6

octahedron. Two phosphates represent this family, $K_2O(MoO)_2(PO_4)_2$ and $Na(MoO)_2(PO_4)_3$. In $K_2O(MoO)_2(PO_4)_2$, the octahedral unit Mo_2O_{11} is built up of two MoO_6 octahedra sharing a corner (Fig. 135).[125] The structure consists of $Mo_2P_2O_{15}$ units built up of one Mo_2O_{11} group sharing four corners with two PO_4 tetrahedra. These $Mo_2P_2O_{15}$ units share their corners along $\vec{\mathbf{c}}$ in such a way that they form $[Mo_2P_2O_{13}]_\infty$ columns, which are themselves linked by corners to give rise to the three-dimensional $[Mo_2P_2O_{11}]_\infty$ framework. Note that the $Mo_2P_2O_{15}$ basic units of this structure (Fig. 128b) have no relation to those in α-$KMo_2P_3O_{13}$ in spite of their identical formulation.

$Na(MoO_2)(MoO)_2(PO_4)_3$ is novel in the sense that there is Mo(V)/Mo(VI) mixed valence and molybdenum is in trigonal-bipyramidal coordination. The formula can therefore be written as $Na(Mo_2^{V})_{oct}(Mo^{VI})_{bip}P_3O_{16}$. The basic structural units are isolated octahedra, tetrahedra, and Mo_2O_{10} units made up of one MoO_6 octahedron and one MoO_5 trigonal bipyramid sharing one corner (Fig. 136a).[126] One Mo_2O_{10} unit alternates with a tetrahedron, forming a $[Mo_2PO_{11}]_\infty$ chain (Fig. 136b). Association of these chains two by two, and through PO_4 tetrahedra, leads to the $[Mo_3P_3O_{16}]_\infty$ framework (Fig. 136c).

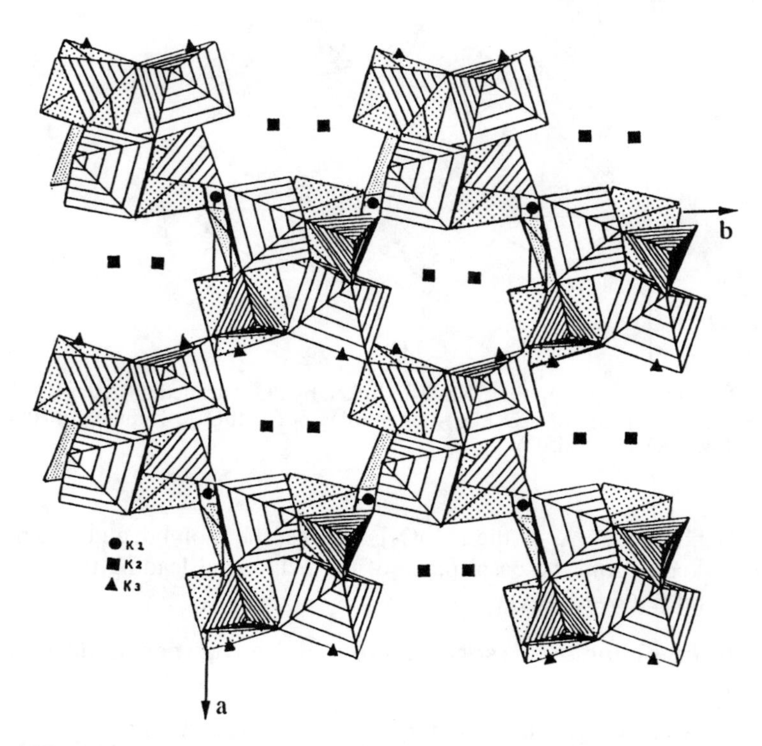

Figure 135. $K_2Mo_2P_2O_{11}$: projection of the structure along $\vec{\mathbf{c}}$. (After Gueho et al.[125])

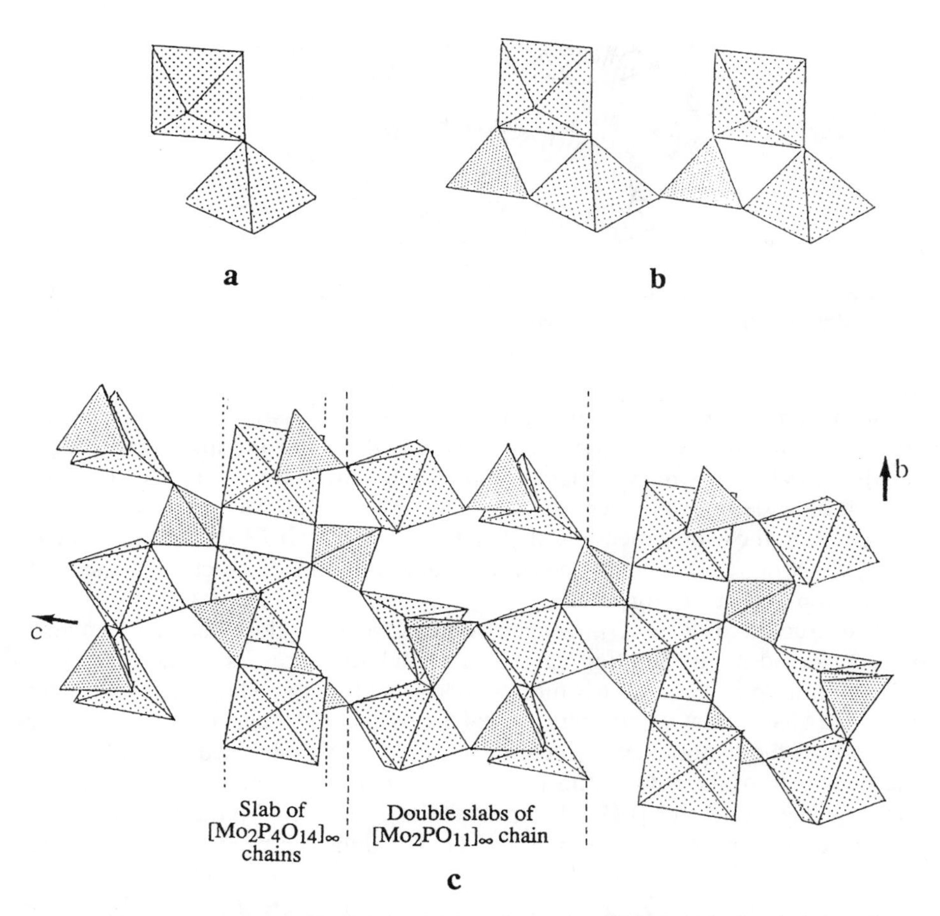

Figure 136. $NaMoO_2(MoO)_2(PO_4)_3$: (a) Mo_2O_{10} units, (b) $[Mo_2PO_{11}]_\infty$ chains, and (c) projection of the structure along $\vec{a}$ showing the stacking of the chains. (After Costentin et al.[126])

9.3 Other Examples of Mixed Frameworks of Octahedra and Tetrahedra

Besides the silicates, germanates, and phosphates, few other oxides with a mixed framework involving tetrahedra and octahedra are known (leaving out the oxygen-deficient perovskites). $Cs_2Nb_4O_{11}$ is an interesting example owing to the great size of its tunnels. The important family of oxides symbolized by $Ga_4O_6(MO_2)_m$ (M = Ti, Ge) is worth describing owing to its analogy with shear structures (Section 11) and the DPTB.

The host lattice of $Cs_2Nb_4O_{11}$ (Fig. 137) is mainly built up of corner- and edge-sharing octahedra.[127] However, a small number of NbO_4 tetrahedra participate in the structure. A striking similarity with the pyrochlore framework can

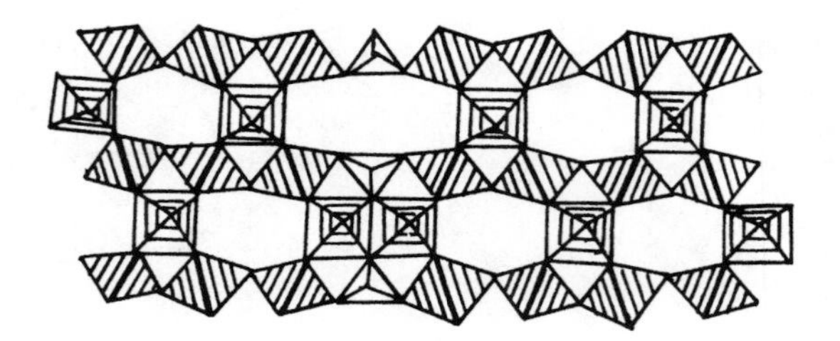

Figure 137. The host–lattice of $Cs_2Nb_4O_{11}$ built up from NbO_6 octahedra and NbO_4 tetrahedra forming large tunnels where the Cs^+ cations are located. (After Gasperin.[127])

be seen from the presence of hexagonal tunnels running in different directions. The main difference with pyrochlores is that there are groups of two edge-sharing octahedra, which associate with NbO_4 tetrahedra, leading to the formation of wide tunnels where the Cs^+ ions are located. These tunnels can be compared to those observed in $CsP_8W_8O_{40}$ and $AM_3P_6Si_2O_{25}$. Like the pyrochlores, $Cs_2Nb_4O_{11}$ is characterized by an intersecting tunnel structure that should exhibit ion exchange properties.

The structure of β-Ga_2O_3, which exhibits an anionic close packing, can also be considered as the limiting case of a tunnel structure. Its framework (Fig. 138a), built up of Ga_4O_{24} structural units (formed by two GaO_4 tetrahedra and two edge-sharing GaO_6 octahedra), gives rise to diamond-shaped tunnels, similar to those of the hypothetical structure BO_2 (Fig. 138b) obtained by a tilting of the octahedra in the TiO_2 rutile form (Fig. 138c).[128,129] The close relationship between the rutile and β-Ga_2O_3 structures is evidenced by the ability of the Ga_4O_{24} units to fit the rutile framework. Ga_4O_{24} units can be introduced into the TiO_2 framework in the form of rows forming empty hexagonal tunnels.

A large family, $Ga_4O_6(BO_2)_m$, can be formed in which m is an odd number characterizing the number of octahedra corresponding to the width of the rutile slab. α-Ga_4GeO_8, $Ga_4Ge_3O_{12}$, and $Ga_4Ti_5O_{16}$ correspond to first three members

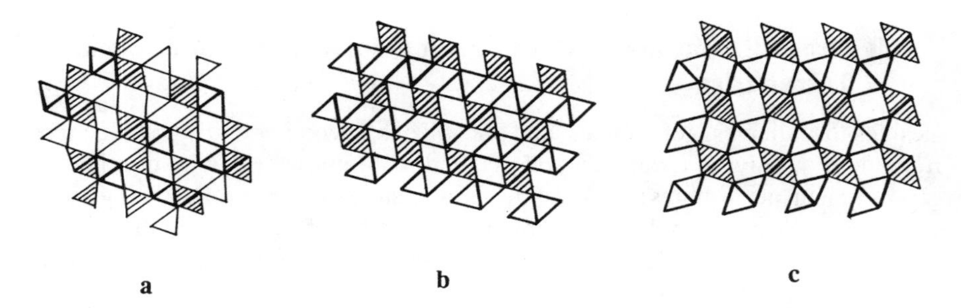

Figure 138. Comparison of (a) the β-Ga_2O_3 structure with (b) the BO_2 oxide derived from (c) the rutile TiO_2 framework by tilting of the octahedra. (After Bursill[128] and Raveau.[129])

of this series (Fig. 139a–c), whereas $Ga_4Ti_{21}O_{48}$ represents the $m = 21$ member.[130,131] The similarity of this structural family with the shear structures and with the diphosphate tungsten bronzes must be pointed out. The $[Ga_4O_6]_\infty$ planes correspond to the phosphate planes or to the shear planes (see Section 11), whereas the rutile slabs can be replaced by the ReO_3-type slabs in the DPTB. The geometry of the Ga_4O_{24} units does not permit an even number of octahedra for the rutile slabs. Nevertheless, the possible existence of even-m members

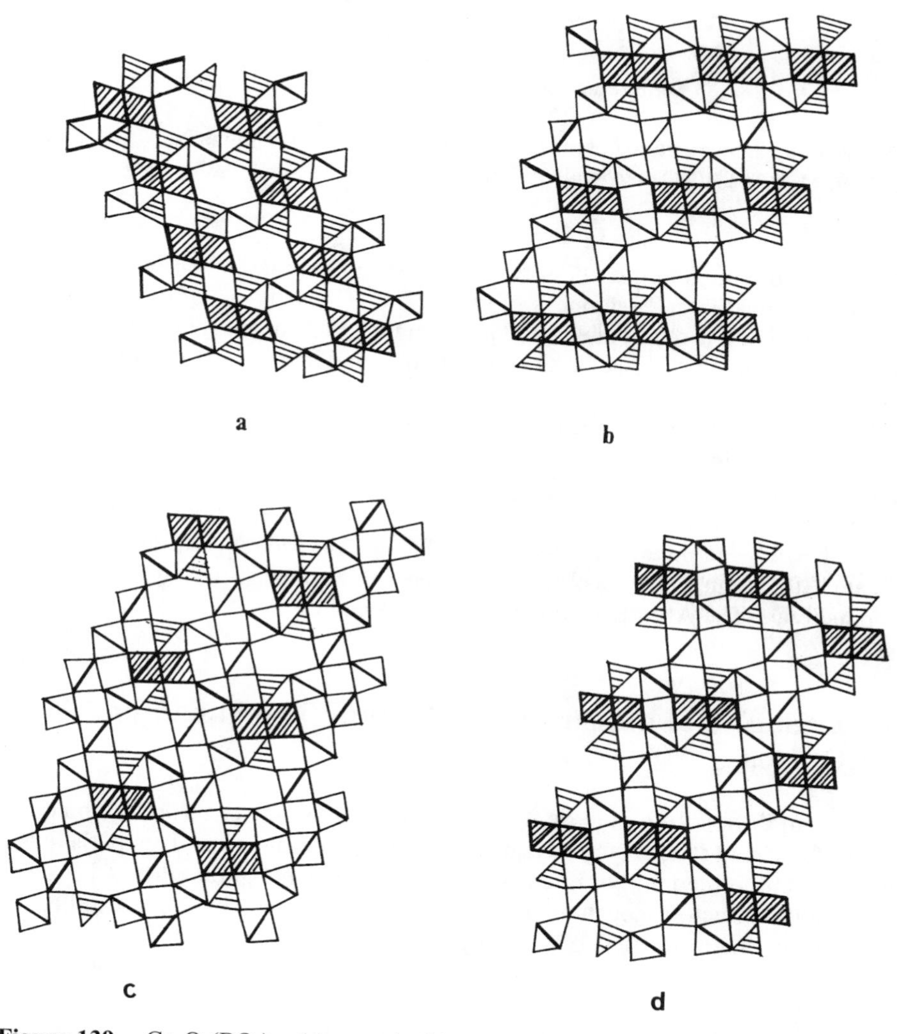

Figure 139. $Ga_4O_6(BO_2)_m$: (a) $m = 1$, (b) $m = 3$, and (c) $m = 5$, and (d) the hypothetical even-m member, $m = 4$, corresponding to the intergrowth of the members $m = 3$ and $m = 5$. (After Agafonov[130] and Lloyd et al.[131])

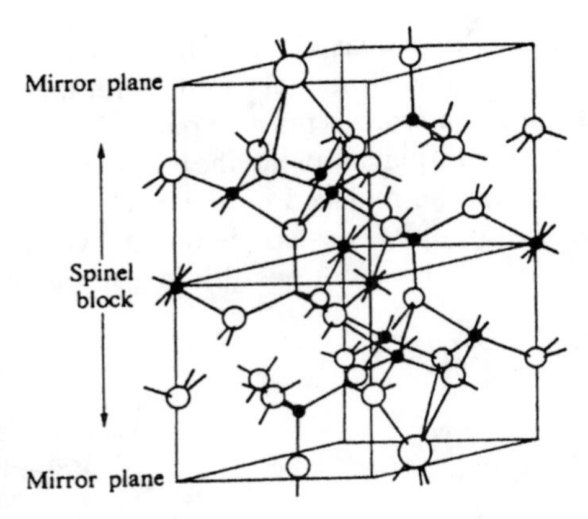

Figure 140. Schematic representation of the structure of β-alumina $NaAl_{11}O_{17}$: large circles, Na; small circles, O; solid circles, Al. (After Beevers and Ross.[132])

resulting from the intergrowth of two odd-*m* members can be considered, as for the hypothetical $m = 4$ member in Figure 139d. The behavior of the first member, α-Ga_4GeO_8, is remarkable; the Ga_4O_{24} units in this oxide share their corners in one direction, forming β-Ga_2O_3-type ribbons, and intergrowth occurs between this member and β-Ga_2O_3.

It is appropriate to examine the structure of β-alumina, which is a well-known superionic conductor, even though it is not a transition metal oxide. This solid electrolyte has the composition $Na_2O \cdot 11Al_2O_3$ and has a structure closely related to the spinel. Its structure (Fig. 140) can be described as spinel blocks linked through AlO_4 tetrahedra.[132] At the level of the AlO_4 tetrahedra, Na^+ ions are located; they can move easily between the spinel blocks, giving rise to high conductivity. The Al in β-alumina can be replaced by Fe.

10 Examples of Unusual Coordination: The Vanadium Oxides

Vanadium differs from other transition elements by virtue of its ability to adopt the trigonal bipyramidal or the square-pyramidal coordination. This is the case in V_2O_5 and in the family of vanadium bronzes, which exhibit either a lamellar or a tunnel structure and have been extensively studied for their metallic or semiconducting properties. Like V_2O_5, the bronzes α-$A_xV_2O_5$ (A = Li, Na, Ag, K, Cu; $0 \le x \le 0.10$) are isostructural. These oxides have an orthorhombic structure ($a \approx 11.5$ Å, $b \approx 3.6$ Å, $c \approx 4.4$ Å) and consist of $[V_2O_5]_\infty$ layers parallel to (001) (Fig. 141a). The coordination of vanadium, which is linked to six near neighbors (1.60–2.45 Å), can be described either as a trigonal bipyramid or as a tetragonal pyramid. The VO_5 pyramids share corners along **b** and their

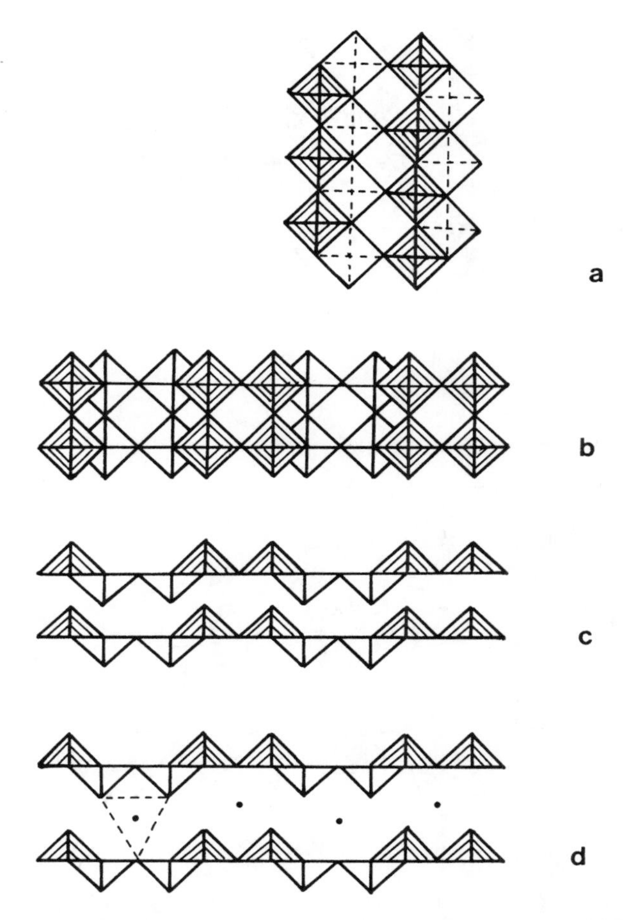

Figure 141. V_2O_5 and α-$A_xV_2O_5$ bronzes: (a) $[V_2O_5]_\infty$ layers, (b) ideal octahedral structure, (c) ideal pyramidal structure equivalent to the octahedral structure, and (d) ideal pyramidal structure after moving apart the V_2O_5 layers from each other, allowing A cations (dots) to be inserted. (After Wells.[4])

edges along $\vec{\mathbf{a}}$, forming the α-$[V_2O_5]_\infty$ layer. In fact, each vanadium has a sixth oxygen neighbor located at 2.8 Å, forming a distorted VO_6 octahedron. The ideal structure of V_2O_5 and of the α-$A_xV_2O_5$ bronzes can thus be obtained from a three-dimensional octahedral lattice built up of double ReO_3-type ribbons, sharing their edges along $\vec{\mathbf{a}}$ (Fig. 141b); in the latter structure, octahedral rows running along $\vec{\mathbf{a}}$ share their corners along $\vec{\mathbf{c}}$ in such a way that the structure can also be described as layers of tetragonal pyramids sharing their edges along $\vec{\mathbf{a}}$ and their corners along $\vec{\mathbf{b}}$ (Fig. 141c). The ideal structure of the α-bronzes is derived from the latter by moving the pyramidal layers apart from each other along $\vec{\mathbf{c}}$. While van der Waals forces are responsible for the cohesion of layers

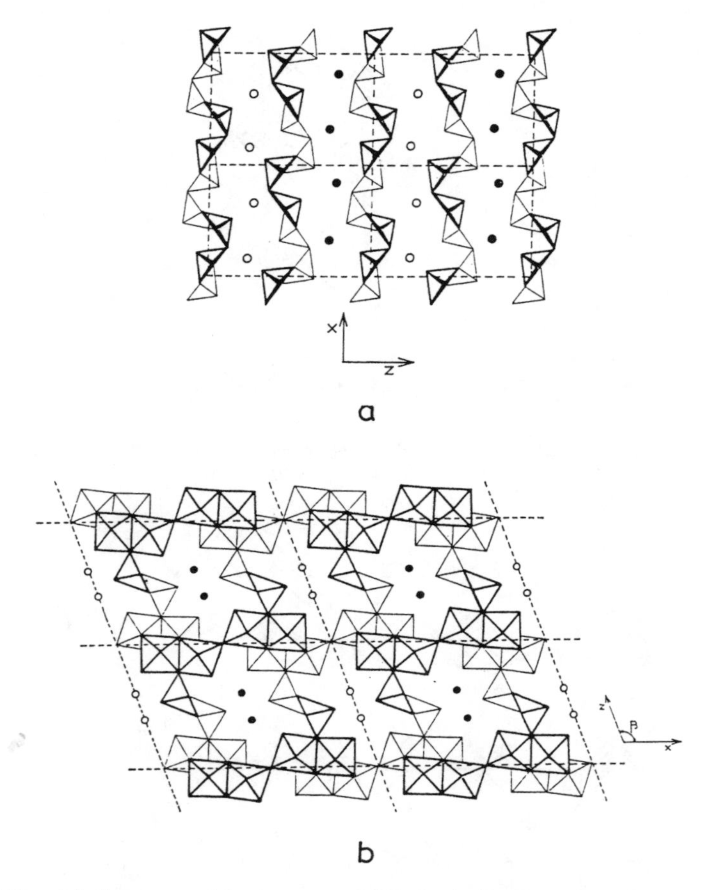

Figure 142. (a) Structure of the bronze γ-$Li_xV_2O_5$, built up of VO_5 pyramids forming layers intercalated with lithium ions (solid and open dots). (b) β-$A_xV_2O_5$ bronzes: view of the structure along $\vec{\mathbf{b}}$. One observes 2 × 2 edge-sharing VO_6 octahedra linked to VO_5 pyramids forming tunnels where the A cations (solid dots) are located. (After Hagenmuller.[133])

in V_2O_5, ionic bonds involving the A^+ ions in trigonal prismatic coordination and the $[V_2O_5]^x_\infty$ layers are important in α-$A_xV_2O_5$ (Fig. 141d).

δ-$Ag_xV_2O_5$ ($0.67 < x \leq 0.86$) bronzes, with a monoclinic structure ($a \approx 11.7$ Å, $b \approx 3.7$ Å, $c \approx 8.8$ Å, $\beta \approx 90.7°$), as well as monoclinic $A_{1+x}V_3O_8$ (A = Li, Na, Ag; $x \approx 0.1$–0.5; $a \approx 6.7$ Å, $b \approx 3.6$ Å, $c \approx 12$ Å, $\beta \approx 107.8°$) are both characterized by a lamellar structure, as described earlier (Section 7.2). The γ-$Li_xV_2O_5$ ($0.88 \leq x \leq 1$) bronzes, with an orthorhombic structure ($a \approx 9.7$ Å, $b \approx 3.6$ Å, $c \approx 10.7$ Å) have such a distorted coordination for vanadium that they cannot be considered as having octahedral layers. The γ-V_2O_5 layers here consist of VO_5 trigonal bipyramids sharing their edges along $\vec{\mathbf{a}}$ and their corners along

$\vec{\mathbf{b}}$ (Fig. 142a).[133] The Li^+ cations have an octahedral coordination. The β-$A_xV_2O_5$ bronzes (A = Li, Na, K, Cu, Ag, Ca, Cd, Pb), which have a variable extent of intercalation ($x = 0.17$–0.67), exhibit a tunnel structure. These monoclinic compounds ($a \approx 10$ Å, $b \approx 3.6$ Å, $c \approx 15.5$ Å, $\beta \approx 110°$) have a three-dimensional host lattice β-$[V_2O_5]_\infty$ (Fig. 142b) built up of highly distorted VO_6 octahedra (V—O distances, 1.6–2.3 Å) and distorted trigonal bipyramids (V—O distances, 1.6–2.2 Å). The octahedral lattice can be described on the basis of 2×2 edge-sharing octahedra (Fig. 142b), sharing their corners along $\vec{\mathbf{b}}$ to form infinite octahedral chains. Along $\vec{\mathbf{a}}$, octahedral chains share their corners to form octahedral layers parallel to (001). The connection between these layers, along $\vec{\mathbf{c}}$, is ensured by trigonal bipyramids, VO_5. The A cations are located in the large tunnels formed by the β-V_2O_5 framework.

Besides the bronzes, the vanadates AV_3O_8 (A = K, Cs), having a monoclinic structure ($a \approx 7.6$ Å, $b \approx 8.2$ Å, $c \approx 8.5$ Å, $\beta \approx 96°$) constitute another type of layered structure that can be described as being built up of distorted VO_6 octahedra. The $[V_3O_8]_\infty$ layers parallel to (100) are interleaved with K^+ or Cs^+ cations.

11 Shear Structures

Many octahedral oxides exhibit a structure derived from a mother framework by the translation of two parts of the crystal with respect to each other in such a way that at the boundary between the two parts, octahedra share their edges or faces instead of their corners. On both sides of the boundary, the original structure of corner-sharing octahedra remains untouched. This phenomenon, first explored by Magnéli, is called *crystallographic shear* (CS), and the plane separating the two shifted parts of the structure is called the crystallographic shear plane (CSP). Two structural types that are particularly suited for the formation of shear structure are the rutile- and ReO_3-type oxides.[6,8,9] High resolution electron microscopy enables direct observation of CSP's in oxides.

11.1 Shear Structure in Rutile-Type Oxides

The rutile family is represented by oxides of general formula B_nO_{2n-1} [B = Ti ($4 \leq n \leq \infty$), V ($3 \leq n \leq \infty$)] and to a limited extent, by elements such as manganese, chromium, and lead. The symmetry of the crystallographic cell varies with the n value, but the cell parameters are related to those of the tetragonal cell of rutile. The different members of the series consist of rutile-type slices that extend infinitely along two directions and are n octahedra thick along the third direction (Fig. 143a). In each microphase, the rutile slice ends in face-sharing octahedra with the next slice (Fig. 143b). These planes of face-sharing octahedar are the CSP's. The successive rutile slices are shifted with respect to each other. For small n values, the shear planes are regularly spaced and each member represents a well-defined phase. For large n values ($n > 10$), the CSP

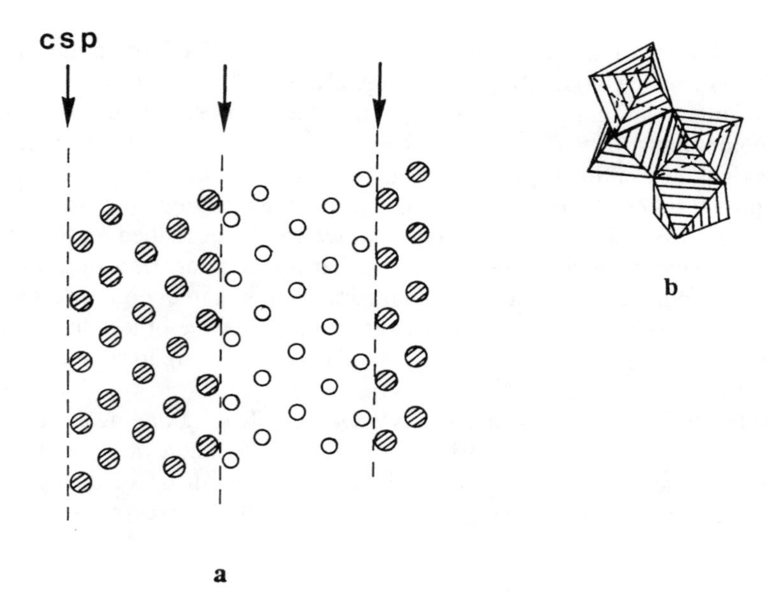

Figure 143. Shear structure oxides Ti_nO_{2n-1}. (a) Rows of Ti atoms of the member $n = 5$, Ti_5O_9, projected in a direction equivalent to [101] for rutile; the vertical dashed lines separate rutile slices, five TiO_6 octahedra thick. (b) The junction between two rutile blocks corresponding to the shear plane; note that in this shear plane the TiO_6 octahedra share their edges and their faces. (After Wadsley.[37])

tends to be disordered; for very large n values ($n > 50$), they appear in the form of two-dimensional defects in the rutile matrix. Such defects, called *Wadsley defects,* are also observed by high resolution electron microscopy for nonintegral n values wherein the rutile slices exhibit different thicknesses and are not distributed in an ordered fashion.

Transition metal oxides with CSPs are associated with the mixed valency of the transition element [Ti(III)/Ti(IV) or V(III)/V(IV)] and generally exhibit interesting transport properties owing to the delocalization of the d electrons.

11.2 Shear Structures in ReO_3-Type Oxides

In the ReO_3-type cubic structure, (see Section 4.1), crystallographic shear occurs along ⟨110⟩ leading to the mode represented here by I (Fig. 144a), in which two ReO_3-type octahedral chains share their edges and are at a distance of $\mathbf{a}/\sqrt{2}$ with respect to each other. The shear can be represented in another way, mode II, involving a translation of $(\vec{\mathbf{c}} - \vec{\mathbf{b}})/2$ (Fig. 144b), in which the edge shared by two octahedra is inclined with respect to the projection plane. All the ReO_3-type shear structures can be described by using these two modes of representation. Three classes of shear structures can be distinguished:

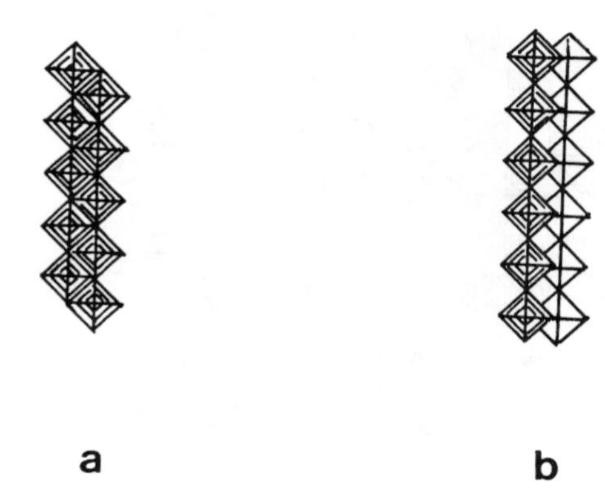

Figure 144. Representation of two ReO_3-type chains sharing edges of their octahedra: (a) mode I and (b) mode II.

1. Suboxides of molybdenum and tungsten of the type B_nO_{3n-1} and B_nO_{3n-2} (B = Mo, W).
2. Oxides of the WO_3/Nb_2O_5 system, which are niobium-rich.
3. Oxides of the TiO_2/Nb_2O_5 system and the vanadium suboxides.

11.2.1 Molybdenum and Tungsten Suboxides Having Formulas B_nO_{3n-1} and B_nO_{3n-2}

The Mo and W oxides described by mode I (Fig. 144) constitute a large family; members of the WO_3/Nb_2O_5 system on the WO_3-rich side also show these structures. Starting from the ReO_3 structure, there are numerous ways to generate shear structures using blocks of $2 \times n$ edge-sharing octahedra, where n is an integral number ($2 \leq n \leq \infty$). The most common shear plane is the {102} CSP, built up of 2×2 edge-sharing octahedra, separated by a double perovskite tunnel. Numerous B_nO_{3n-1} phases are generated by a regular spacing of these CSPs, as shown in the cases of Mo_8O_{23} and Mo_9O_{26} in Figure 145.[134] These two oxides correspond to the $n = 8$ and 9 members, respectively. The thickness of the ReO_3-type slices is characterized by the number of octahedra (n) running along $\vec{\mathbf{a}}_p$, between two successive CSPs; it corresponds to $n/2$ octahedra in the perpendicular direction. These numbers are 8 and 4, respectively, for the even-n member Mo_8O_{23}. For the odd-n member, Mo_9O_{26}, $n = 9$: if one counts nine octahedra along $\mathbf{a}_p$, one counts four and five octahedra alternately in the perpendicular direction. The odd-n members form strings of $(n + 1)/2$ and $(n - 1)/2$ octahedra in the second direction, whereas the even-n members form strings of $n/2$ octahedra. These {102} shear structures are mainly encountered in the suboxides Mo_nO_{3n-1} and $(Mo_{1-x}W_x)_nO_{3n-1}$.

The shear plane {103} corresponds to the association of two ReO_3-type lattices through the edges of their octahedra along the {103} plane. This CSP is

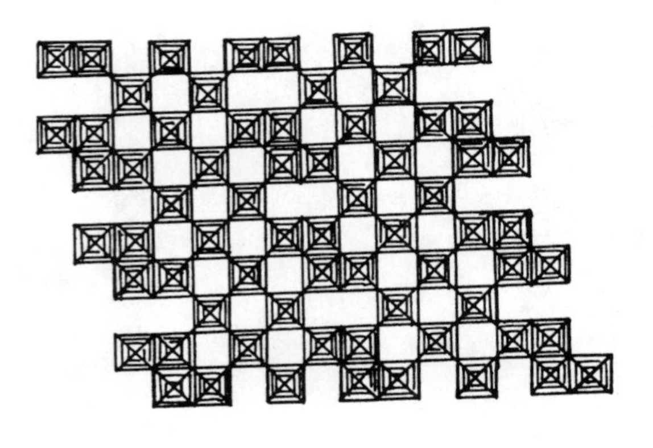

a

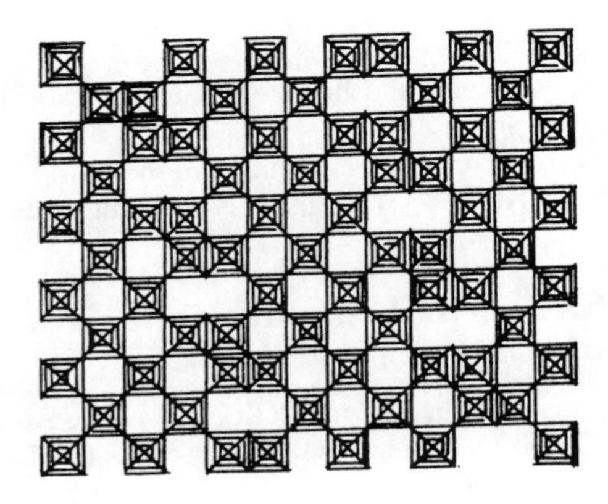

b

Figure 145. Structures of the two shear structure oxides Mo_nO_{3n-1}: (a) Mo_8O_{23} (n = 8) and (b) Mo_9O_{26} (n = 9). (After Magnéli.[134])

built up of blocks of 2 × 3 edge-sharing octahedra separated by double empty perovskite tunnels (Fig. 146).[135] The regular spacing of these shear planes leads to the B_nO_{3n-2} family, whose smallest member is $W_{20}O_{58}$ (Fig. 146), with n = 20. The number of oxides of this family is limited to the tungsten suboxides, W_nO_{3n-2} (n = 20, 24, 25, 40), and $Ta_4W_{35}O_{115}$ (n = 39). Partial replacement of W by Mo favors the formation of {102} CSPs instead of the {103} CSPs, as shown in $(W_{1-x}Mo_x)_{10}O_{29}$, which is identical to $W_{20}O_{58}$ with respect to the formula but is in fact the n = 10 member of the B_nO_{3n-1} series. The width of the ReO_3-type slices is determined by the number of octahedra between two successive CSP's in the {103} family as well.

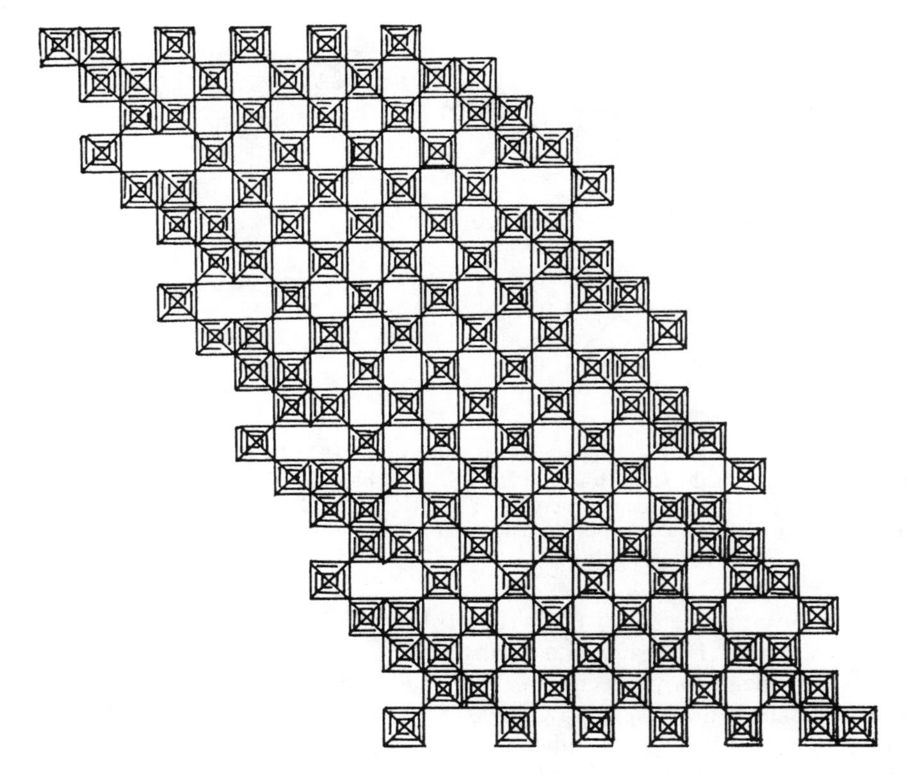

Figure 146. Structure of the oxide $W_{20}O_{58}$, $n = 20$, member of the shear structure series W_nO_{3n-2}. (After Magnéli.[135])

The {102} and {103} CSPs are also encountered in the WO_3/B_2O_5 (B = Nb, Ta) and Ti/W/O systems. However, they are generally not spaced regularly, especially on the WO_3-rich side; rather, they are distributed in a disordered fashion and occur in combination with other shear planes. The formation of such extended defects explains the nature of changes in oxygen stoichiometry observed during the reduction of WO_3 to WO_{3-x}. All these defects have been extensively studied by high resolution electron microscopy. High resolution electron microscopy of Nb_2O_5/WO_3, on the WO_3-rich side has shown that besides these shear planes, CSP's such as {104}, {105}, and {001} also occur. The {104} CSP's correspond to blocks of 2 × 4 edge-sharing octahedra (Fig. 147a), whereas the {105} CSP's correspond to blocks of 2 × 5 edge-sharing octahedra (Fig. 147b).[136] No pure {104} or {105} shear structures have been characterized, but some (e.g., $Nb_6W_{54}O_{177}$, $n = 60$, of B_nO_{3n-3}) have been observed in microcrystals by electron microscopy.

The {001} shear plane corresponds to infinite chains of edge-sharing octahedra (2 × n octahedra, $n \rightarrow \infty$). The corresponding formulas are identical to

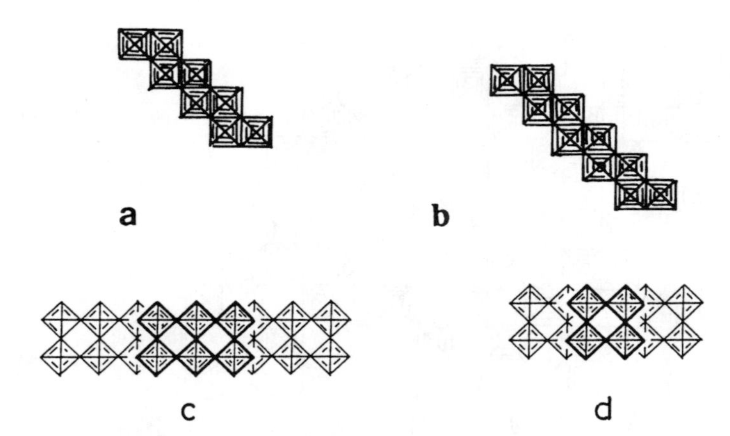

Figure 147. Structural units characteristic of (a) the {104} CSP (i.e., blocks of 2 × 4 edge-sharing octahedra) and (b) the ⟨105⟩ CSP (i.e., blocks of 2 × 5 edge-sharing octahedra). The {001} ideal shear structures of B_nO_{3n-1} of (c) Nb_3O_7F or V_2MoO_8 ($n = 3$) and (d) R-Nb_2O_5 ($n = 2$). (After Wadsley and Andersson.[136])

those involved in the {102} CSPs (i.e., B_nO_{3n-1}). The oxyfluoride Nb_3O_7F, as well as V_2MoO_8 (Fig. 147c) and R-Nb_2O_5 (Fig. 147d), correspond to the $n = 3$ and $n = 2$ members, respectively. The other members have been obtained as single crystals in the system Nb_2O_5/WO_3 for n ranging from $n = 8$, $(Nb,W)O_{2.88}$, to $n = 16$, $(Nb,W)O_{2.94}$. Another family of B_nO_{3n-1} oxides resulting from a combination of shearing modes I and II should be considered. Here, two kinds of CSP's, called $\{001\}^a$ and $\{001\}^b$, respectively, alternate in the structure (Fig. 144). The second member of this series, P-Nb_2O_5 (Fig. 148a) is tetragonal ($a \approx$ 3.9 Å, $c \approx$ 25.4 Å); higher members have not been observed till now. The first member of the series (Fig. 148b) corresponds to a close-packed structure built up of edge-sharing octahedra (TiO_2 in anatase form). Table 14 lists the different homologous series of oxides involving shear planes.

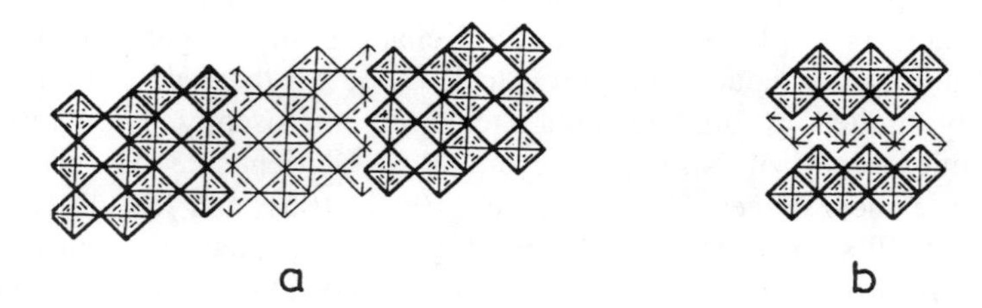

Figure 148. (a) Ideal structure of P-Nb_2O_5: $n = 2$ member of the $\{001\}_{a,b}$ shear structure series B_nO_{3n-1}. (b) Ideal structure of anatase TiO_2: $n = 1$ member of $\{001\}_{a,b}$ shear structures series B_nO_{3n-1}. One observes ribbons of edge-sharing TiO_6 octahedra at two levels. (After Wadsley and Andersson.[136])

Table 14 Typical Homologous Series of Oxides Involving CS[a]

Parent Structure	CS Planes	Series Formula	Approximate Composition Range
WO_3	$\{\bar{1}20\}$ ReO_3	W_nO_{3n-1}	WO_3–$WO_{2.93}$
	$\{\bar{1}30\}$ ReO_3	W_nO_{3n-2} ($n = 20, 24, 25, 40$)	$WO_{2.93}$–$WO_{2.87}$
MoO_3	$\{\bar{1}20\}$ ReO_3	Mo_nO_{3n-1} ($n = 8, 9$)	
	$\{311\}$ MoO_3	Mo_nO_{3n-2} ($n = 18$)	
TiO_2	$\{\bar{1}21\}$ rutile	Ti_nO_{2n-1} ($4 < n < 10$)	$TiO_{1.75}$–$TiO_{1.90}$
	$\{\bar{1}32\}$ rutile	Ti_nO_{2n-1} ($16 < n < \sim 37$)	$TiO_{1.90}$–$TiO_{1.9375}$

[a] Crystallographic shear with one set of nonintersecting shear planes.

11.2.2 Block Structures Derived from M-, N-, and H-Nb_2O_5

Let us consider the oxide B_4O_{11}, the $n = 4$ member of the series B_nO_{3n-1}. Its structure can be represented in two ways. In the first representation (mode I, Fig. 149a), the shear is performed in the projection plane, whereas in mode II (Fig. 149b) it takes place in the perpendicular direction. Starting from the second representation (Fig. 149b), a second shear can be performed along the same direction, perpendicular to the projection plane (mode II), on octahedral planes located at 90° of the CSP of B_4O_{11}, every n' octahedra. The resulting structure would then be built up of rectangular ReO_3-type blocks of $4 \times n'$ octahedra. This is exactly what is found in tetragonal M-Nb_2O_5 ($a \approx 20$ Å, $c \approx 3.8$ Å) whose structure (Fig. 150a) consists of blocks of 4×4 octahedra connected together by their edges. A series of phases with $n \times n'$ ReO_3-type blocks can, in principle, occur. However, M-Nb_2O_5 appears to be the only member of this series.

11.2.3 $B_{3n'}O_{8n'-3}$-Type Oxides Derived from the Nb_2O_5 Structure

Starting from the B_4O_{11} structure ($n = 4$ member of B_nO_{3n-1}), with the mode II representation (Fig. 149b), a second shear can be performed in the perpendicular direction (mode I), every n' octahedra. The structure (Fig. 150b) of the monoclinic form of N-Nb_2O_5 ($a \approx 28.5$ Å, $b \approx 3.8$ Å, $c \approx 17.5$ Å, $\beta \approx 120.8°$) results from such an operation. It is built up of ReO_3-type blocks of 4×4 octahedra ($n' = 4$) connected through their edges. A large family of oxides corresponding to this double shear can be predicted. One subgroup can be for-

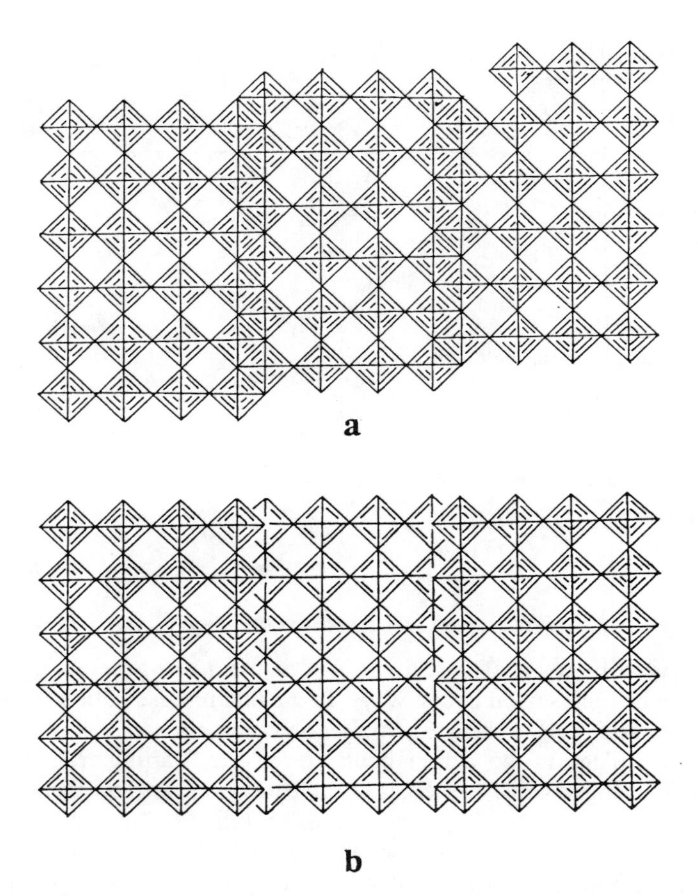

Figure 149. The ideal structure of the member $n = 4$ of the B_nO_{3n-1} {001} series represented in two different ways according to mode I (a) and mode II (b), respectively.

mulated as $B_{3n'}O_{8n'-3}$, represented by the $n' = 2$ member, V_6O_{13}, involving ReO_3-type blocks of 2 × 3 octahedra (Fig. 151a),[137] and the $n' = 3$ member, $TiNb_2O_7$ (or $TiTa_2O_7$), involving blocks of 3 × 3 octahedra (Fig. 151b).[136] $Ti_2Nb_{10}O_{29}$ is the $n' = 4$ member, involving ReO_3-type blocks of 4 × 3 octahedra (Fig. 152a); the oxyfluoride $MgNb_{14}O_{35}F_2$ is the $n' = 5$ member, with blocks of 5 × 3 octahedra (Fig. 152b). Note that in these oxides $n = 3$ and n' varies from 2 to 5; N-Nb_2O_5 is the only oxide of the series with $n = 4$ (n' also is 4), $Ti_2Nb_{10}O_{29}$ being classified in the subgroup $B_{3n'}O_{8n'-3}$ with $n = 3$ and $n' = 4$. The orthorhombic forms of $Ti_2Nb_{10}O_{29}$ and $Nb_{12}O_{29}$ must also be mentioned here. Their structure is also built up from ReO_3-type blocks of 3 × 4 octahedra with CSP's at 90°. There are regular stacking faults, however, which render each CSP the mirror image of an adjacent CSP.

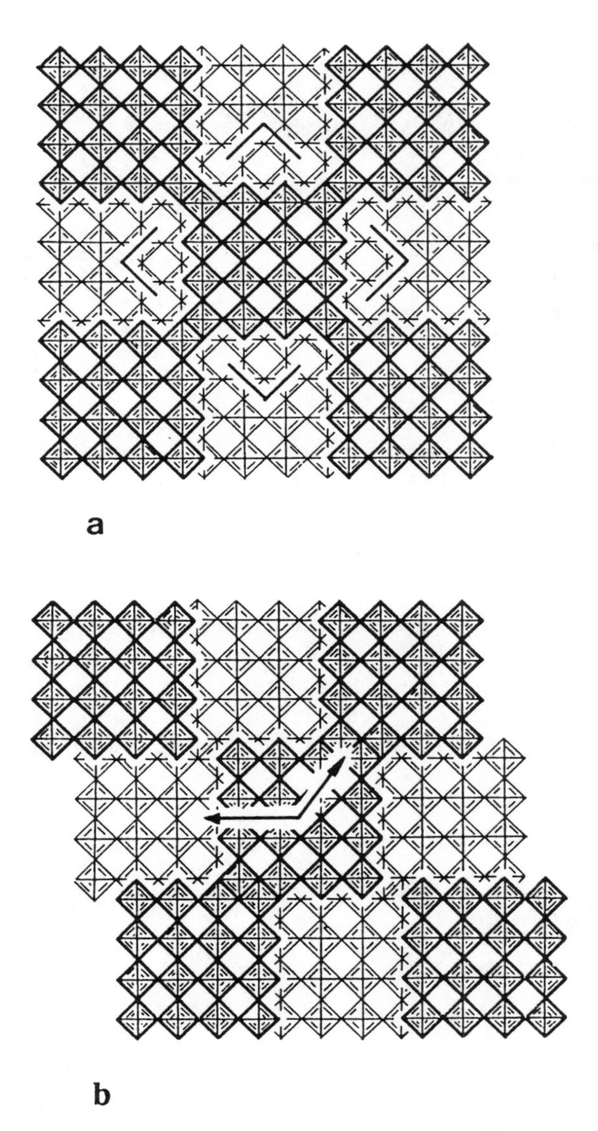

Figure 150. Ideal structures of (a) M-Nb_2O_5; (b) N-Nb_2O_5. Note that in both structures the blocks are 4 × 4 octahedra. (After Wadsley and Andersson.[136])

11.2.4 $B_{3n+1}O_{8n-2}$-Type Oxides Derived from H-Nb_2O_5

It is possible to describe this series by simple shear operations in spite of their close relationship with the oxides described earlier. Nevertheless, they can be described in terms of ReO_3-type blocks. Two groups can be distinguished according to the parity of n. For even-n values, blocks of $3 \times n/2$ octahedra are

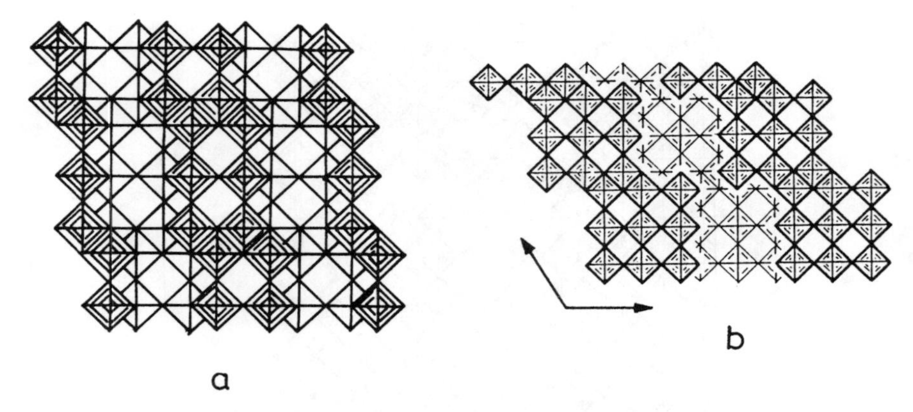

Figure 151. (a) The ideal structure of V_6O_{13}, $n' = 2$, member of the shear structure family $B_{3n'}O_{8n'-3}$. (After Wilhelmi et al.[137]) (b) The ideal structure of $TiNb_2O_7$, $n' = 3$, member of the shear structure family $B_{3n'}O_{8n'-3}$. The blocks are 3 × 3 corner-sharing octahedra. (After Wadsley and Andersson.[136])

obtained, whereas for odd-n values, there are blocks of two sorts, $3 \times (n + 1)/2$ and $3 \times (n - 1)/2$, which coexist in the structure. $TiNb_{24}O_{62}$ represents the first member of this series, corresponding to $n = 8$. In this oxide, ReO_3-type blocks of 3 × 4 octahedra form octahedral layers by sharing their edges in the projection plane (mode I), as shown in Figure 153a. Along the perpendicular

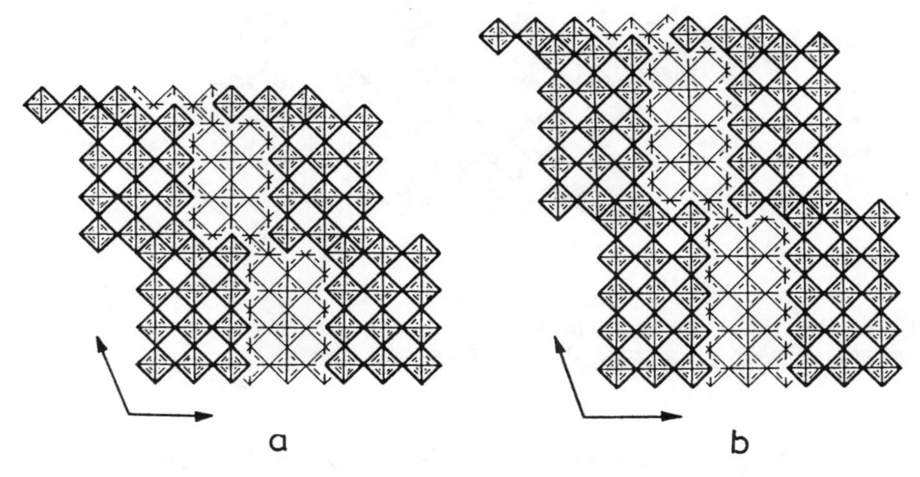

Figure 152. (a) The ideal structure of monoclinic $Ti_2Nb_{10}O_{29}$, $n' = 4$, member of the shear structure family $B_{3n'}O_{8n'-3}$. The blocks are 3 × 4 corner-sharing octahedra. (b) The ideal structure of $MgNb_{14}O_{35}F_2$, $n' = 5$, member of the $B_{3n'}O_{8n'-3}$ shear structure family. The blocks are 3 × 5 corner-sharing octahedra. (After Wadsley and Andersson.[136])

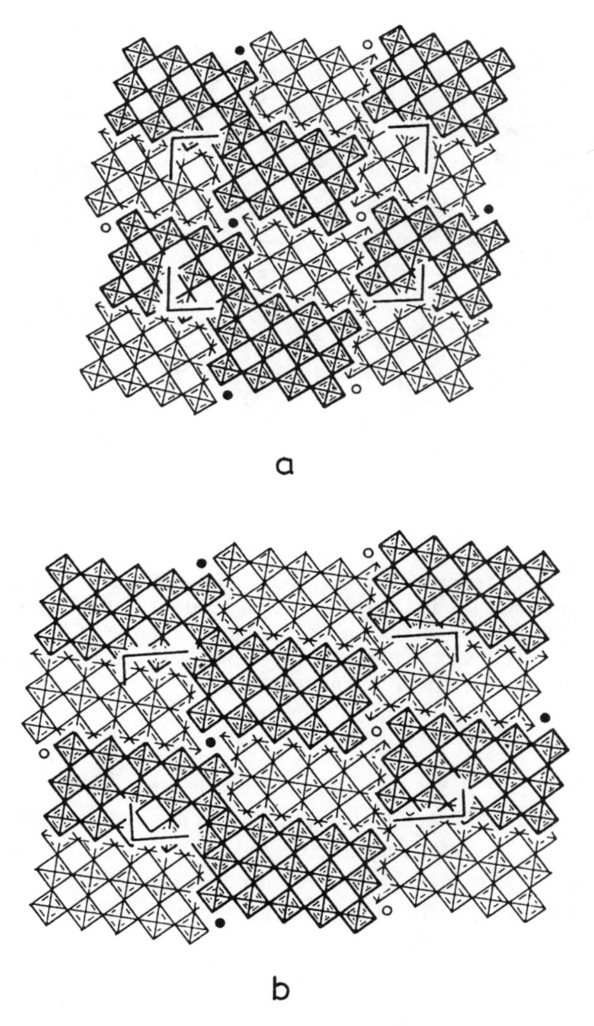

Figure 153. (a) The ideal structure of $TiNb_{24}O_{62}$, $n = 8$, member of the $B_{3n+1}O_{8n-2}$ shear structure family. Blocks of 3 × 4 corner-sharing octahedra to form tetrahedral sites where additional Nb atoms (solid and open dots) are located. (b) The ideal structure of $Nb_{31}O_{77}F$, $n = 10$, member of the $B_{3n+1}O_{8n-2}$ shear structure family. The blocks are 3 × 5 corner-sharing octahedra to form tetrahedral sites occupied by Nb atoms (dots). (After Wadsley and Andersson.[136])

direction, these layers share the edges of their octahedra (mode II), forming tetrahedral sites at the intersection of four octahedral blocks, where niobium is located. $Nb_{25}O_{62}$ and other substituted phases $(B, Nb)_{25}O_{62}$ (B = Al, V, Fe, Cs) adopt this structure. The $n = 10$-member, characterized by blocks of 3 × 5 octahedra is known in the oxyfluoride, $Nb_{31}O_{77}F$ (Fig. 153b).

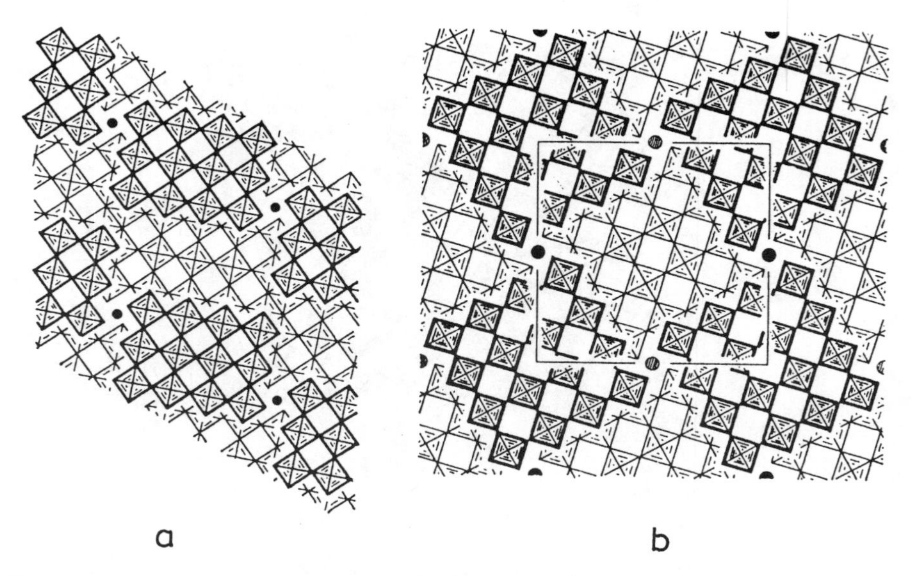

Figure 154. (a) Ideal structure of H-Nb_2O_5 involving two kinds of blocks of "3 × 5" and "3 × 4" octahedra, located at two different levels respectively and forming tetrahedral sites occupied by Nb atoms (dots). (b) Ideal structure of $W_3Nb_{14}O_{44}$, showing blocks of 4 × 4 octahedra located at two different levels and forming tetrahedral sites occupied by Nb or W. (After Wadsley and Andersson.[136])

H-Nb_2O_5 represents the $n = 9$ member of the series, built up of ReO_3-type blocks of 3 × 4 and 3 × 5 octahedra. The blocks are distributed in two different layers parallel to the projection plane (Fig. 154a). In the first layer, octahedral blocks of the larger size (3 × 5) share their edges along one direction (mode I), forming octahedral ribbons that are three octahedra wide and shifted with respect to each other. The smaller blocks (2 × 3) form the second layer, which is similarly built. The two layers share octahedral edges (mode II) to form the three-dimensional structure, which gives rise to tetrahedral sites partially occupied by niobium. $Nb_{22}O_{54}$ and $Nb_{34}O_{84}F_2$ correspond to the $m = 7$ and $m = 11$ members of this series, respectively.

11.2.5 Block Structures Derived from P-Nb_9O_{25}

Although closely related to H-Nb_2O_5 type-oxides, the block structures derived from P-Nb_9O_{25} have a simpler structure. One observes only one kind of octahedral junction formed by edges (mode II). P-Nb_2O_5, described earlier (Fig. 148a), also belongs to this family; it consists of ReO_3-type blocks of 3 × 3 octahedra sharing their edges and forming tetrahedral sites, where the phosphorus atoms are located. Other oxides exhibiting this structure are MNb_9O_{25} (M = V, As) and $Nb_{6.67}Ta_{3.33}O_{25}$. Several structures in the WO_3/Nb_2O_5 and MoO_3/

Nb_2O_5 systems, on the niobium-rich side, are derived from this structure type. This is the case in $B_3Nb_{14}O_{44}$ (B = W, Mo) involving blocks of 4×4 octahedra (Fig. 154b). $W_5Nb_{16}O_{55}$ is built up of blocks of 4×5 octahedra, while $W_5Nb_{18}O_{69}$ is characterized by blocks of 5×5 octahedra.

12 The Highly Complex Structural Behavior of Transition Metal Oxides

The structure of transition metal oxides is often more complicated than described in the preceding sections. Some of the oxides possess incommensurately modulated structures, whereas others have extended defects and exhibit order–disorder phenomena, leading to local modifications of the structure extending over several tens or hundreds of angstroms. In some instances, single crystals of a given composition do not consist of a single phase at the microscopic scale, but instead form *phasoids*. Phasoids have been defined by Magnéli as the coexistence within the same single crystal of several closely related structures having different compositions and different lattice periodicities (as in intergrowths). All such phenomena have been extensively studied in the last 20 years, thanks to the progress made in high resolution electron microscopy. We shall discuss a few illustrative examples of such oxides in Sections 12.1–12.4.

12.1 Modulated Structures

Several transition metal oxides show extra reflections in the X-ray and electron diffraction patterns in incommensurate positions with respect to the basic lattice. These incommensurate satellites are due either to a pseudoperiodic displacement of the atoms with respect to the ideal positions in the solids (displacive modulation) or to a pseudoperiodic distribution of two or more kinds of atom on the same crystallographic site (modulated distribution). Modulation may appear in tunnel structures—for instance, in the hollandite $La_{1.16}Mo_8O_{16}$, which exhibits a one-dimensional displacive modulation acting on La, Mo, and O atoms and a modulation wave governing the occupancy probability of La sites inside the tunnels. These phenomena give rise to the formation of La—La pairs in the tunnels and the formation of Mo_3 triangular clusters.

A recent example of modulation is that of the layered high T_c superconducting cuprates, $Bi_2Sr_2Ca_{n-1}Cu_nO_{2n+4}$.[49,138] For example, the structure of the 22 K superconductor $Bi_2Sr_2CuO_6$ (Fig. 38a) is more complex than described in Section 4.5. Electron diffraction patterns of this cuprate (Fig. 155) show incommensurate satellites along the a^*-axis. The corresponding HREM images (Fig. 156), clearly show modulation of the contrast. In the (001) images (Fig. 156a,b), we see the coexistence of blocks of 4 $(\sqrt{2}/2)a_p$, 5 $(\sqrt{2}/2)a_p$, and 6 $(\sqrt{2}/2)a_p$ in a nonperiodic fashion. The modulations along $\vec{\mathbf{a}}$ are clearly visible on the right-hand side of the micrograph of Figure 156c. The modulation is associated with

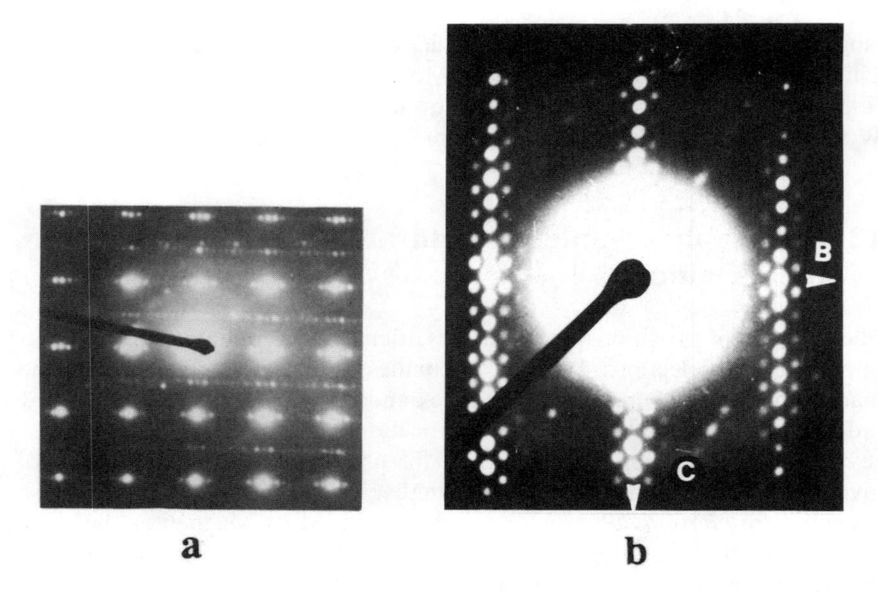

Figure 155. (001) and (010) electron diffraction patterns of the bismuth cuprate $Bi_2Sr_2CuO_6$. Satellites lie in incommensurate positions along $\vec{a}$. (After Raveau et al.[49])

the displacement of the metal and oxide ions. In fact, the BiO, SrO, and CuO layers have a wavy pattern in the structure. Single crystal X-ray diffraction studies have shown that there is an anharmonic modulation of all the atoms with regard to both the displacements and the site occupancies. On the left-hand side of the Figure 156c, modulation is not observed because of the (100) orientation of the lamellae; that is, in this part of the crystal, because the layers are turned by 90°, their undulation is rendered invisible.

12.2 Extended Defects and Order–Disorder Phenomena

The presence of extended two-dimensional defects, a common feature in transition metal oxides, is at the root of the wide ranges of stoichiometry found in many of the oxides. The phenomenon of intergrowth was discussed earlier, in Section 4.5, where we showed how new structures are created by the juxtaposition of perovskite and rock salt layers. This property can be generalized to many other structures. To have intergrowths, the two participating structures should have an identical atomic arrangement along one crystallographic plane. We discussed earlier the intergrowth in intergrowth tungsten bronzes (ITB's between HTB's and ReO_3-type structures),[139] between perovskite layers and Bi_2O_2 layers in the Aurivillius phases,[140,141] or between oxygen-deficient perovskites and distorted rock salt layers in layered cuprates.[49,138] When one of the two participating structures in the intergrowth is proportionally smaller than the

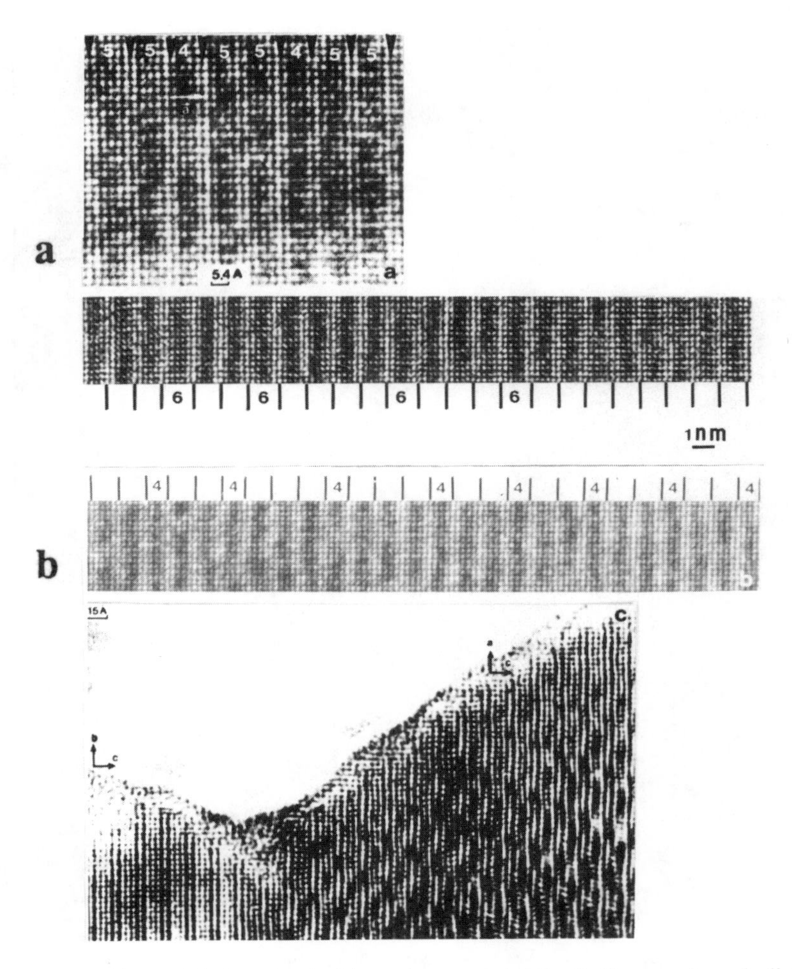

Figure 156. (a) Modulations in the bismuth cuprate $Bi_2Sr_2CuO_6$: pseudoperiodicity of blocks corresponding to $4(\sqrt{2}/2)a_p$ and $5(\sqrt{2}/2)a_p$ in a (001) image. (b) (001) projections of $Bi_2Sr_2CaCuO_8$ crystals, the insertion of blocks of $\times 4$ and $\times 6(\sqrt{2}/2)a_p$ is not perfectly periodic. (c) HREM image of 90°-oriented domains in $Bi_2Sr_2CuO_6$. The undulations of the bismuth layers are clearly visible at the right-hand side, in agreement with a (010) orientation of the crystals. The left part of the micrograph exhibits a contrast that agrees with a (100) orientation. Bismuth positions are correlated with the black spots of the image. Note the perfect interface. (After Raveau et al.[49])

other one, it forms isolated layers distributed in the matrix of the second structure in a statistical manner.[140] Such layers are identified as intergrowth defects. When the intergrowth is substantial and the participating units are distributed at random, one considers them to be disordered intergrowths. This is the case in disordered Cs_xWO_3 (ITB). An HREM micrograph (Fig. 157) shows different rows

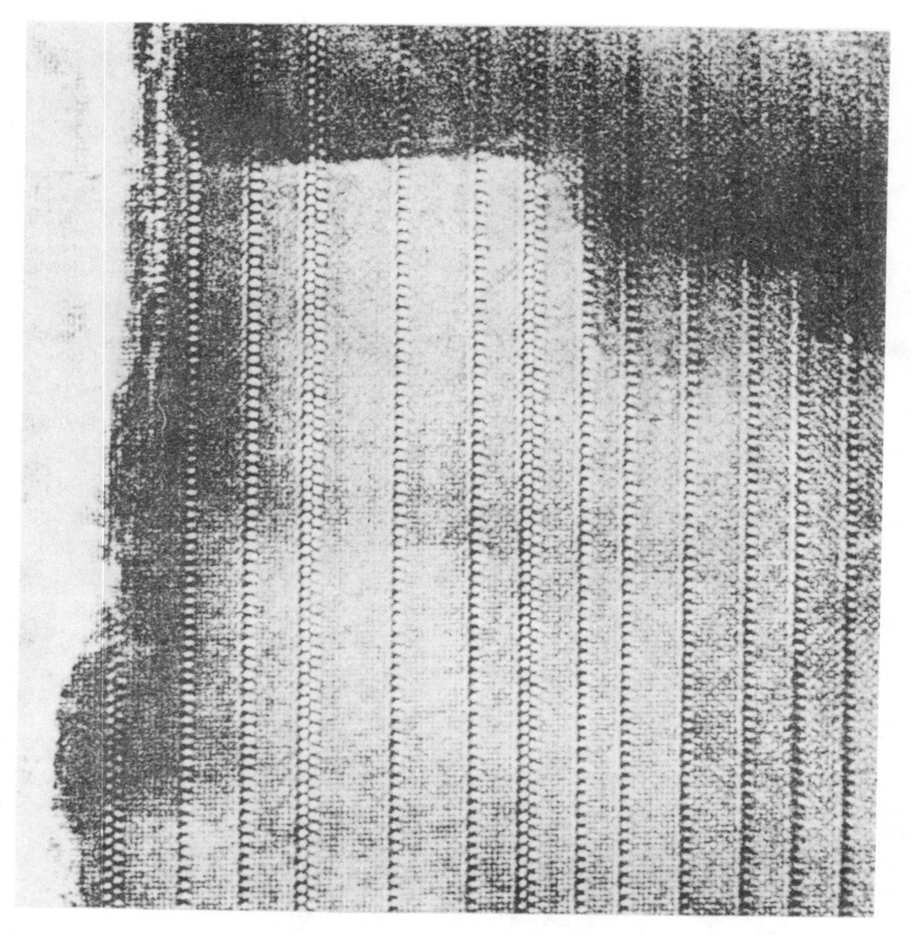

Figure 157. Disordered Cs_xWO_3 showing different widths of WO_3 slabs. (After Kihlborg[139])

of HTB tunnels distributed at random in a WO_3-type matrix, revealing a dramatic variation of the WO_3 slabs.[139] HREM images of two other disordered intergrowths appear in Figures 158 and 159. The micrograph in Figure 158 presents a disordered intergrowth in an Aurivillius phase of nominal composition $Bi_9Ti_3Fe_5O_{27}$.[141] One observes that the Bi_2O_2 layers (dark rows) are distributed at random, causing the perovskite slabs (white dots) to exhibit a variable thickness. The micrographs in Figure 159 are related to intergrowth defects that appear in superconducting thallium cuprates.[49] In $Tl_2Ba_2Ca_3Cu_4O_{12}$, which corresponds to the $m = 4$ member of the series $Tl_2Ba_2Ca_{m-1}Cu_mO_{2m+4}$, one observes extended defects of the larger perovskite slabs corresponding to the $m = 5$ and $m = 7$ members (see white dots, Fig. 159a). In the same way, Figure 159b shows the

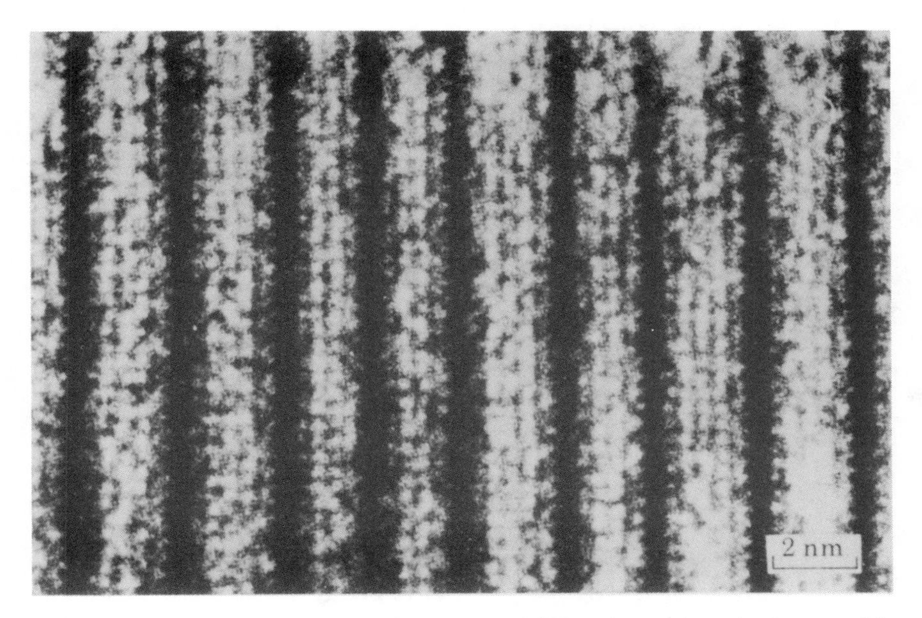

Figure 158. Disordered intergrowth in the Aurivillius phase of nominal composition $Bi_9Ti_3Fe_5O_{27}$. (After Hutchison et al.[141])

existence of extended defects corresponding to double rock salt layers involving thallium monolayers in a matrix of composition $Tl_2Ba_2CaCu_2O_8$, characterized by triple rock salt layers.

Shear defects or Wadsley defects (see Section 11) appear when the density of crystallographic shear planes in the structure is sufficiently low that the defects are distributed at random. Such defects appear in tungsten suboxides having different orientations with respect to the cubic perovskite subcell (i.e., {001}, {102}, and {103}). HREM images (Figs. 160 and 161)[142] show the formation of such crystallographic shear planes as extended defects in the WO_3 slabs of the diphosphate tungsten bronze $KP_4O_8(WO_3)_{30}$. In the first image (Fig. 160), besides the regularly spaced rows of hexagonal tunnels (rows of bright dots), one observes three rows of dark dots [see arrows parallel to (102)]; these rows correspond to the existence of {102} shear planes in the WO_3 slabs, built up of 2×2 edge-sharing octahedra. In the second image (Fig. 161), the sequence of hexagonal tunnels (rows of bright dots) is regularly spaced in the crystal, except for one WO_3 slab, which is much larger than the others. There is a zigzag row of dark dots that corresponds to a mixed shear plane {001}/{103}. This shear plane consists of two branches; the larger one, which is inclined with respect to the rows of tunnels, is an element of the {103} shear plane, whereas the shorter one, which is almost transverse, ensures the junction between two {103} segments. A combination of these elements of shear planes gives rise to block

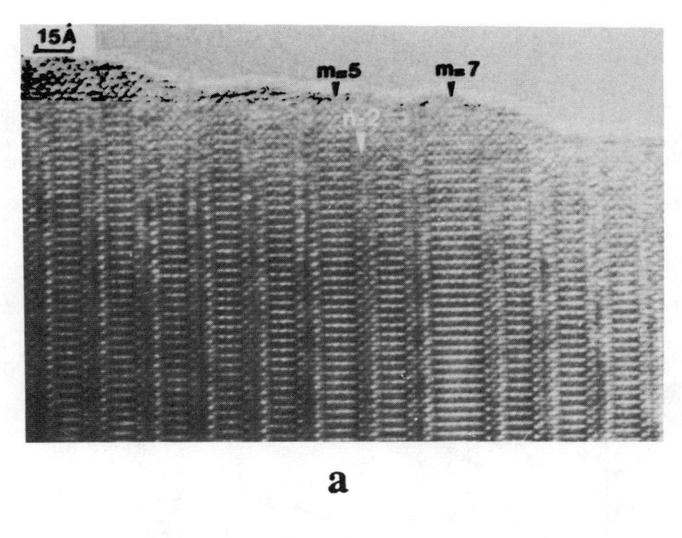

a

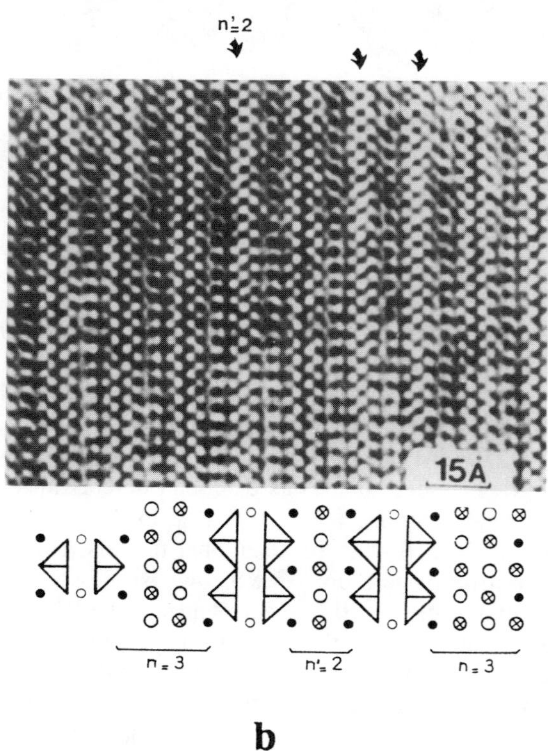

b

Figure 159. Intergrowth defects in thallium cuprates. (a) The $m = 5$ and $m = 7$ defective members appearing in the matrix of the $m = 4$ member of composition $Tl_2Ba_2Ca_3Cu_4O_{12}$. (b) Defective rock salt layers in a matrix of composition $Tl_2Ba_2CaCu_2O_8$. (After Raveau et al.[49])

Figure 160. HREM micrograph of a thin crystal fragment of $KP_4O_8(WO_3)_{30}$, where {102} CS planes appear as localized defects (black arrows). (From Hervieu and Raveau.[142])

structures (see Section 11.2) and leads to complex microstructures containing octahedral blocks of different sizes, as evidenced from the fault bands in HREM: see Figure 162[143] representing the block structure of $TiNb_{24}O_{62}$.

Chemical twinning, or microtwinning, is another type of extended defect that is observed in transition metal oxides. Microtwinning is systematically observed in the orthorhombic superconductor $YBa_2Cu_3O_7$, as seen in electron diffraction and microscopic images (Fig. 163). The phenomenon is easily explained in terms of the ability of copper to exhibit various coordinations so that the twin boundary (Fig. 164) is built up of CuO_6 octahedra, CuO_5 pyramids, or CuO_4 tetrahedra. This flexibility, in turn, allows the rows of CuO_4 groups on both sides of this boundary to be oriented at 90° with respect to each other. Another example of

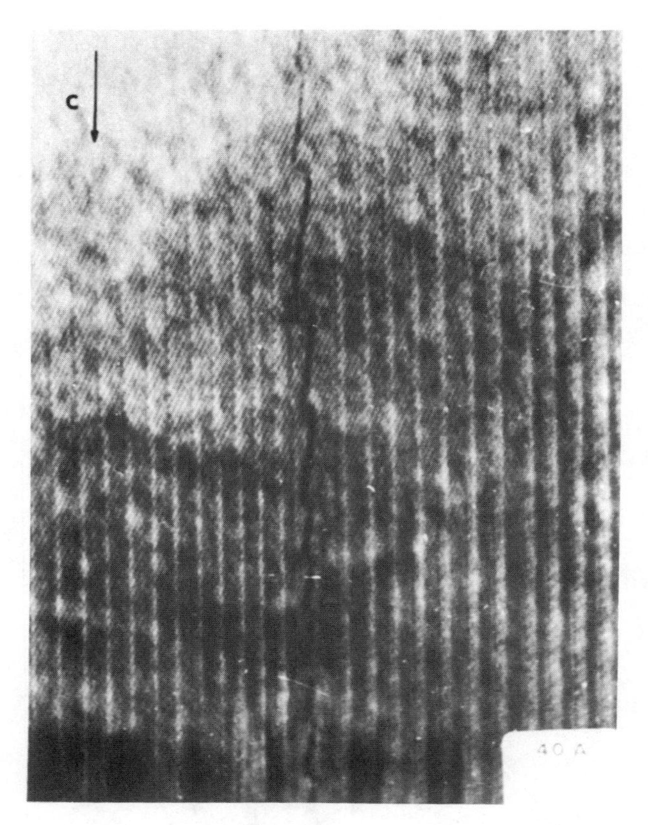

Figure 161. HREM micrograph of a crystal of overall composition $KP_4O_8(WO_3)_{22}$. In the widest ReO_3-type slab ($m = 17$), mixed {001}/{103} CS planes arose. The contrast of the {101} Cs branches is sometimes diffuse as a result of misalignment of the crystal. (From Hervieu and Raveau.[142])

chemical twinning is found in diphosphate tungsten bronzes with pentagonal tunnels DPTB's (see Section 4.6). Since pure phases corresponding to a regular periodicity cannot be observed in these oxides, they can be considered as phasoids. HREM images of these crystals (Fig. 165) show twinned ReO_3-type slabs, whose junction is parallel to (102) ReO_3[144]; the spacing of the twin boundaries that corresponds to rows of pentagonal tunnels is variable, and therefore no periodicity is obtained in the direction perpendicular to the chemical twin boundary.

Order–disorder phenomena leading to local superstructures are observed in ion-deficient transition metal oxides. This is the case, for instance, in the oxygen-deficient perovskite $YBa_2Cu_3O_{7-\delta}$ ($0 < \delta < 1$), which exhibits many local superstructures simultaneously in the same crystal, as indicated in recently published HREM images.[49] The image in Figure 166 corresponds to tripling of the

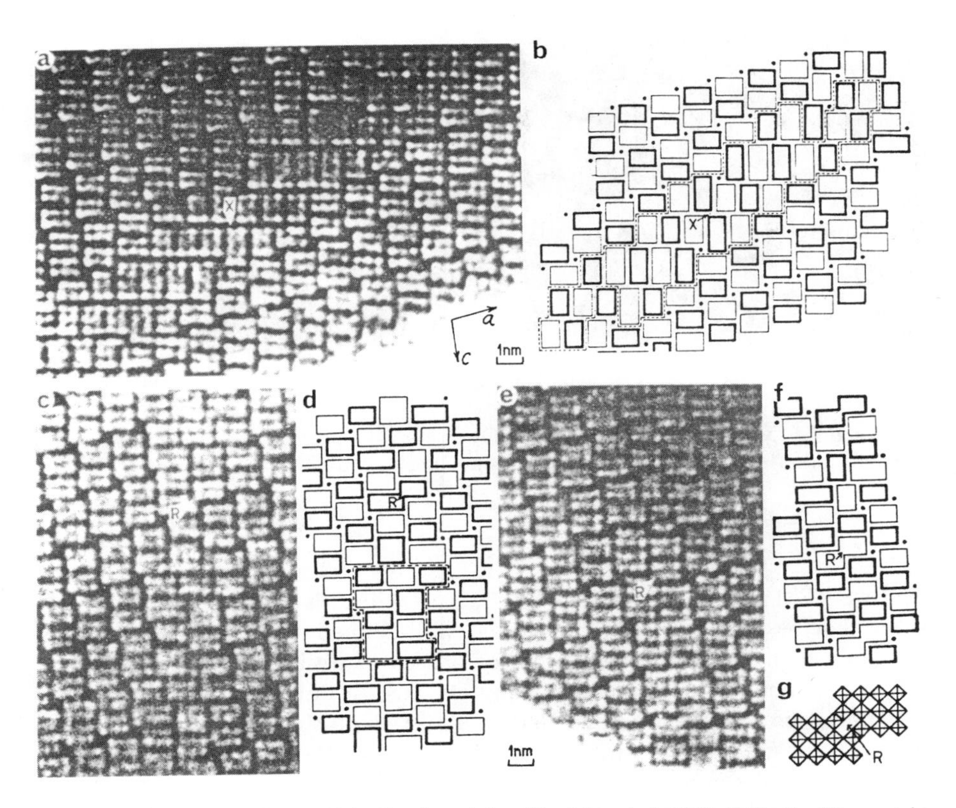

Figure 162. The lattice and idealized models of fault bands in $TiO_2 \cdot 7Nb_2O_5$. The matrix is $TiNb_{24}O_{62}$, and the magnification and matrix axes in (a) apply to all the micrographs (a, b). The band contains 4 × 3 and 5 × 3 blocks whose long axes lie perpendicular to those of blocks in the matrix. The point X marks the center of symmetry of the outlined region of the band in (b) and (d). The band contains 4 × 3, 4 × 4, and 5 × 3 blocks, and the repeating unit of fault structure is outlined in (d). This repetition is interrupted in the vicinity of R, which is a new type of block junction. (e)–(g) Row of R-type junctions occurs ion this band, and the model in (g) shows details of the proposed structure of these junctions. (From Iijima.[143])

a parameter of the perovskite, whereas that in Figure 167 exhibits different local superstructures along the ⟨210⟩*, ⟨110⟩*, and ⟨310⟩* directions of the perovskite subcell. These local superstructures are explained on the basis of the ordering of oxygen and anionic vacancies in the structure.

12.3 Phasoids

The notion of phasoids would not have come about without the help of HREM. It is well established that phasoids are not anomalies but are present in many oxide systems where two types of structures can coexist in the same crystal.

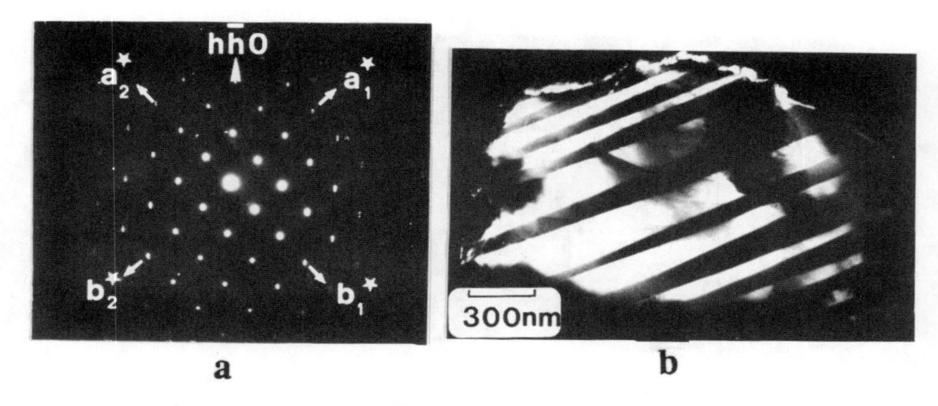

Figure 163. Microtwinning of $YBa_2Cu_3O_7$: (a) typical electron diffraction pattern and (b) dark field (001) image of a crystal. (From Raveau et al.[49])

Thus, HREM observations of $DPTB_P$'s, Aurivillius phases, and $YBa_2Cu_3O_{7-\delta}$ show that besides the well-ordered and well-defined phases, phasoids corresponding to disordered intergrowths of phases of different compositions and different structures are present in many crystals. In this respect, an interesting example is provided by reduced Ba(Sr)/Nb/O, which shows a disordered intergrowth structure built of NbO and $BaNbO_3$ blocks. A typical HREM image of such a phasoid is shown in Figure 168[145] for the gross composition $BaNb_6O_8$. In this image, the square dark spots correspond to the NbO_6 octahedra, which form NbO blocks of different sizes that are inserted in a perovskite $BaNbO_3$ matrix as shown in Figure 169.

12.4 Infinitely Adaptive Structures

In the homologous series Ti_nO_{2n-1}, crystallographic shear occurs on $(\bar{1}21)$ for $n < 10$ and on $(\bar{1}32)$ for $n > 16$ (Table 14). A remarkable feature of the compositions in which n is in the 10–14 range is that the $(\bar{1}21)$ plane swings continuously to $(\bar{1}32)$ through all possible orientations [(253), (374), (495), etc.]. As a consequence, almost any composition in the range of $TiO_{1.900}$ to $TiO_{1.937}$ can remain structurally ordered. Such continuous series of ordered structures may have apparently irrational CS planes. These structures have been called infinitely adaptive structures, a concept put forward by Anderson.[146] In the relevant composition range, the crystallographic shear plane swings continuously, being pivoted in its own zone from one stable direction to another.

An example of infinitely adaptive structures not involving CS is found in L-Ta_2O_5 and its solid solutions, with WO_3 up to the composition $11Ta_2O_5{\cdot}4WO_3$. The structure of L-Ta_2O_5 is related to that of U_3O_8 and consists of edge-shared

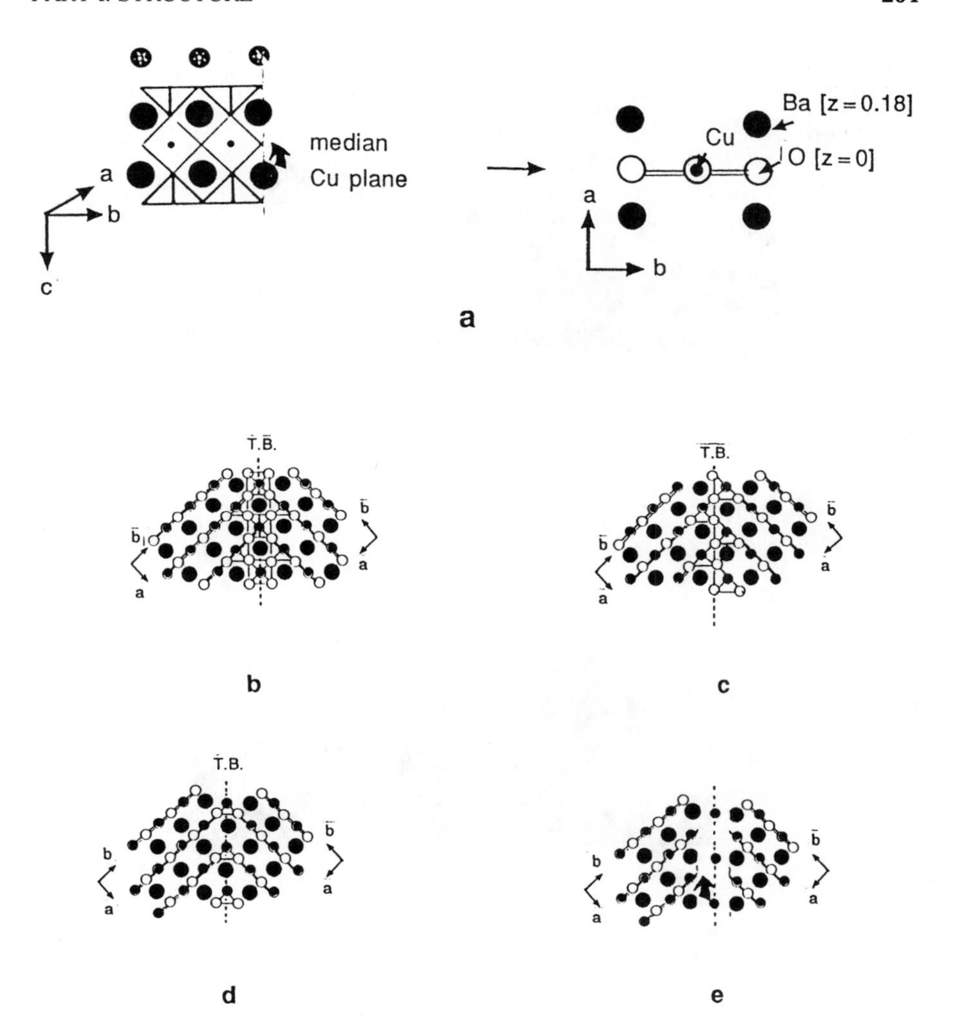

Figure 164. (a) Schematic drawings of the twin boundaries (TBs). (b)–(e) Four idealized models of junctions between the twinning domains in the $[CuO_2]_\infty$ layers: (b) junction in a mirror plane through CuO_6 octahedra and CuO_5 pyramids, (c) junction through CuO_5 pyramids, (d) junction in a mirror plane through CuO_4 tetrahedra, and (e) junction through twofold coordinated copper, which implies a threefold coordination for the adjacent copper atoms (arrowed) at the boundary; such arrangements between octahedra and pyramids are frequently observed in oxygen-deficient perovskites. (After Raveau et al.[49])

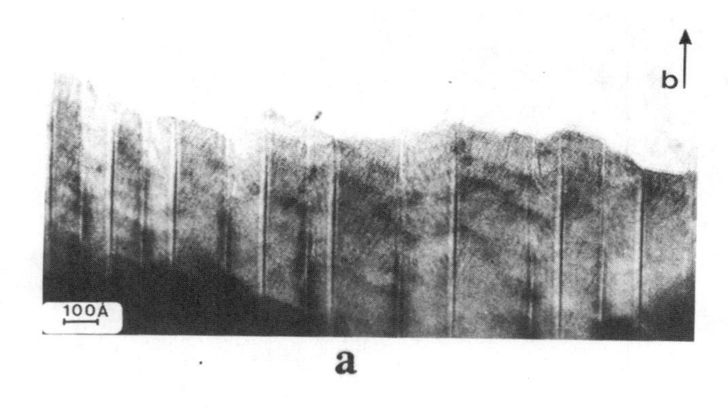

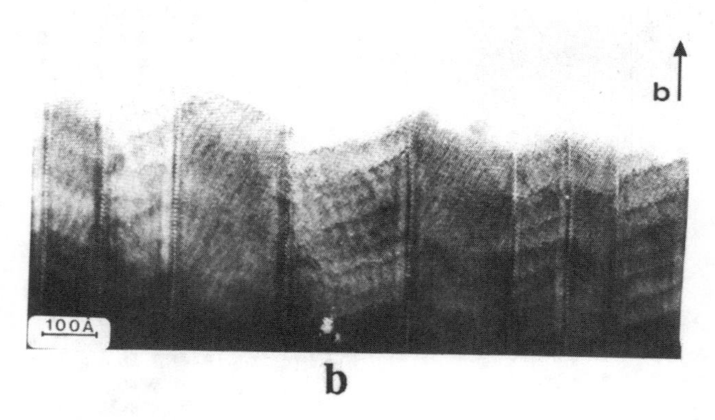

Figure 165. Diphosphate tungsten bronzes with pentagonal tunnels HREM images projected down [010]. (a) Twin domains of various widths are observed. (b) Enlargement of a micrograph showing the WO_3 structure of each domain. The domains are inclined by 65° with respect to the twin boundary, which appears as large white dots. (From Hervieu et al.[144])

pentagonal bipyramids and distorted octahedra. Addition of WO_3 to L-Ta_2O_5 results in unique superstructures for every composition. The resulting homologous series corresponds to $M_{2m}O_{(16m-2)/3}$ ($M_{10}O_{26}$, $M_{22}O_{58}$, etc.). An infinite series of structures can be built by an ordered stacking of these units in varying proportions. Any composition between $MO_{2.625}$ and $MO_{2.636}$ in this system can be built by using *a* units of $M_{10}O_{26}$ and *b* units of $M_{22}O_{58}$ so that the formula ($M_{10a+22b}O_{26a+58b}$) may be made to correspond to any composition in the range by an appropriate choice of *a* and *b*.

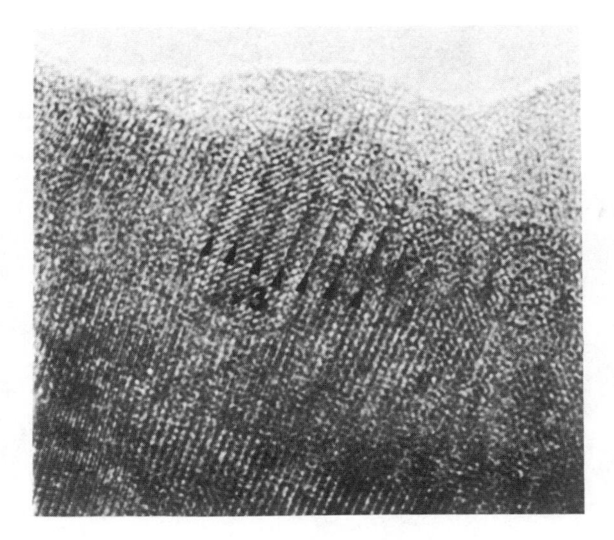

Figure 166. $YBa_2Cu_3O_{7-\delta}$: [001] HREM image of a local superstructure corresponding to a tripling of the a parameter $a' = 3a$. (After Raveau et al.[49])

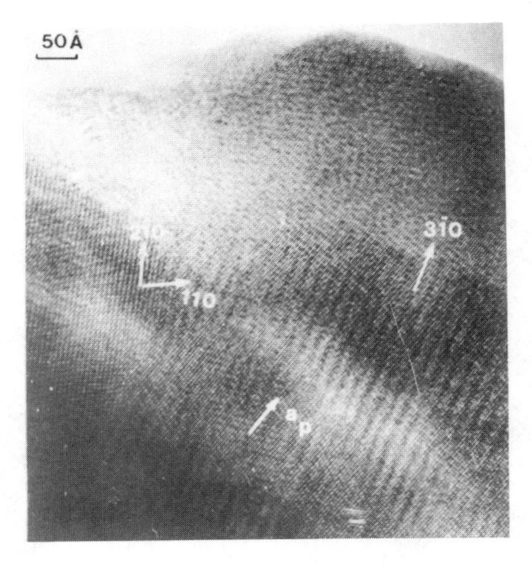

Figure 167. Complex modulations observed in an orthorhombic crystal $YBa_2Cu_3O_{6.66}$. New periodicities appear along the ⟨210⟩* (area 1) and ⟨110⟩* and ⟨310⟩* (area 2) directions of the perovskite subcell. (After Raveau et al.[49])

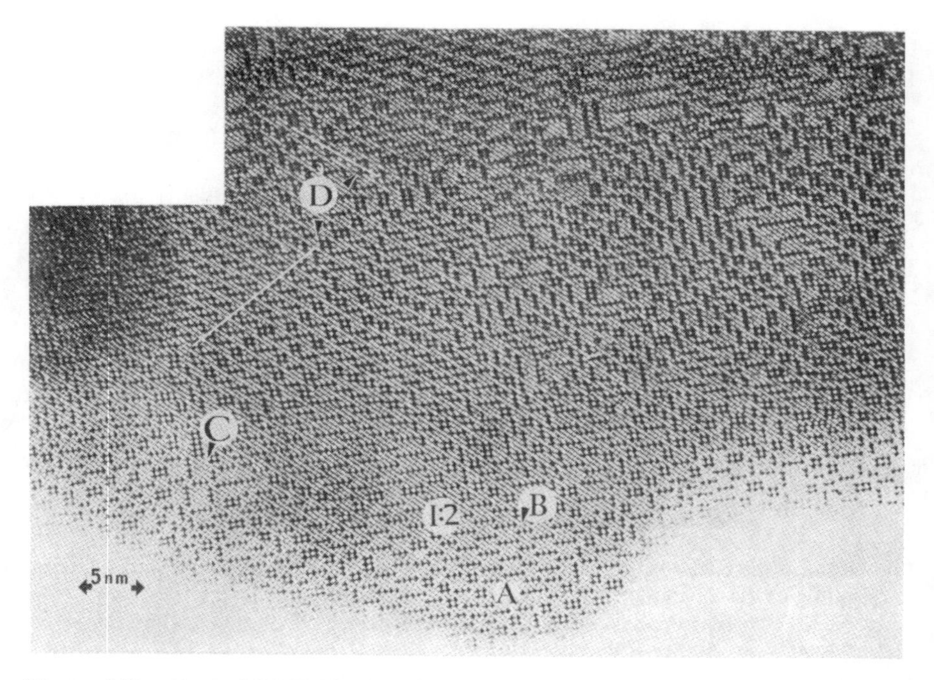

Figure 168. Typical HREM image of a phasoid called α, observed for a crystal formed in a sample with the gross composition $BaNb_6O_8$. (From Svensson.[145])

Figure 169. Model of an NbO block inserted in a $BaNbO_3$ matrix. The NbO_6 octahedra with Nb atoms, which are boundary atoms between the perovskite matrix and the NbO block, are shown with broken lines. (From Svensson.[145])

References

1. D. M. Adams, *Inorganic Solids, An Introduction to Concepts in Solid State Structural Chemistry,* Wiley, London, 1974.
2. C.N.R. Rao (ed.), *Solid State Chemistry,* Dekker, New York, 1974.
3. C.J.M. Rooymans and A. Rabenau (eds.), *Crystal Structure and Chemical Bonding in Inorganic Chemistry,* North-Holland, Amsterdam, 1975.
4. A. F. Wells, *Structural Inorganic Chemistry,* Clarendon Press, Oxford, 1975.
5. A. R. West, *Solid State Chemistry and Its Applications,* Wiley, Chichester, 1975.
6. C.N.R. Rao and J. Gopalakrishnan, *New Directions in Solid State Chemistry,* Cambridge University Press, New York, 1980.
7. B. G. Hyde and S. Andersson, *Inorganic Crystal Structures,* Wiley, New York, 1989.
8. L. Eyring and M. O'Keeffe (eds.), *The Chemistry of Extended Defects in Nonmetallic Solids,* North-Holland, Amsterdam, 1970.
9. J. S. Anderson and R.J.D. Tilley, in *Surface and Defect Properties of Solids* (M. W. Roberts and J. M. Thomas, eds.), The Chemical Society, London, 1974.
10. M. P. Tosi, in *Solid State Physics,* Vol. 16 (F. Seitz and D. Turnbull, eds.), Academic Press, New York, 1964, p. 1.
11. C.N.R. Rao, *Modern Aspects of Solid State Chemistry,* Plenum Press, New York, 1970.
12. M. O'Keeffe and A. Navrotsky, *Structure and Bonding in Crystals,* Vol. I, Academic Press, New York, 1981.
13. R. D. Shannon and C. T. Prewitt, *Acta Crystallogr., B25,* 925 (1969); also see *Acta Crystallogr., A32,* 751 (1976).
14. C.R.A. Catlow and W. C. Mackrodt (eds.), *Computer Simulation of Solids,* Lecture Notes in Physics, Springer-Verlag, Berlin, 1982.
15. M. J. Gillian and P.W.M. Jacobs, *Phys. Rev., B28,* 758 (1983).
16. A. M. Stoneham, *Theory of Defects in Solids,* Clarendon Press, Oxford, 1975.
17. F. A. Kröger, *The Chemistry of Imperfect Crystals,* North-Holland, Amsterdam, 1974.
18. O. T. Sørensen (ed.), *Nonstoichiometric Oxides,* Academic Press, New York, 1981.
19. B. Andersson and J. Gjonnes, *Acta Chem. Scand., 24,* 2250 (1970).
20. M. Morinaga and J. B. Cohen, *Acta Crystallogr., A35,* 745, 975 (1979).
21. W. L. Roth, *Acta Crystallogr., 13,* 140 (1960).
22. F. Koch and J. B. Cohen, *Acta Crystallogr., A37,* 307 (1981).
23. A. K. Cheetham, B.E.F. Fender, and R. J. Taylor, *J. Phys. C, 4,* 2160 (1971).
24. J. S. Anderson, *Proc. Indian Acad. Sci. (Chem. Sci.), 93,* 861 (1984).
25. M. R. Thornber and D.J.M. Bevan, *J. Solid State Chem., 1,* 536 (1970).
26. B.T.M. Willis, *J. Phys., 25,* 431 (1964).
27. R.J.D. Tilley, in *Chemical Physics of Solids and Surfaces,* Vol. 8, (M. W. Roberts and J. M. Thomas, eds.) The Chemical Society, London, 1980.

28. C.R.A. Catlow and R. James, in *Chemical Physics of Solids and Surfaces,* Vol. 8, (M. W. Roberts and J. M. Thomas, eds.) The Chemical Society, London, 1980.

29. J. L. Hutchison, J. S. Anderson, and C.N.R. Rao, *Nature, 255,* 541 (1975).

30. C.N.R. Rao and J. M. Thomas, *Acc. Chem. Res., 18,* 113 (1985).

31. D. A. Jefferson, M. K. Uppal, and C.N.R. Rao, *Mater. Res. Bull., 19,* 1403 (1984).

32. C.N.R. Rao, *Acc. Chem. Res., 18,* 113 (1985).

33. C.N.R. Rao and K. J. Rao, *Phase Transitions in Solids,* McGraw-Hill, New York, 1978.

34. D. Watanabe, in ref. 8.

35. R. Collongues, in *La nonstoichiométrie,* Masson, Paris, 1971.

36. R.W.G. Wycoff, *Crystal Structures,* Vols. 1 and 2, Wiley-Interscience, New York, 1964.

37. A. D. Wadsley, in *Nonstoichiometric Compounds* (L. Mandelcorn, ed.), Academic Press, New York, 1964.

38. L. Eyring in ref. 8.

39. F. S. Gallasso, *Structure, Properties and Preparation of Perovskite Type Compounds,* Pergamon Press, Oxford, 1969.

40. J. B. Goodenough and J. M. Longo, *Landolt–Börnstein Tabellen,* New Series, III/4a, Springer-Verlag, Berlin, 1970.

41. S. Nomura, *Landolt–Börnstein Tabellen,* New Series, III/12a, Springer-Verlag, Berlin, 1978.

42. F. Jona and G. Shirane, *Ferroelectric Crystals,* Pergamon Press, New York, 1962.

43. A. M. Chippindale, P. G. Dickens, and A. V. Powell, *Prog. Solid State Chem., 21,* 133 (1991).

44. P. Labbé, *Diffusion Phase Transitions and Related Structures in Oxides* (C. Bonlesteix, ed.), Trans Tech Publications, Switzerland, 1992.

45. C.N.R. Rao, J. Gopalakrishnan, and K. Vidyasagar, *Indian J. Chem., 23A,* 265 (1984); M. T. Anderson, J. T. Vaughey and K. R. Poeppelmeier. *Chem. Mater., 5,* 151 (1993).

46. B. Raveau, *Proc. Indian Natl. Sci. Acad., 52A,* 67 (1986).

47. J. C. Grenier et al., *Mater. Res. Bull., 11,* 1219 (1976).

48. A Reller, J. M. Thomas, D. A. Jefferson, and M. K. Uppal, *Proc. R. Soc. London, A394,* 224 (1984); also *J. Phys. Chem., 87,* 913 (1983).

49. B. Raveau, C. Michel, M. Hervieu, and D. Groult, *Crystal Chemistry of High* T_c *Superconducting Copper Oxides,* Springer-Verlag, Berlin, 1991.

50. P. Ganguly and C.N.R. Rao, *J. Solid State Chem., 53,* 193 (1984).

51. S. N. Ruddlesden and P. Popper, *Acta Crystallogr., 10,* 538 (1957); *11,* 54 (1958).

52. M. Dion, M. Ganne, and M. Tournoux, *Mater. Res. Bull., 16,* 1429 (1981); A. J. Jacobson, J. W. Johnson, and J. T. Lewandowski, *Inorg. Chem., 24,* 3727 (1985).

53. V. Fornichev et al., *Superconductivity, 3,* 126 (1990).

54. M. Huvé, C. Michel, A. Maignan, M. Hervieu, C. Martin, and B. Raveau, *Physica C, 205,* 219 (1993).

55. D. Pelloquin, M. Caldes, A. Maignan, C. Michel, M. Hervieu, and B. Raveau, *Physica C, 208,* 121 (1993).

56. B. Raveau, M. Huvé, A. Maignan, M. Hervieu, C. Michel, B. Domengés, and C. Martin, *Physica C,* 209, 163 (1993).

57. C.N.R. Rao et al., *Solid State Commun., 88,* 757 (1993). R. Nagarajan, S. Ayyappan, and C.N.R. Rao, *Physica C, 220,* 373 (1994); R. Mahesh, R. Nagarajan, and C.N.R. Rao, *Solid State Commun., 90,* 435 (1994).

58. A. Maignan, M. Hervieu, C. Michel and B. Raveau. *Physica C, 208,* 116 (1993).

59. C. Greaves, *Chem Bri.,* 30, 743 (1994).

60. S. Andersson and A. D. Wadsley, *Acta Crystallogr., 15,* 194 (1962).

61. A. Magnéli, *Acta Chem. Scand., 7,* 315 (1953).

62. M. Goreaud, P. Labbé, J. C. Monier, and B. Raveau, *J. Solid State Chem., 30,* 311 (1979).

63. L. Kihlborg and R. Sharma, *J. Micros. Spectros. Electron., 7,* 387 (1982).

64. J. Galy and M. Darriet, *Rev. Chim. Miner., 11,* 513 (1974).

65. G. K. Layden, *Mater. Res. Bull., 3,* 349 (1968).

66. A. A. Awadalla and B. M. Gatehouse, *J. Solid State Chem., 24,* 183 (1978).

67. L. Jahnberg, *Acta Chem. Scand., 17,* 2548 (1963).

68. G. D. Fallon and B. M. Gatehouse, *J. Solid State Chem., 49,* 59 (1983).

69. A. Magnéli, *Ark. Kem., 1,* 213 (1949).

70. A. W. Sleight, *Acta Chem. Scand., 20,* 1102 (1966).

71. L. Parmentier, *Rev. Chim. Miner., 9,* 519 (1979).

72. C. Michel, M. Hervieu, R.J.D. Tilley, and B. Raveau, *J. Solid State Chem., 52,* 281 (1984).

73. D. Groult, M. Hervieu, and B. Raveau, *J. Solid State Chem., 53,* 184 (1984).

74. A. D. Wadsley and L. Kihlborg, in *Nonstoichiometric Compounds* (L. Mandelcorn, ed.), Academic Press, New York, 1964.

75. A. Magnéli, *Ark. Kem., 1,* 223 (1949).

76. L. Kihlborg, *Acta Chem. Scand., 14,* 1612 (1960).

77. A. D. Wadsley, *Acta Crystallogr., 17,* 623 (1964).

78. A. F. Reid, A. D. Wadsley, and M. J. Sienko, *Inorg. Chem., 7,* 112 (1968).

79. K. Okada, H. Morikawa, F. Marumo, and S. Iwai, *Acta Crystallogr., B32,* 1522 (1976).

80. M. Sleeborg, *Chem. Comm.,* 1126 (1967).

81. B. M. Gatehouse and M. C. Nesbit, *J. Solid State Chem., 33,* 153 (1980). B. M. Gatehouse, *Natl. Bur. Stand (U.S.) Spec. Publ., 364,* 15 (1972). B. M. Gatehouse, *J. Less Common Metals, 36,* 53 (1974).

82. M. A. Subramanian, G. Aravamudan, and G. V. Subba Rao, *Prog. Solid State Chem., 15,* 55 (1983).

83. B. Raveau, *Rev. Inorg. Chem., 1,* 81 (1979).

84. M. Goreaud, G. Desgardin, and B. Raveau, *J. Solid State Chem., 27,* 145 (1979).

85. G. Desgardin, C. Robert, D. Groult, and B. Raveau, *J. Solid State Chem., 22,* 101 (1977).

86. C. Michel, A. Guyomarc'h, and B. Raveau, *J. Solid State Chem., 22,* 393 (1977). *J. Solid State Chem., 25,* 251 (1978).

87. M. Hong, *Acta Cryst., B30,* 945 (1974).

88. M. Nanot, F. Queyroux, J. C. Gilles, A. Carpy, and J. Galy, *J. Solid State Chem., 11,* 272 (1974).
89. B. Raveau, *Rev. Inorg. Chem., 9,* 37 (1987).
90. R. Ruh and A. D. Wadsley, *Acta Crystallogr., 21,* 974 (1966).
91. W. G. Mumme and J. A. Watts, *J. Solid State Chem., 3,* 319 (1971).
92. P. M. Gasperin, *Acta Crystallogr., B38,* 2024 (1982).
93. J. Graham and A. D. Wadsley, *Acta Crystallogr., 20,* 93 (1966).
94. C. Fouassier, G. Matejka, J. M. Reau, and P. Hagenmuller, *J. Solid State Chem., 6,* 532 (1973); C. Delmas, C. Fouassier, and P. Hagenmuller, *J. Solid State Chem., 13,* 165 (1975).
95. S. Lacorre, M. Hervieu, and B. Raveau, *Rev. Inorg. Chem., 6,* 195 (1984).
96. L. O. Hagman and P. Kierkegaard, *Acta Chem. Scand., 22,* 1822 (1968).
97. N. Nguyen, J. Choisnet, and B. Raveau, *J. Solid State Chem., 34,* 1 (1980).
98. J. Shannon, *Inorg. Chem., 17,* 158 (1978).
99. J. Shannon and L. Katz, *Acta Crystallogr., B26,* 105 (1970).
100. A. Leclaire, J. M. Monier, and B. Raveau, *Acta Crystallogr., B40,* 180 (1984).
101. A. Leclaire, J. M. Monier, and B. Raveau, *Z. Kristallogr., 184,* 247 (1988).
102. A. Leclaire, M. M. Borel, A. Grandin, and B. Raveau, *J. Solid State Chem., 76,* 131 (1988).
103. K. H. Lii and C. C. Wang, *J. Solid State Chem., 77,* 117 (1988).
104. R. C. Haushalter, *J. Solid State Chem., 76,* 218 (1988).
105. R. C. Haushalter, *J. Chem. Soc., Chem. Commun.,* 1566 (1987).
106. K. H. Lii, R. C. Haushalter, and C. J. O'Connor, *Angew. Chem., Int. Ed. Engl., 26,* 549 (1987).
107. A. Leclaire, M. M. Borel, A. Grandin, and B. Raveau, *J. Solid State Chem., 80,* 12 (1989).
108. A. Leclaire, A. Benabbas, M. M. Borel, A. Grandin, and B. Raveau, *J. Solid State Chem., 83,* 245 (1989).
109. P. M. Parmentier, C. Gleitzer, A. Courtois, and J. Protas, *Acta Crystallogr., B35,* 1963 (1979).
110. A. Leclaire, A. Benmoussa, M. M. Borel, A. Grandin, and B. Raveau, *J. Solid State Chem., 78,* 227 (1989).
111. J. W. Johnson, D. C. Johnston, H. E. King Jr., T. R. Halbert, J. F. Brody, and D. P. Goshorn, *Inorg. Chem., 27,* 1646 (1988).
112. C. S. Lee and K. H. Lii, *J. Solid State Chem., 92,* 363 (1991). K. H. Lii et al. *Inorg. Chem. 29,* 3298 (1990).
113. L. Benhamada, A. Grandin, M. M. Borel, A. Leclaire, and B. Raveau, *J. Solid State Chem., 97,* 131 (1992). P. Crespo, A. Grandin, M. M. Borel, A. Leclaire and B. Raveau, *J. Solid State Chem., 105,* 307 (1993).
114. L. Benhamada, A. Grandin, M. M. Borel, A. Leclaire, and B. Raveau, *J. Solid State Chem., 94,* 274 (1991).
115. A. Grandin, J. Chardon, M. M. Borel, A. Leclaire, and B. Raveau, *J. Solid State Chem., 99,* 297 (1992).
116. A. Grandin, J. Chardon, M. M. Borel, A. Leclaire, and B. Raveau, *J. Solid State Chem., 104,* 226 (1993).

117. A. Leclaire, M. M. Borel, A. Grandin, and B. Raveau, *Z. Kristallogr., 188,* 77 (1989).

118. K. H. Lii and R. C. Haushalter, *J. Solid State Chem., 69,* 320 (1987).

119. G. Costentin, Thesis, Université de Caen, 1992.

120. G. Costentin, M. M. Borel, A. Grandin, A. Leclaire, and B. Raveau, *J. Solid State Chem., 89,* 31 (1990).

121. G. Costentin, M. M. Borel, A. Grandin, A. Leclaire, and B. Raveau, *Z. Kristallogr., 201,* 53 (1992).

122. K. H. Lii, D. C. Johnston, D. P. Goshorn, and R. C. Haushalter, *J. Solid State Chem., 71,* 131 (1987).

123. P. Kierkegaard and M. Westerlund, *Acta Chem. Scand., 18,* 2217 (1964).

124. A. Leclaire, *Z. Kristallogr., 190,* 135 (1990).

125. C. Gueho, M. M. Borel, A. Grandin, A. Leclaire, and B. Raveau, *J. Solid State Chem., 104,* 202 (1993).

126. G. Costentin, M. M. Borel, A. Grandin, A. Leclaire, and B. Raveau, *J. Solid State Chem., 95,* 168 (1991).

127. P. M. Gasparin, *Acta Crystallogr., B37,* 641 (1981).

128. L. Bursill, *Acta Crystallogr., B35,* 530 (1979).

129. B. Raveau, *Proc. Indian Acad. Sci. (Chem. Sci.), 96,* 419 (1986).

130. V. Agafonov, Thesis, Université de Paris VI, 1985.

131. D. J. Lloyd, I. E. Grey, and L. A. Bursill, *Acta Crystallogr., B32,* 1756 (1976).

132. C. A. Beevers and M.A.S. Ross, *Z. Kristallogr., 97,* 59 (1937).

133. P. Hagenmuller, in *Progress in Solid State Chemistry,* Vol. 5 (H. Reiss, ed.), Pergamon Press, New York, 1971.

134. A. Magnéli, *Acta Chem. Scand., 2,* 501 (1948).

135. A. Magnéli, *Ark. Kem., 6,* 133 (1953).

136. A. D. Wadsley and S. Andersson, in *Perspectives in Structural Chemistry,* Vol. 3, (J. Dunitz and H. Ibers, eds.), Wiley, New York, 1970.

137. K. A. Wilhelmi, K. Waltersson, and L. Kihlborg, *Acta Chem. Scand., 25,* 2675 (1971).

138. C.N.R. Rao (ed.), *Chemistry and High-Temperature Superconductors,* World Scientific, Singapore, 1991.

139. L. Kihlborg, *Chem. Scripta, 14,* 187 (1978).

140. C.N.R. Rao, *Bull. Mater. Sci., 7,* 155 (1985).

141. J. L. Hutchison, J. S. Anderson, and C.N.R. Rao, *Proc. R. Soc. London, A355,* 301 (1977).

142. M. Hervieu and B. Raveau, *Chem. Scripta, 22,* 123 (1983).

143. S. Iijima, *J. Solid State Chem., 7,* 94 (1973).

144. M. Hervieu, B. Domengés, and B. Raveau, *J. Solid State Chem., 58,* 233 (1985).

145. G. Svensson, Thesis, University of Stockholm, 1989.

146. J. S. Anderson, *J. Chem. Soc., Dalton Trans.,* 1107 (1973).

PART

II

Properties and Phenomena

PART

II

Properties and Phenomena

1 Introduction

Transition metal oxides exhibit such a variety of properties and phenomena that a review of these topics would serve to promote an understanding of most aspects of inorganic materials.[1–10] By virtue of the diverse structures adopted by transition metal oxides, these materials are ideal for demonstrating relations between structure and properties and for showing how such relations help in designing materials. In Part II we shall briefly examine the electronic structure of oxides, to provide the background necessary to understand the properties; a discussion of the properties of oxides of different structures follows. We shall also see how the results of empirical theoretical calculations have helped investigators to understand the properties of transition metal oxides. We shall examine the results obtained by combined use of electron spectroscopy and empirical theory to demonstrate how we are now able to characterize oxides in terms of their electronic structure.

Physical properties of metal oxides and other solids arise as responses to external stimuli (physical forces). The important physical forces are mechanical stress, electric field, magnetic field, and temperature. A magnetic field causes magnetization, an electric field gradient produces a current flow, and stress produces strain. Physical properties can be equilibrium, steady state, hysteretic, and irreversible. Equilibrium properties are measured when the solid is in thermal equilibrium with its surroundings. During the measurement of a steady state property, on the other hand, the solid is not in thermal equilibrium with the surroundings, but it does not change with time. An example of a steady state property is electrical resistivity, which relates the potential gradient to charge flow. Hysteretic properties are found when the relation between two variables is irreversible and depends on whether the intensive variable is increased or decreased. A typical example is the magnetization–magnetic field relationship in a ferromagnet. Thus in Part II we discuss the electronic, magnetic, dielectric, optical, and catalytic properties of transition metal oxides, along with important phenomena such as metal–nonmetal transitions and superconductivity.

2 Electrons in Transition Metal Oxides

Correlation of the structure and physical properties of transition metal oxides requires a description of the valence electrons that bind the atoms in the solid state.[9,11,12] The two limiting descriptions of atomic outer electrons in solids are the localized electron theory (or the *ligand field theory*) and the *band theory*. Localized electron theory is applicable when interatomic interactions are weak and the electrons are tightly bound to the atomic core. Localized electrons are characterized by a large value of the energy U required to transfer a valence electron from one site to an occupied orbital on another equivalent site (of the order of the atomic ionization energy minus the electron affinity in the zero-order approximation) and a small band with W ($U \gg W$). When there is appreciable overlap between orbitals of neighboring atoms, the *band theory* of Bloch and Wilson becomes applicable. That is, we assume that valence electrons are shared equally by all the like atoms in the solid. In this model, $U \ll W$; in the extreme case, $U = 0$. When $U \sim W$, we have the third possibility of strongly *correlated electrons* in solids. Outer s and p electrons that are weakly bound to the atomic core and interact strongly with neighboring atoms are described by the band model. On the other hand, $4f$ electrons of the rare earths, which are screened from the neighboring atoms by the $5s^2 5p^6$ electrons, are always localized in solids and are described by ligand field theory. In coordination complexes of transition metals, the d electrons are localized, being described by ligand field theory. In transition metals, the d electrons are itinerant, being in narrow d bands that overlap the broad s–p bands. In transition metal oxides, we come across both localized and itinerant d-electron behavior. The electronic structure of transition metal oxides has been discussed in detail by many authors[1,4,8,10] and we shall examine the bare essentials here.

2.1 Band Model

Band theory assumes that the electrons in a solid are distributed among a set of available stationary states following the Fermi–Dirac statistics. The states are given by solutions of the Schrödinger equation,

$$\mathcal{H}\psi_{nk} = \mathcal{E}\psi_{nk}$$

Here, the Hamiltonian operator $\mathcal{H}$ includes the kinetic energy term $(p^2/2m)$ as well as the crystal potential $V(x)$, which accounts for the interaction of an electron with other particles in the crystal lattice. $V(x)$ has the translational periodicity of the crystal structure, namely, $V(x + a) = V(x)$ where a is the lattice constant. The wave functions *(Bloch functions)* in one dimension are of the form, $\exp(ikx)u_{nk}(x)$, where u_{nk} is periodic with the same periodicity as the crystal potential. The wave function is modulated by $u_{nk}(x)$.

Each wave function is characterized by an integer, the band index n, and the wave vector **k**. Whereas in the simple free electron model the quadratic dependence of the energy on the wave-number k is continuous, the energy dependence on k in the band model is discontinuous, showing breaks at k equal to π/a, $2\pi/a$, $3\pi/a$, etc. The dependence of the energy eigenvalue, E_{nk}, on the crystal momentum (hk) for different values of n describes the electronic band structure of a crystalline solid. The discontinuities in energy that give rise to allowed (energy bands) and forbidden (band gaps) regions of energy arise from the periodic potential. Thus, at $k = \pi/a$, the solutions to the Schrödinger equation are $\exp[i(\pi/a)x] \pm \exp[-i(\pi/a)x]$, which correspond to two different allowed energies, giving rise to the energy gap. This range of k is known as the first *Brillouin zone.* At values of k far from $\pm n\pi/a$, the energy shows the free electron quadratic dependence on k. In calculating simple band structures of transition metal oxides, one employs the linear combination of the metal atomic d and oxygen p orbitals (LCAO), just as in molecular orbital calculations (tight-binding scheme), and the periodic nature of the system helps to solve the problem.

The number of allowed electron states $N(E)dE$ per unit cell in the energy range between E and $E + dE$ is called the *density of states.* Application of the Fermi–Dirac distribution shows that all the states lower in energy than the *Fermi energy* E_F are occupied by electrons, while those above E_F are unoccupied at 0 K. In a metal, E_F lies within a band (the highest occupied band is partly filled). In an *insulator* or a *semiconductor,* E_F lies between bands with an energy gap separating the highest-lying filled band *(valence band)* and the lowest-lying empty band *(conduction band).* Thus Bloch–Wilson band theory classifies crystalline solids into metals and insulators depending on whether their electronic structures involve partially filled bands. Many of the physical properties of solids are determined by the electron states close to the *Fermi surface,* which is the locus of k values corresponding to the boundary between the occupied and the empty levels. Figure 1 presents energy band schemes for oxides of different types. Table 1 lists different metal oxide types on the basis of their electronic properties. Electronic excitation in insulators and semiconductors gives rise to an electron in the conduction band and a hole in the valence band. Electrostatic interaction between electrons and holes gives rise to *excitons.*

The simple picture just described does not depict the real behavior of most of the transition metal oxides. Typical examples of such oxides are MnO, CoO, and NiO, possessing the rock salt structure. The cation d orbitals in the rock salt structure would be split into t_{2g} and e_g sets by the octahedral crystal field of the anions. In transition metal monoxides, TiO-NiO ($3d^2$–$3d^8$), the d levels would

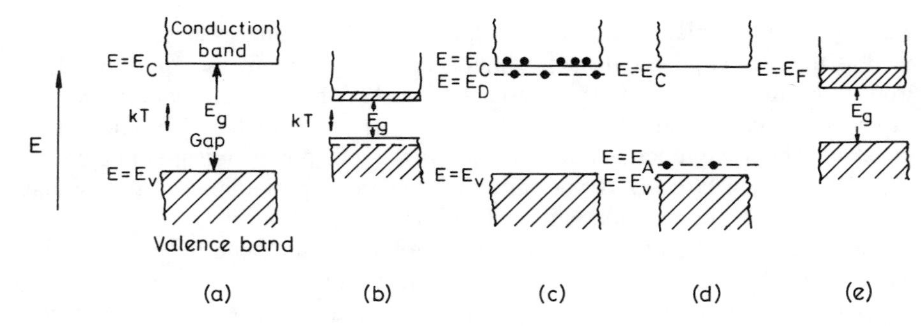

Figure 1. Schematic illustrations of five band structures of solids. (a) Insulator ($kt \ll E_g$), where E_g is the energy gap). (b) Intrinsic semiconductor ($kt \sim E_g$). (c) and (d) Extrinsic semiconductors with donor (D) and acceptor (A) impurities in n- and p-type semiconductors, respectively. (e) Metal. In oxides, n-doping (electron doping) can be done by substitution by a cation of a higher oxidation state (e.g., Pb^{4+} in place of Tl^{3+}; Ce^{4+} in place of Nd^{3+}) or by an anion of lower charge (F^- in place of O^{2-}). Hole-doping can be done by increasing the oxygen content.

be partially filled; hence the simple band theory predicts them to be metallic. The prediction is true in the case of TiO and to some extent in the case of VO. Stoichiometric MnO, CoO, and NiO are however all good insulators showing antiferromagnetism. The insulating nature of FeO can be understood by assuming that the t_{2g} subband is completely filled for the $3d^6$ configuration (i.e., Fe^{2+} is in the low spin state), but the insulating nature of MnO, CoO, and NiO cannot be understood in terms of simple band theory. Similar problems arise with transition metal oxides of other structures. Sesquioxides of corundum structure with 1, 3, and 5 d electrons per cation can in principle be insulators; the electrical properties of Ti_2O_3 ($3d^1$), Cr_2O_3 ($3d^3$), and Fe_2O_3 ($3d^5$) conform to such a pre-

Table 1 Electronic Properties of Transition Metal Oxides of Different Types

Class of Materials	Examples
Metals	
$3d$ oxides	TiO, CrO_2
$4d$ oxides	MoO_2, RuO_2
$5d$ oxides	ReO_3, A_xWO_3
Semiconductors or insulators	
$3d$	MnO, CoO, NiO, Fe_2O_3, Cr_2O_3
$4d$ and $5d$	MoO_3, WO_3, Nb_2O_5, Rh_2O_3
$4f$	Pr_2O_3, Pr_6O_{11}, PrO_2
Nonmetal (semiconductor)-to-metal transition	
$3d$	V_2O_3, VO_2, V_3O_5, V_4O_7, V_6O_{13}, Ti_2O_3, Ti_3O_5
$4d$	NbO_2
$4f$	EuO
Superconducting transition	TiO, $BaPb_{0.75}Bi_{0.25}O_3$, $LiTi_2O_4$, $YBa_2Cu_3O_7$

diction; Ti_2O_3, however, shows metallic conductivity above ~400 K. The other oxides V_2O_3 ($3d^2$), Mn_2O_3 ($3d^4$), and Co_2O_3 ($3d^6$), if they existed in the corundum structure, most likely would be metallic. In reality, only V_2O_3 occurs in corundum structure above ~150 K, and this phase is metallic.

It would appear that the simple band picture is independent of the electron spin and employs the same wave function for both spin-up and spin-down states. This statement is however not correct. By invoking antiferromagnetic ordering, one can account for the insulating behavior of some oxides at $T = 0$ K; magnetic order leads to a doubling of the size of the primitive cell and consequently to an exchange splitting of all the bands. This may be valid for NiO and MnO. To explain the insulating behavior of CoO, however, one must assume that crystallographic distortion to a low symmetry (i.e., below the antiferromagnetic ordering temperature) introduces an energy gap. The concept of *spin density waves* (spiral spin structures) is employed to describe magnetically ordered insulators. The behavior of the monoxides at finite temperatures is not explained, since for these materials, metallic conductivity at temperatures higher than the antiferromagnetic ordering temperature is predicted (MnO, CoO, and NiO are insulating at all temperatures). Thus, elementary band theory fails to account even qualitatively for the properties of these oxides. The reason for this failure is the neglect of electron correlation (arising from large U and small W). Band theory modified by considering electron correlations would predict a low temperature insulating state of some of these oxides, as we will see later. It is to be noted that although the simple LCAO approximation provides a simple method for calculating the band structures of oxides and other solids, it is more convenient to employ other variants such as the augmented plane wave (APW) method, which properly approximates the three-dimensional potential experienced by the electrons.

Other than magnetic perturbations and electron–lattice interactions, certain situations present instabilities of other types, and these must be taken into account. One such situation is that related to low dimensionality. In fact, one-dimensional solids cannot be metallic because a periodic lattice distortion *(Peierls distortion)* destroys the Fermi surface in these systems. The perturbation of the electron states results in *charge-density waves* (CDW), involving a periodicity in electron density in phase with the lattice distortion. Blue molybdenum bronze, $K_{0.3}MoO_3$, shows such features. In two- or three-dimensional solids, however, one observes *Fermi surface nesting* due to the presence of parallel Fermi surface planes perturbed by periodic lattice distortions. Certain molybdenum bronzes seem to exhibit this behavior.

It was mentioned earlier that electron correlation plays an important role in determining the electronic structures of transition metal oxides. Hubbard treated the correlation problem in terms of the parameter U. Figure 2 shows how U varies with the bandwidth W, resulting in the overlap of the upper and lower Hubbard states (or in the disappearance of the band gap). In NiO, since $W < U$, there is a splitting between the upper and lower Hubbard bands. Thus the relative values of U and W determine the electronic structure of oxides. Unfortunately,

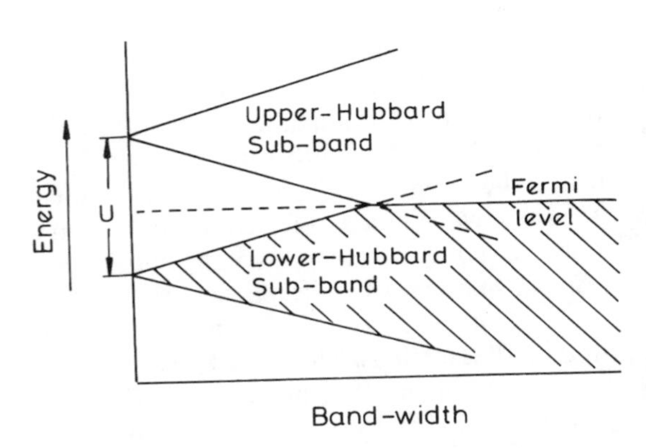

Figure 2. Hubbard model showing variation of U with bandwidth W.

it is difficult to obtain reliable values of U. The Hubbard model takes into account only the d orbitals of the transition metal (single band model). One has to include the mixing of the oxygen p and metal d orbitals in a more realistic treatment. Presence of mixed valence of the metal, as commonly encountered in superconducting cuprates (Cu^{2+}, Cu^{3+}), must also be taken into account.

Magnetic insulators have two important features: they are nonmetallic (i.e., there is a band gap), and they possess unpaired electrons. They also show crystal field transitions due to the presence of open-shell d^n ions. Many years ago, Mott proposed that electron repulsion can be responsible for the breakdown of the normal band properties of transition metal oxides and related materials. This model gives a simple interpretation of the properties and systematics found in magnetic insulators (e.g., 3d transition metal oxides with small bandwidths) and metallic systems (e.g., some of the sulfides). These ideas are in fact incorporated in the Hubbard model. The point to note is that in the so-called *Mott insulators,* the band gap involves the split Hubbard states. The realization that the top-filled band in many of the insulating transition metal oxides has considerably more oxygen p character than metal d character has given rise to the concept of *charge-transfer insulators.* Whether one should consider oxides like NiO Mott or charge-transfer insulators is a problem of interest, which we shall examine in Section 11 while dealing with the results from photoelectron spectra of oxides. It is interesting that in the superconducting cuprates, the oxygen–copper charge transfer energy is very small and charge transfer therefore plays an important role (see Section 8).

A transition between the magnetically insulating state to the metallic state (which occurs with increase in the W/U ratio) is called the *Mott transition.* V_2O_3 is supposed to undergo such a transition, but there are doubts. Many other transition metal oxides have been suspected of undergoing a Mott transition. The problem is that a real Mott transition should occur at a critical lattice constant when the Coulombic interaction between electrons and holes becomes very weak, giving rise to a large free carrier concentration. In principle, a Mott transition should occur upon application of high pressures, inasmuch as the lattice parameter of the oxides goes through a critical value. In real crystals, however, the lattice constant can be varied only through a limited range.

A concept that has become important is that of *Anderson localization* in disordered solids. The electronic structure of highly disordered or doped oxides (with high impurity concentration) cannot be described in terms of the normal band structure because disorder can localize electron states. The criterion for localization is that the energy difference between localized sites be greater than the bandwidth. Mott pointed out that the boundary between localized and extended states is sharp *(mobility edge)* and that electrons in localized states move by a phenomenon known as *variable-range hopping.* Several transition metal oxides exhibit Anderson localization and variable-range hopping.

2.2 Localized Electron Model

The localized electron model assumes that a crystal is composed of an assembly of independent ions fixed at their lattice sites and that the overlap of atomic orbitals is small. Crystal field theory or ligand field theory describes how the d^n electron configuration of transition metal ions is perturbed by the chemical environment. When interatomic interactions are weak, intra-atomic exchange *(Hund's rule splitting)* and electron–phonon interactions favor the localized behavior of electrons. This tendency increases the relaxation time of charge carriers from about 10^{-15} s in an ordinary metal to $\sim 10^{-12}$ s, which is the order of time required for a lattice vibration in a polar crystal.

Localized d or f electrons retain their one-atom manifolds except that states arising from different d^n or f^n are split by crystal field (CF) and spin-orbit coupling (SO). The magnitude of the CF splitting determines whether the transition metal ion occurs in the low spin or the high spin configuration. Multiplet splittings due to spin-orbit coupling are larger than crystal field splittings for $4f^n$ levels; the converse is the case for $3d^n$ levels. The difference in energy between $d^n(f^n)$ and $d^{n+1}(f^{n+1})$ manifolds corresponds to free atom U that is decreased as a result of interatomic interaction in the solid. In the crystal field approach, the Hamiltonian employed is $\mathcal{H} = \mathcal{H}_{CF} + \mathcal{H}_{SO} + \mathcal{H}_{es}$, where $\mathcal{H}_{es}$ is the electrostatic repulsion between the d electrons. In the localized model of solids, the band gap is interpreted in terms of the Madelung potential and the effects of polarization and orbital overlap.

The localized electron model readily predicts the insulating ground state of solids. At finite temperatures, electron–phonon and electron–electron interactions become important, especially when the bands are narrow, as in the case of d bands. Considerable work has been carried out on the effect of electron–phonon interactions on the transport properties of oxide materials. The strength of the *electron–phonon interaction* (nature of the *polaron*) is represented by Frohlich's coupling constant. If the interaction is sufficiently large, charge carriers move along with the associated polarization (small polaron); the mobility is small and is thermally activated. The problem of charge transport is treated in terms of the classical diffusion theory. In the presence of an electric field, preferential diffusion of electrons occurs through the crystal, giving rise to a net current. Such uncorrelated hopping of polarons is mostly encountered in compounds having the same cation in more than one valence state (e.g., Fe_3O_4, Pr_7O_{12}, $Na_xV_2O_5$). If the electron–lattice interaction is small, the electron movement is affected only slightly and we have the large polaron regime.

2.3 Cluster Model

Between the localized electron model and the band model, we have cluster models, which in their simplest form take into account the interaction of a metal atom with the surrounding ligand atoms (oxygens in the case of oxides). Basically, cluster calculations are molecular orbital calculations carried out on a cluster. Figure 3 shows the molecular orbital diagram of a transition metal ion

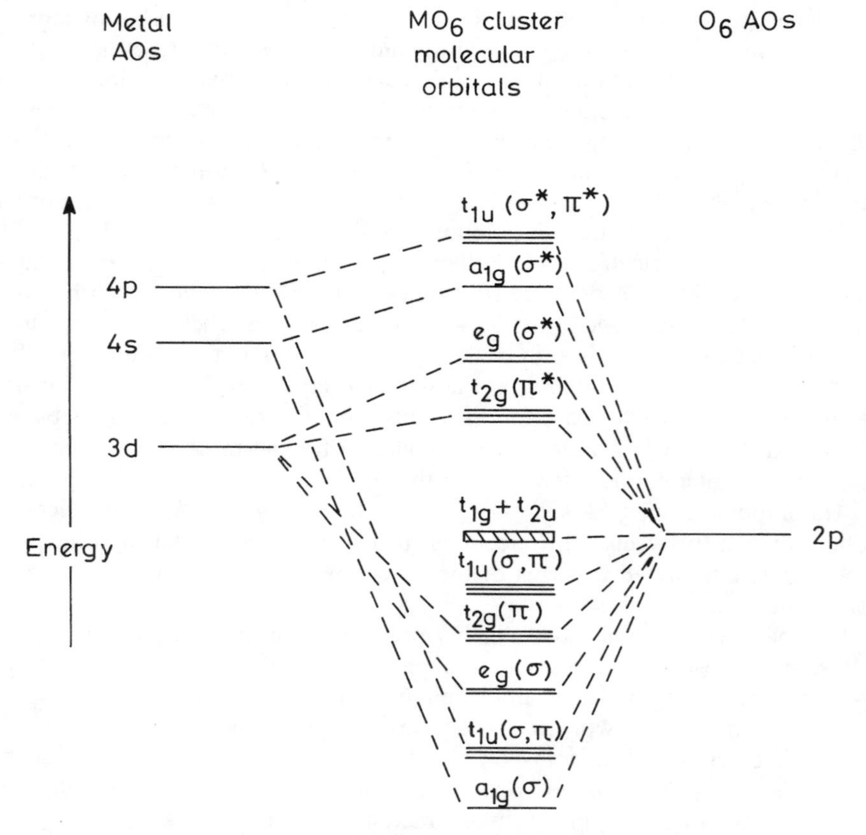

Figure 3. Molecular orbital diagram for an octahedral MO_6 unit.

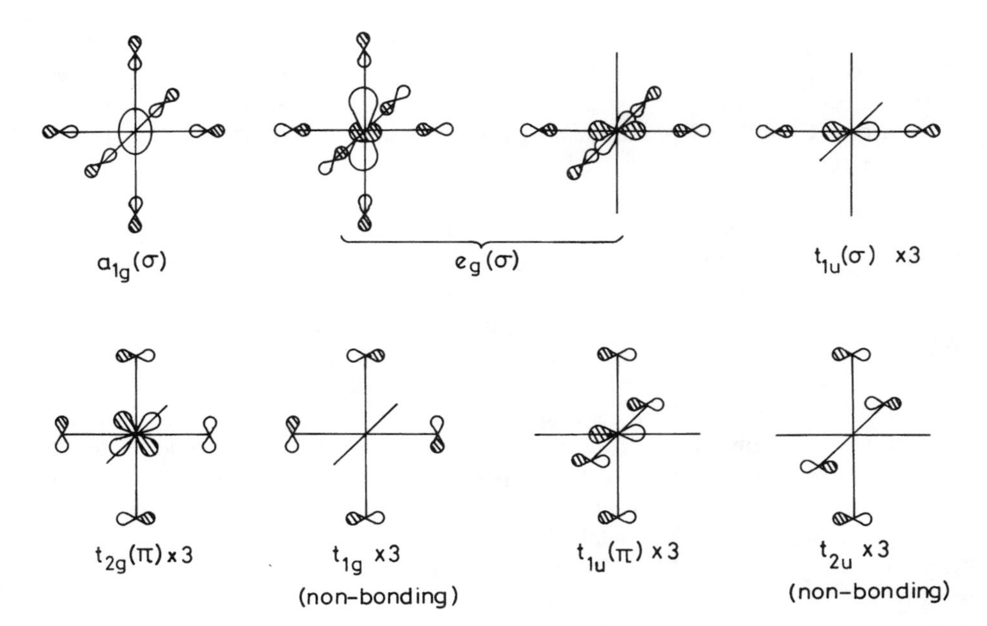

Figure 4. Molecular orbitals arising from combinations of metal *d* orbitals and oxygen *p* orbitals in octahedral MO_6.

octahedrally surrounded by six oxygens (O_h point group). The different orbitals arising from combinations of metal *d* orbitals and oxygen *p* orbitals are shown in Figure 4. In principle, one can deal with larger clusters (instead of the smallest cluster corresponding to the first coordination sphere in oxides) as well. Molecular orbital calculations of varying degrees of sophistication have been carried out on a variety of transition metal–oxygen clusters. Besides ab initio methods, approximate methods such as the Xα method method (which employs an approximation for the electron repulsion term) have been employed. Configuration interaction (CI), involving mixtures of different electron configurations of the cluster, is included in most recent calculations on clusters. Especially significant are the so-called charge-transfer impurity models (the impurity hole or electron is assumed to be localized on a specific ion), which take into account states corresponding to d^n, d^{n+1}L, d^{n+2}L^2, and so on, where L is the hole in the oxygen 2*p* band states. We shall examine results of such calculations on transition metal oxides in Section 11.

2.4 Chemical Bond Approach

Goodenough has employed a semiempirical approach based on crystal chemistry to explain the electronic properties of transition metal compounds.[1,3,4] With empirically derived criteria for the overlap of cation–cation and cation–anion–

cation orbitals, Goodenough rationalizes the nature of d electrons in transition metal compounds. This approach appeals to chemical intuition and has enabled an understanding of the conditions that give rise to localized and itinerant d electrons. Conceptual phase diagrams are constructed in terms of the *transfer energy* b_{ij}:

$$b_{ij} = (\psi_i H \psi_j) \simeq \epsilon_{ij}(\psi_i \psi_j)$$

where H is the interaction term and ϵ_{ij} is the one-electron energy, while b_{ij} measures the strength of interaction between the localized orbitals ψ_i and ψ_j on the neighboring like atoms i and j (b_{ij} is proportional to $\psi_i \psi_j$, the overlap integral). Although it is not possible to get good estimates of b_{ij}, one can predict its variation in a series of isostructural oxides, since the relative magnitudes of the overlap integral and orbital energies can be estimated.

In oxides exhibiting significant cation–cation interaction, b_{ij} can be related to R^{-1}, where R is the cation–cation separation. Where the cation–anion–cation interaction is important, b_{ij} is related to λ, the covalent mixing parameter of the cation–anion orbitals. For octahedrally coordinated cations as in rock salt and perovskite structures, the relevant mixing parameters are λ_σ and λ_π in the following "molecular" wave functions:

$$\psi_e = N_e(f_e - \lambda_\sigma \phi)$$

$$\psi_{t_2} = N_t(f_t - \lambda_\pi \phi_\pi)$$

where N_e and N_t are normalization constants, f_e and f_t are the cation d orbitals of e_g and t_{2g} symmetry, respectively, and ϕ and ϕ_π are the anion sp_σ and $p\pi$ orbitals of appropriate symmetry. In general, $\lambda_\sigma > \lambda_\pi$ and $b_{ij} \sim \lambda^2$. For small values of b, the outer d electrons are localized, and for large values they are itinerant. In a series of isostructural oxides, there is a critical value of transfer energy b_c, separating the itinerant electron regime from the localized electron regime. Since b is related to R and λ, expressions for the critical values R_c and λ_c are expressed in terms of the position of the metal in the periodic table, the principal quantum number of the d orbital, the oxidation state, and the total spin of the cation. Calculated values of R_c in oxides of rock salt, rutile, and corundum structures separate members with localized electron properties from those exhibiting itinerant electron properties. Critical transfer energies b_m, b_g, and b_{cs} relevant to the changes in various physical properties have also been defined. For $b > b_m \sim b_g$, spontaneous magnetism disappears and metallic conductivity sets in. Other important electronic properties that can be related to b are the antiferromagnetic ordering temperature, (Néel temperature) $T_N \sim b^2/U$, for $b < b_c$ and the d bandwidth, $W = 2zb$, where z is the number of nearest neighbors.

A representative phase diagram plotting transfer energy b versus temperature T for the case of a single d electron per interacting orbital ($n_l = 1$) is shown in Figure 5a. The physical significance of the phase diagram is as follows. The magnetic interaction in a system with $n_l = 1$ is antiferromagnetic. Since T_N is proportional to b^2, T_N increases up to the critical value, b_c. For $b < b_c$, the

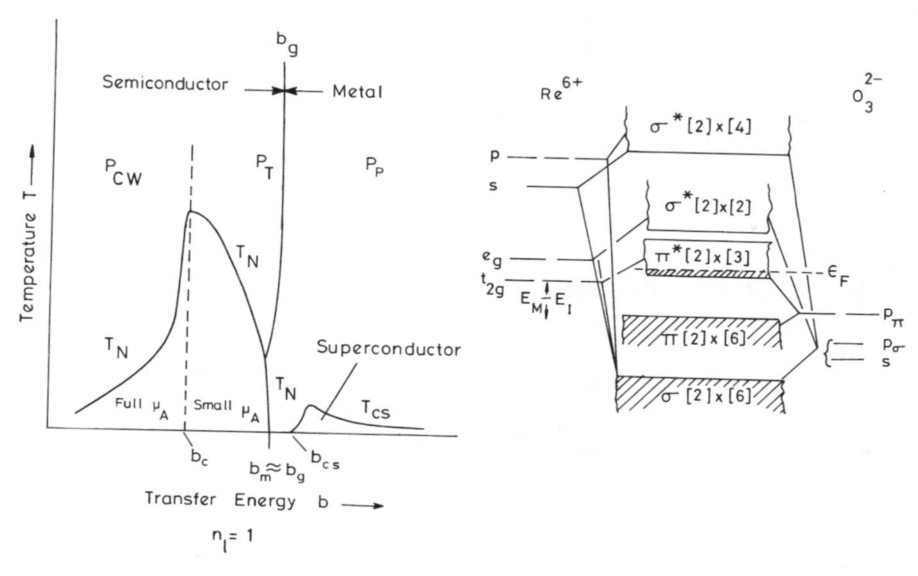

Figure 5. (a) Conceptual diagram relating b, the transfer energy, with temperature. (b) The schematic energy level diagram for ReO_3. (After Goodenough.[3,4])

localized electrons exhibit Curie–Weiss paramagnetism above T_N. For $b > b_c$, the states are bandlike and at b_m, spontaneous magnetism disappears ($T_N = 0$ at $b = b_m$) and a first-order semiconductor-to-metal transition is predicted. For $b > b_m$, the d electrons are truly collective, exhibiting Fermi surface-dependent electronic properties and Pauli paramagnetism. A superconducting state may be stabilized for $b > b_{cs}$ at low temperatures. The ratio W/U is somewhat equivalent to b_{ij}.

To apply conceptual phase diagrams to interpret electronic properties of isostructural series of transition metal oxides, Goodenough constructs one-electron energy level diagrams assuming the most probable hybridization of the cationic and anionic orbitals. These diagrams are similar to the one-electron molecular orbital energy level diagrams of isolated molecules except that discrete energy levels are allowed to become band states where necessary, depending on the cation–cation and the cation–anion–cation interactions in the oxides. In Figure 5b, a typical one-electron energy level diagram applicable to ReO_3, we see close similarity to the diagram of a discrete octahedral complex ion (see Fig. 3) such as $Ti(H_2O)_6^{3+}$. In the diagram for ReO_3, the filled bonding states (which are primarily anionic $2s$, $2p$) and the empty antibonding states (which are mainly cationic $6s$, $6p$) are split by a large gap (~5 eV) due to the difference in electronegativity between the constituent atoms. The $5d$ states of the cation, which are antibonding with respect to the anionic $2s$ and $2p$ states, lie in the gap. The electronic properties of the oxide are determined by the nature of the d states, (i.e., localized or itinerant). In ReO_3, both t_{2g} and e_g orbitals are itinerant because of large λ_π and λ_σ; the t_{2g} band is partially occupied, making the material metallic. A similar diagram is applicable to ABO_3 perovskites. Whether the t_{2g} and e_g states are localized or itinerant in a perovskite depends on the magnitude of covalent mixing, λ_π and λ_σ, of the B and O orbitals. One-electron energy level diagrams can be conveniently constructed for different oxides in rock salt, rutile, and other structures.

3 Properties of Oxide Materials

Sections 3.1–3.4 discuss briefly the more important properties of oxides that are of interest, including magnetic, electrical, dielectric and optical properties[8,9,11,12]. Table 2, which presents the electrical and magnetic properties of some binary transition metal oxides, indicates the variety that can be observed.

3.1 Magnetic Properties

Oxides with positive magnetic susceptibility are called paramagnetic, while those with negative magnetic susceptibility are called diamagnetic. Paramagnetic oxides generally follow the *Curie law,*

$$\chi_M = \frac{M}{H} = \frac{C}{T}$$

where M is the magnetization, H the magnetic field, χ_M the molar susceptibility, and C the Curie constant. Because of the interaction between atomic moments,

Table 2 Electrical and Magnetic Properties of Binary Transition Metal Oxides

Metal Oxides	Properties
d^0	
TiO_2, V_2O_5, ZrO_2, Nb_2O_5, HfO_2, Ta_2O_5, WO_3, MoO_3	Diamagnetic semiconductors or insulators when pure, but exhibit n-type extrinsic conduction when doped or slightly reduced
d^n	
TiO, NbO CrO_2, RuO_2, OsO_2, MoO_2, RhO_2, WO_2, IrO_2, ReO_3	Metallic and Pauli paramagnetic (CrO_2 is ferromagnetic)
Ti_2O_3, Ti_3O_5, Ti_4O_7, V_2O_3, V_3O_5, VO_2, NbO_2	Exhibit temperature-induced metal-to-nonmetal transition
MnO, FeO, CoO, NiO, Cr_2O_3, Fe_2O_3, Mn_3O_4	Insulators
f^n	
PrO_2, Ln_2O_3 (Ln = rare earth), Pr_nO_{2n-2}, Tb_nO_{2n-2}, EuO	Insulators or hopping semiconductors; paramagnetism characteristic of f^n configuration; EuO shows a metal-to-nonmetal transition under pressure

giving rise to an internal field in the oxide, the susceptibility is generally described by the *Curie–Weiss law,*

$$\chi_M = \frac{C}{T - \Theta}$$

The sign of Θ depends on the nature of magnetic interaction (ordering).

Following the Curie–Weiss law, rare earth compounds are typical paramagnets at ordinary temperatures. Experimental values of the effective magnetic moment generally agree well with those calculated from the ground state J values as expected of the localized $4f$ electrons. Samarium and europium compounds show discrepancies with theory that arise from the presence of higher multiplets within energies comaprable to kT. Eu^{3+} deserves special mention; the $J = 0$ ground state ($^7F_0$) of this ion would predict zero effective magnetic moment, and the nonzero moment found experimentally is therefore due to the population of higher multiplets.

The paramagnetism of transition metal compounds is different from that of rare earth compounds. This is because of the involvement of d electrons to varying degrees in crystal binding. For compounds of first-row transition metals, where definite electronic configurations can be assigned to the cations, agreement between the calculated and experimental moments is poor. However, if one assumes that the orbital moment plays no part in magnetism and that the moment is entirely due to spin, one obtains good agreement between experiment and theory in certain cases. The reason for the quenching of the orbital moment is the inhomogeneous crystal field, which strongly interacts with the cation d orbitals. Itinerant electrons of simple metals show a weak temperature-independent (Pauli) paramagnetism.

Magnetism in transition metal oxides is considerably more complex than that of isolated atoms because of the interaction (coupling) between atomic moments. Coupling between moments is responsible for cooperative magnetism. A pair of electrons of like spin, localized on an atom, is lower in energy than a pair with opposite spin by an amount called the intra-atomic exchange energy. Consequently, there is a statistical correlation for electrons of like spin, with each surrounded by a void due to the local depletion of parallel spin electrons. This phenomenon is called exchange. One can distinguish two classes of exchange. *Direct exchange* occurs between moments on atoms that are close enough to have significant overlap of their wave functions; exchange coupling is strong but decreases rapidly with increasing interatomic distance. *Indirect exchange,* on the other hand, couples moments over relatively large distances. It can act through an intermediary nonmagnetic ion (superexchange) or through itinerant electrons (RKKY). Superexchange generally occurs in insulators while RKKY (Ruderman, Kittel, Kasuya, and Yoshida) coupling is important in metals.

Superexchange, or interaction between localized moments of ions in insulators that are too far apart to interact by direct exchange, operates through the intermediacy of a nonmagnetic ion (e.g. metal–oxygen–metal). Superexchange is able to occur when localized electron states as described by the formal valences are stabilized by an admixture of excited states involving electron transfer between the cation and the anion. A typical example is the 180° cation–anion–cation interaction in oxides of rock salt structure, where the antiparallel orientation of spins on the neighboring cations is favored by covalent mixing of the anion *p* orbital with the cation *d* orbitals on each side. Different types of superexchange interaction are possible, depending on the structure of the oxide and the electronic configuration of the cations.

Two important superexchanges are correlation superexchange and delocalization superexchange. Delocalization superexchange involves transfer from one cation to another, as a result of cation–cation or cation–anion–cation interaction. Correlation superexchange is restricted to cation–anion–cation interaction. Anderson, Goodenough, and Kanamori have described rules governing the sign and magnitude of superexchange. For example, 180° cation–anion–cation interaction between half-filled orbitals is antiferromagnetic. A 90° cation–anion–cation interaction between half-filled orbitals is ferromagnetic, provided the orbitals are bonded to orthogonal anion orbitals. Superexchange involving σ bonds is stronger than that involving π bonds. Thus, the ordering temperature T_N of $LaCrO_3$ is lower than that of $LaFeO_3$. In the 3*d* transition metal monoxides, T_N increases in the order MnO < FeO < CoO < NiO because the σ interaction increases in that order. For cations of the same electronic configuration, superexchange is stronger for the higher valency cation (e.g., $Fe^{3+} > Mn^{2+}$).

Diamagnetism and paramagnetism are not cooperative phenomena. *Ferromagnetism,* on the other hand, is a cooperative phenomenon: there is a long-range collinear order of the moments in the solid and the *exchange parameter* $J > 0$. (The exchange parameter J is defined by the equation, $H_{ex} = \Sigma_{ij} J r_{ij} S_i S_j$, where H_{ex} is the exchange energy of atoms *i* and *j* separated by a distance r_{ij} and having spins S_i and S_j.) A ferromagnetic solid is spontaneously magnetized even in the absence of the field. To maximize its magnetostatic energy, a crystalline ferromagnet divides into domains that are spontaneously magnetized nearly to saturation; the moment of each domain, however, is oriented so as to produce a zero net moment. An external field changes the size of the domains, enlarging those of favorable orientation at the expense of others. Thus, in ferromagnetism, the external field acts as an agent to show the ordering that exists microscopically. Typically, magnetization rises sharply at lower fields, as the domains with more favorable alignments expand at the expense of others, and saturates when the maximum domain alignment is reached. The important characteristic of ferromagnets is the hysteresis in magnetization (with respect to the magnetic field). With increasing temperature, the thermal energy increases, becoming comparable to and eventually exceeding the exchange energy. Spontaneous magnetization thus decreases with temperature and indeed disappears at the Curie temperature T_c. Above T_c, an ideal ferromagnet becomes a paramagnet obeying the Curie–Weiss law. Examples of ferromagnetic oxides are CrO_2, $SrRuO_3$, $La_{0.5}Sr_{0.5}CoO_3$, $La_{0.5}Sr_{0.5}MnO_3$, and $LaMn^{3+}_{0.6}Mn^{4+}_{0.4}O_{3+\delta}$.

Antiferromagnetism, like ferromagnetism, is characterized by long-range ordering of identical spontaneous moments. But since the exchange parameter J is negative, the moments of neighboring atoms are exactly opposed; there is no overall spontaneous magnetization. Below T_N, an antiferromagnet can be regarded as consisting of two identical interpenetrating sublattices, with the result that the spins of one sublattice are opposed to those of the other. Most antiferromagnets are insulating solids (e.g., MnO, NiO, $LaCrO_3$, $CaMnO_3$), but antiferromagnetism is known among metals and alloys as well. Above T_N, antiferromagnetic materials become paramagnetic.

Ferrimagnetism requires two or more chemically different magnetic species, which occupy two kinds of lattice site, producing two sublattices, A and B. The moments of ions in each sublattice are ferromagnetically coupled, but the coupling between the moments of A and B is antiferromagnetic. Since the net moments of A and B are different, there is a resultant spontaneous magnetization. The temperature dependence of ferrimagnetism is similar to that of ferromagnetism, except that the spontaneous magnetization decreases more rapidly with increasing temperature. In the paramagnetic state, there is deviation from the Curie–Weiss law, particularly close to T_c. Ferrimagnetism can be observed in Fe_3O_4 and spinel ferrites.

Certain oxides comprising two (or more) antiferromagnetic lattices that are canted at an angle leave a net magnetization. They are called *weak ferromagnets* or *canted antiferromagnets.* Weak ferromagnetism can arise from differences in single ion anisotropy, from Dzyaloshinsky–Moriya interaction, or from both. Dzyaloshinsky–Moriya interaction, which results in an asymmetric arrangement of anion moments, arises from a strong spin-orbit coupling of the anion, which upsets the superexchange interaction. A typical example of a weak ferromagnet arising from this interaction is α-Fe_2O_3. When the particle size is very small, ferromagnetic and ferrimagnetic materials become *superparamagnetic.* Superparamagnetic materials do not show hysteresis in magnetization.

Another important magnetic behavior found in some of the spinel oxides (e.g., $Ga_{0.8}Fe_{0.2}NiCrO_4$) and other systems is that of the *spin glass,* arising from the freezing of randomly oriented moments. These materials show a cusp in the ac susceptibility. A spin-glass state exists in crystalline materials within a limited range of concentrations of the magnetic solute. The concentration must be high enough to give mutual interactions but low enough to avoid cluster formation. For concentrations below the dilute limit, the *Kondo regime* obtains, and magnetic atoms are screened from interaction with others. When the concentration of magnetic atoms is large enough, clusters are formed. A *mictomagnet* is similar to spin glass, except that local correlations of magnetic atoms are dominant because of clustering.

Besides the types of ordering of moments discussed hitherto, various *magnetic excitations* can occur in the different magnetic states. One of the elementary excitations is a quantized spin wave in a classical ferromagnet. In the ground state of a ferromagnet, all the spins are oriented parallel. We can form an excited state by reversing one of the spins. Alternatively, we can let all the spins share the reversal, in which case the spins take on a wavelike arrangement. The elementary excitations of a spin system having a wavelike form are called *magnons.* Magnons are analogous to phonons and can be similarly treated to derive magnon dispersion relations. The magnon spectrum of a solid can be determined by inelastic neutron scattering.

It is instructive to recount some of the applications of magnetic oxide materials. Spinel ferrites are used as thermistor materials. Thus, $NiFe_2O_4$ and $MgFe_2O_4$ have negative temperature coefficients of resistance. In contrast, $MgAl_2O_4$ and Zn_2TiO_4 show a positive temperature coefficient of resistance. Hard ferrites such as $BaFe_{12}O_{19}$ and soft ferrites such as $MnFe_2O_4$ find applications in memory devices, recording tapes, and transformer cores. Orthoferrites and hexaferrites, along with materials such as $GdCo_5$, can support *magnetic bubbles* for memory applications. Rare earth ion garnets $Ln_3Fe_5O_{12}$ (Ln = rare earth) provide the flexibility required in such devices. In these cubic materials, the Ln ion is in the dodecahedral site while the Fe^{3+} ions, distributed between octahedral (2) and tetrahedral (3) sites, are coupled in antiparallel fashion. In yttrium iron garnet, $Y_3Fe_5O_{12}$ (YIG), the magnetization is much too large (to sustain bubbles of 1–8 μm diameter) and chemical manipulation becomes necessary to control the bubble diameter. Typical examples are $Sm_{0.4}Y_{2.6}Ga_{1.2}Fe_{3.8}O_{12}$ (8 μm), $Sm_{0.18}Lu_{0.19}Y_{1.73}Ca_{0.9}Ge_{0.9}Fe_{4.1}O_{12}$ (6 μm), and $Sm_{0.51}Ln_{0.42}Y_{1.21}Ca_{0.86}Ge_{0.70}Si_{0.16}Fe_{4.14}O_{12}$ (2–3 μm). Garnet films are grown on solid substrates by liquid phase epitaxy; Fe_3O_4 dispersed in oil (with some surfactant) acts as a ferrofluid.

3.2 Electrical Properties

Electrical conductivity, thermal conductivity, the Seebeck effect, and the Hall effect are some of the common electron transport properties of solids that characterize the nature of charge carriers. On the basis of electrical properties, solid materials may be classified into metals, semiconductors, and insulators, wherein the charge carriers move in band states. In certain semiconductors and insulators, charge carriers are localized; their motion involves a diffusive process.

The expression for the electrical conductivity of a metal is given by

$$\sigma = \left(\frac{8\pi}{3n}\right)^{1/3} ne^2 \frac{l}{m}$$

where l is the mean free path and m the effective mass of the electrons. Above the Debye temperature, the density of phonons varies as T and $\sigma \propto T^{-1}$; below the Debye temperature, the density of phonons goes as T^3 and $\sigma \propto T^5$ at low temperatures. In *intrinsic semiconductors* with both electrons and holes as charge carriers,

$$\sigma = n_i e(u_n + u_p)$$

$$n_i = (N_c N_v)^{1/2} \exp\left(\frac{-E_g}{2kT}\right)$$

where N_c and N_v are the effective density of states in the conduction and valence bands, respectively, and u_n and u_p are electron and hole mobilities. In the presence of donors and acceptors, the conduction is extrinsic at low temperatures ($kT \ll E_g$) and $\sigma \propto \exp(-E_d/2kT)$ if the donor concentration is higher than that of the acceptor; at high temperatures, conduction becomes intrinsic. In the case of localized charge carriers (as in *hopping semiconductors*), $\sigma \propto \exp-(E_t + E_u)/kT$, where E_t is the energy required to ionize charge carriers and E_u is the activation energy associated with mobility.

A gradient in the electrochemical potential caused by a temperature gradient in a conducting material causes *thermoelectric effect.* The *Seebeck coefficient,* α, is the constant of proportionality between the voltage and the temperature gradient that causes it when there is no current flow. It is defined as $\Delta V/\Delta T$ as $\Delta T \rightarrow 0$, where ΔV is the thermo-emf caused by the temperature gradient ΔT; it is related to the entropy transported per charge carrier ($\alpha = -S^*/e$). The expression for Seebeck coefficient of metals is,

$$\alpha = \frac{k}{e}\left[\frac{\pi^2}{3}\right]\left[\frac{kT}{E_F}\right]$$

Metals are characterized by a small $|\alpha|$ with temperature. In intrinsic semiconductors, α is given by

$$\alpha = \pm\frac{k}{e}\left[\frac{S^*}{k} - \frac{E_F}{kT}\right]$$

where S^* is the entropy transported by charge carriers ($S^*/k < 10\ \mu V/K$ in metal oxides) and E_F is the Fermi energy relative to the appropriate band edge. The sign of α is generally (but not always) positive for hole conduction and negative for electron conduction. We can also write

$$\frac{E_F}{kT} = \ln\left[\frac{c}{1-c}\right] = \ln\left[\frac{n}{N}\right]$$

where $1 - c$ or N represents the density of unoccupied states in the relevant band and c or n, the density of charge carriers in the band. In extrinsic semiconductors, α is generally large (0.1–1 mV/K) and decreases with increasing temperature. The maximum value of α occurs for small c, and there is a change of sign for α as c varies from 0 to 1. In localized extrinsic semiconductors (small polaron hopping), c is generally constant and α is nearly temperature independent.

The *Hall coefficient, R*, of a solid with a single type of charge carrier is given by

$$R = \pm\frac{C_r}{ne}$$

where $C_r \sim 1$. If R and σ are known, then the product C_rU_d is a measure of the *drift mobility.*

Electrical transport properties of oxides provide useful criteria for distinguishing localized and itinerant electrons in solids. Thus, the temperature dependence of drift mobility u for collective electrons ($b > b_c$) is different from the behavior for small polarons ($b < b_c$). For collective electrons, u goes as $T_{-3/2}$, and when the bands are narrow, mobility becomes thermally activated $u \sim e^{-E_a/kT}$ where E_a is the activation energy for hopping. Mobility is small (<0.1 $cm^2/V{\cdot}s$) for localized semiconductors exhibiting hopping conduction and large (>1 $cm^2/V{\cdot}s$) in the band limit. Experimentally, the product of the Hall coefficient and the electrical conductivity gives a measure of the drift mobility. The Seebeck coefficient also provides a convenient means of characterizing charge carriers.

Motion of ions through solids results in charge as well as mass transport. Whereas charge transport manifests as ionic conductivity in the presence of an applied electric field, macroscopic mass transport (diffusion) occurs in a concentration gradient. Ionic conductivity as well as diffusion arise from the presence of point defects in solids. For a solid showing exclusive *ionic conduction,* conductivity is written as follows:

$$\sigma = \sum_i n_i q_i e u_i$$

where the summation is taken over all the charged species i, n_i denotes the concentration of the i th type ion bearing a net charge $q_i e$ and possessing mobility u_i. If Schottky defects give rise to electrical conductivity, n refers to the concentration of vacancies in the cation and anion sublattices, n_c and n_a. The expression for the electrical conductivity of an oxide where the conductivity is dominated by the motion of cations is given by

$$\sigma = \left(\frac{\sigma_0}{T}\right) \exp\left[\frac{-E_A}{kT}\right]$$

where σ_0 is a constant for a given crystal, $E_a = \frac{1}{2}E_S + E_m$. ($E_S$ is the *Schottky defect* formation energy and E_m is the activation required for ion migration). Comparing the expressions for the ionic conductivity and the diffusion coefficient, D, we obtain the *Nernst–Einstein relation,*

$$\frac{\sigma}{D} = \frac{Ne^2}{kT} \quad \text{or} \quad \frac{u}{D} = \frac{e}{KT}$$

In the early work on ionic conductivity, the data were presented as log σ versus T^{-1} plots, which showed two linear regions joined by a "knee," the position of the "knee" being dependent on the purity of the sample. Since the mobility of ions due to point defects depends not only exponentially on T^{-1} but also inversely on T, plotting $\log(\sigma T)$ versus T^{-1} is more appropriate. Diffusion coefficients determined by the tracer technique D_T and those obtained from electrical conductivity using the Nernst–Einstein relation D_σ show subtle differences because ionic motion in tracer experiments is correlated. The *Haven ratio,* D_T/D_σ, is proportional to the correlation factor. Electrically neutral defects contribute to D_T, but not to D_σ. The presence of electronic conduction in an oxide results in an abnormally small Haven ratio. Mechanisms of diffusion and ionic conductivity in terms of point defects in ionic solids have been discussed extensively in the literature. In certain oxides, conductivity of one of the ions is unusually large, being comparable to that of aqueous electrolytes (e.g., ZrO_2 doped with Ca^{2+} or Y^{3+}). Such superionic conductors have many applications.

Several oxide systems exhibit *fast ion conduction* (superionic conduction). In some cases, the ionic conductivity is accompanied by electron conduction. Typical superionic oxides are calcia-stabilized zirconia, $Zr_{1-x}Ca_xO_{2-x}$, β-alumina, $Na_{1+x}Al_{11}O_{17+x/2}$ and nasicon $Na_3Zr_2PSi_2O_{12}$. In $LiAlSiO_4$ (β-encryptite), ionic conduction is one-dimensional, while in nasicon, it is three-dimensional. Lisicon is formed by solid solutions of Li_2ZnGeO_4 and Li_4GeO_4. Goodenough has discussed framework (skeleton) structures of oxides for fast ion conduction of Na^+ and other ions in one-, two-, or three-dimensions. $NaZr_2(PO_4)_3$ has a three-dimensional interstitial space (Fig. 6), where along the *c*-axis, the vacant trigonal-prismatic sites, *p*, the octahedral Zr^{4+} sites, *Z*, and the octahedral sites available to the Na^+ ions, *M*, are ordered as *-Z-p-Z-M-Z-*. This phosphate is not an Na^+ conductor, and neither is $Na_4Zr_2(SiO_4)_3$ with the same framework, but their solid solutions (nasicon) are good Na^+ conductors. β-Alumina has received much attention because of possible use in the sodium–sulfur battery; partial substitution of Al by Mg in β-alumina makes it a still better ionic conductor. $Zr_{1-x}Ca_xO_{2-x}$ has been employed as an oxygen sensor in vehicle exhausts, furnaces, and similar applications. Some of the oxide materials also act as proton conductors.

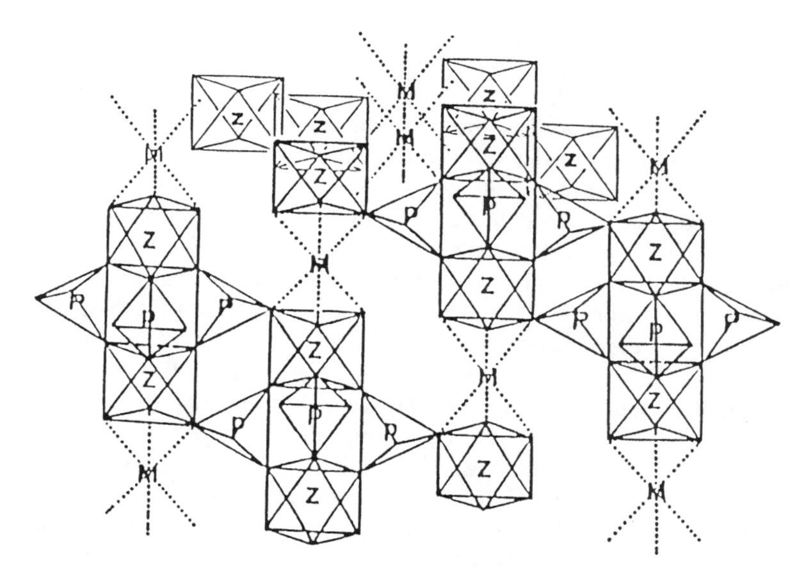

Figure 6. Structure of $NaZr_2(PO_4)_3$.

Incorporation of a small quantity of Na in WO_3 to give the bronze, Na_xWO_3, is accompanied by a change in color from pale yellow to deep blue. Since this process (intercalation) is reversible and fast, the material can be used in electrochromic displays. H_xWO_3 can similarly be used as an electrochromic material. Note that the bronze will have both electronic and ionic (electrolytic) conduction.

Photoelectrolysis can be carried out by the use of a semiconductor electrode and a metal counter electrode. Thus, photoelectrolysis of water has been carried out by using n-type $SrTiO_3$ as the anode, to produce H_2 and O_2. Unfortunately, the band gap of $SrTiO_3$ is too large (3.2 eV). Efficient solar energy conversion requires band gaps of the order of 1.3 eV. Photoelectrolysis cells making use of Mg^{2+}- and Si^{4+}-doped Fe_2O_3 have been examined. Photoelectrolysis of water has been achieved in n-TiO_2/p-GaP cells as well. The efficiency of n-TiO_2 is increased by making the compound oxygen-deficient or by alloying with VO_2. Visible light-induced decomposition of water has been possible by using RuO_2 and Pt-loaded TiO_2 as catalysts.

3.3 Superconductivity

When the electrical resistance of a material vanishes below a critical temperature $T_c > 0$ K, we refer to it as a superconducting transition. The transition was first observed in mercury by Kamerlingh Onnes in 1911. A sharp discontinuity in resistivity is found in pure, homogeneous materials, and a broad transition is seen in inhomogeneous materials. Typical resistivities of materials in the superconducting state are of the order of 10^{-23} Ω·cm compared to the lowest resistivity of $\sim 10^{-13}$ Ω·cm in metals. The temperature interval ΔT_c over which the transition between the normal and the superconducting state occurs, may be as little as 10^{-4} K or several degrees in width, depending on the material. In the superconducting state, materials exhibit perfect diamagnetism, excluding magnetic field below a critical field H_c *(Meissner effect)*. When $H > H_c$, superconductivity is destroyed and the material reverts to the normal state. The Meissner effect is a fundamental property of the superconducting state and is used as a means of detecting superconducting transitions.

Two kinds of superconductor are distinguished depending on their magnetic behavior. Materials in which superconductivity is destroyed abruptly at a critical field H_c are called type I superconductors (H_c is a function of temperature). In type II superconductors, the magnetic flux starts to penetrate the material at a field of H_{c1} lower than the thermodynamic critical field H_c. The magnetization then gradually decreases with increasing field strength until at $H = H_{c2}$ the superconducting state is completely destroyed. Between H_{c1} and H_{c2}, the material is said to be in a *vortex state* consisting of rod-shaped regions of normal conductivity within the superconducting bulk substance. Because of the possibility of many technical applications (magnets, power generators, etc.), superconductivity has been widely investigated and the number of materials showing superconductivity is continually growing. Oxides are some of the most important superconducting materials, showing the highest T_c values known to date. We shall discuss these materials later in Section 8. Some of the important properties of superconductors are the H_c values, the energy gap Δ and the penetration depth λ.

Bardeen, Cooper, and Schrieffer (BCS) explained the nature of the superconducting state as resulting from the formation of bound electron pairs *(Cooper pairs)* whose energy is lower than the free electron energy in a normal metal. Pairing occurs through electron–lattice interaction in such a way that a spin-up electron of wave vector $+\mathbf{k}$ is paired with a spin-down electron of wave vector of $-\mathbf{k}$. BCS theory does not tell us which substances can become superconducting and which cannot. Matthias suggested that the occurrence of superconductivity strongly depends on the crystal structure and on the average number of valence electrons per atom. For instance, in alloys of Nb_3Ge and BCC structures, T_c is maximum when the electron concentration is between 4 and 5 or 6 and 7. The observation of high temperature superconductivity (T_c 30–155 K) in cuprates has changed the entire picture, as we shall see later in Section 8.

3.4 Dielectric and Optical Properties

The molar polarization of a solid is given by the Clausius–Mossotti equation,

$$\frac{\epsilon_r - 1}{\epsilon_r + 2}\frac{M}{\rho} = \frac{N_0\alpha}{3\epsilon_v}$$

where ϵ_r and ϵ_v are the dielectric constants of the material (relative to vacuum) and of vacuum respectively, M is the molecular weight, N_0 the Avogadro number, and α the polarizability. In dielectric materials, the randomizing effect of temperature is balanced by the orienting effect of the internal field. The dielectric constant is then given by

$$\epsilon_r = \epsilon_\infty + \frac{C}{T - T_C}$$

where T_C is the Curie temperature, ϵ_∞ the optical (high frequency) dielectric constant, and C the Curie constant. In ferroelectric materials ϵ goes to infinity when $T = T_C$.

The dielectric constant is separated into real and imaginary parts, as given by $\epsilon_r(\omega) = \epsilon_{r'} + i\epsilon_{r''}$ where $\epsilon_{r'} = n^2 - k^2$ and $\epsilon_{r''} = 2nk$, where n and k are the real and imaginary parts of the refractive index. The reflection coefficient of a solid is determined by the values of n and k. The real and imaginary parts of ϵ become equal at the *plasma frequency,* ω_p which is equal to $(ne^2/\epsilon_0 m)^{1/2}$.

The dielectric constant of a transparent oxide varies as a function of the frequency of the oscillating electric field in the nonabsorbing region of the electromagnetic spectrum. The refractive index similarly varies with the frequency, generally increasing with ω. Such dispersion behavior of materials is of importance in the choice of materials for prisms and other purposes. Oxides absorb electromagnetic radiation by different modes depending on the nature of bonding.

Tightly bound electrons and ions in oxides give rise to narrow resonance absorption, electrons in the ultraviolet region, and ions in the infrared region *(restrahlen absorption).* Absorption of radiation due to loosely bound electrons is generally attributable to *interband transitions.* Such absorption bands in semiconductors are broad and featureless, but with a sharp absorption edge. Transition metal ions give absorption bands because of the presence of *d–d* transitions which are Laporte-forbidden. Both donor and acceptor impurities in semiconducting materials give absorption bands owing to photoionization at energies lower than the gap energies. Excitons (electron–hole pairs) in insulators give characteristic bands with the energy depending on how tightly bound the electron–hole pair is.

Electrons in metallic and semiconducting materials give rise to free carrier absorption, the absorption coefficient being proportional to square of the incident wavelength (hence high in the infrared region for most metals). The reflectivity of metals (e.g., ReO_3, TiO) which is related to the plasma frequency is almost total up to high frequencies.

Ineleastic scattering in solids is typified by the *Raman effect.* If a single phonon is involved, the scattering is referred to as first-order Raman effect. In second-order Raman effect, two phonons are involved. In *Brillouin scattering,* a special case of Raman scattering, the phonons involved are from the acoustic branch.

The electron–hole pair produced in photoexcitation processes reverts to the original state by releasing energy to the lattice through creation of phonons or radiative recombination, giving rise to *luminescence.* Such a de-excitation can be induced at impurity sites called activators. Absorption and emission processes in *photoluminescence* need not occur at the same impurity sites. For example, in $Ca_3(PO_4)_2$ doped with Ce and Mn, absorption occurs at Ce sites (sensitizer) and recombination at Mn sites (activator). This process involves energy transfer between two centers, and the luminescence intensity therefore depends on the average distance between impurity centers. The cation Eu^{2+} in barium magnesium aluminate gives blue emission, while Eu^{3+} in Y_2O_3 gives red emission. $Na_3Ce_{0.65}Tb_{0.35}(PO_4)_2$ is an efficient phosphor, since the emission of Ce^{3+} overlaps with the excitation of Tb^{3+}.

Laser (light amplification through stimulated emission of radiation) action is seen in materials such as ruby (0.01 atom % Cr^{3+} in Al_2O_3). This occurs because of population inversion, achieved by optical pumping to higher energy levels ($^4F_1$ and $^4F_2$) from which de-excitation occurs through a nonradiative process to an intermediate level (^2E), from which lasing occurs to the ground state ($^4A_2$), giving 6943 Å radiation. The stimulated laser light has the special property of coherence; it has the same phase and direction of propagation as the incident photon. When Nd^{3+} is doped in $CaWO_4$, there is laser emission at 1.06 μm ($^4F_{3/2}$ to $^4I_{11/2}$). The commonly employed Nd-YAG laser is Nd^{3+} in $Y_3Al_5O_{12}$ (yttrium aluminum garnet).

The polarization produced in a solid is generally proportional to the electric field of the radiation. Electric fields produced can be very high in laser beams and nonlinear effects therefore become significant. Polarization would then contain multiples of frequencies. The number of photons with frequencies 2ω, 3ω . . . , will however be small. Generation of second and higher harmonics is accomplished by using noncentrosymmetric crystals such as KH_2PO_4, $LiNbO_3$, and $KTiOPO_4$.

4 Electronic and Magnetic Properties of Oxides in Relation to Structure

Electronic properties of oxides of nontransition metals (e.g., Na_2O, MgO, Al_2O_3, SiO_2) consist of a filled valence band (derived mainly from oxygen $2p$) and an empty conduction band (derived from the outer shells of the metal atoms), separated by a large gap. Such oxides are therefore diamagnetic insulators under ordinary conditions. Since the intrinsic activation energy for electronic conduction is higher than the energy required for the creation and migration of point defects, ionic conduction predominates over electronic conduction in many of these oxides at moderately high temperatures.

Two classes of transition metal oxides can be distinguished: those in which the metal ion has a d^0 electronic configuration and those in which the d shell is partially filled (Table 2). The former class of oxides has a filled oxygen $2p$ valence band and an empty metal d conduction band; the energy gap is around 3–5 eV. High purity oxides of this class exhibit intrinsic electronic conduction only at high temperatures. With the d^0 cations at octahedral sites, the oxides exhibit spontaneous ferroelectric distortions (e.g., WO_3). Many lose oxygen at high temperatures, becoming nonstoichiometric. Oxygen loss or insertion of electropositive metal atoms into these oxides promotes electrons to the conduction band. The nature of electronic conduction in these materials depends on the strength of electron–phonon coupling and the width of the conduction band derived from metal d states. When the coupling is large and the band is narrow, small polarons are formed and such materials (e.g., $Na_xV_2O_5$) exhibit hopping conduction. When the conduction band is broad, the material (e.g., Na_xWO_3) exhibits metallic properties.

Transition metal oxides with partially filled d bands can be metallic or insulating. Some of them exhibit temperature-induced nonmetal-to-metal transitions (Table 2). Magnetic properties vary anywhere from Curie–Weiss paramagnetism to Pauli paramagnetism (through spontaneous magnetism). Rare earth metal oxides containing localized $4f^n$ electrons are generally insulators or hopping semiconductors and exhibit paramagnetism.

Transition metal oxides with the d^n configuration exhibit metallic properties when the overlap between orbitals of the valence shells of constituent atoms is large. Two kinds of metallic behavior can be distinguished: one due to strong cation–cation interaction arising from a small cation–cation separation, and the other due to strong cation–anion–cation interaction arising from a large covalent mixing of oxygen $2p$ orbitals with cation d orbitals. Isostructural series of transition metal oxides, possessing rock salt, corundum, rutile, and perovskite structures, exhibit systematic changes in electronic properties, wherein at least one member of the series shows properties characteristic of itinerant electrons while others exhibit properties due to localized electrons. Electronic properties of transition metal oxides are conveniently discussed in terms of Goodenough's arguments [1–7]. Literature references to the properties of various transition metal oxides may be found in refs. 6, 7 and 8.

4.1 Monoxides

Monoxides of $3d$ transition metals, TiO to NiO, possess the rock salt structure and exhibit interesting properties (Table 3). While TiO and VO exhibit properties characteristic of itinerant d electrons, MnO, FeO, CoO, and NiO show localized electron properties. The properties can be understood in terms of cation–cation and cation–anion–cation interactions in the rock salt structure. Direct cation–cation interaction can occur through the overlap of cationic t_{2g} orbitals across the face diagonal of the cubic structure. When such an interaction is strong ($R < R_c$ and $b > b_c$), cationic t_{2g} orbitals are transformed into a cation sublattice t_{2g}^* band; if this band is partially occupied, the material becomes metallic (e.g., TiO). VO is less metallic, and its magnetic susceptibility becomes temperature dependent at low temperatures because the increased nuclear charge contracts the radial extension of the $3d$ orbitals of vanadium, thereby decreasing the overlap of the t_{2g} orbitals. The increasing nuclear charge across the TiO–NiO series has the effect of lowering the $3d$ manifold energy relative to the top of the anionic $2p$ states. The radial extension of the $3d$ wave functions therefore decreases, resulting in a decrease in the cation–cation transfer energy b_{cc} and rendering the t_{2g} electrons localized beyond VO. Furthermore, the cation–anion–cation transfer energy b_{ca}, involving the e_g orbitals of cations, increases across the series because λ, the covalent mixing parameter between the cation and anion orbitals, is inversely proportional to the energy difference between the cationic d states and the anionic p states. This energy difference decreases across the series,

Table 3 Transition Metal Monoxides with Rock Salt Structure

Oxide	R (Å)	R_c (Å)	Properties
TiO	2.94	3.02	Metallic, Pauli paramagnetic
VO	2.89	2.92	Semimetal, weak temperature dependence of susceptibility
MnO	3.14	2.66	Semiconductor, Curie–Weiss, T_N = 122 K
FeO	3.03	2.95	Semiconductor, Curie–Weiss, T_N = 198 K
CoO	3.01	2.87	Semiconductor, Curie–Weiss, T_N 293 K
NiO	2.95	2.77	Semiconductor, Curie–Weiss, T_N = 523 K

becoming a minimum at NiO. The rock salt structure permits 180° cation–anion–cation superexchange interaction through the metal e_g orbitals (Fig. 7a). This in turn is responsible for the antiferromagnetic behavior of MnO, FeO, CoO, and NiO. The increase in T_N in the series is explained in terms of the increase in the cation–anion transfer energy b_σ across the series. Figure 7b shows the energy band diagram for TiO.

Ternary oxides of the type $LiMO_2$ (M = V, Cr, Mn, Fe, Co, Ni) crystallize in ordered rock salt structures. When M is V, Cr, Co, and Ni, the M^{3+} and Li^+ ions are ordered in alternate (111) cation planes, introducing a unique [111] axis and rhombohedral symmetry. In $LiVO_2$, the V—V distance is 2.84 Å $< R_c$ (V^{3+}) = 2.95 Å. The susceptibility is Curie–Weiss type at high temperatures and shows a sharp drop below 460 K, indicating cation–cation bonding in the (111) planes. $NaVO_2$, which is isostructural with $LiVO_2$, does not show a similar magnetic transition because the V—V distance is larger than R_c (V^{3+}). $ACrO_2$ (A = Li, Na, K) compounds are isostructural with $LiVO_2$ and exhibit two-dimensional antiferromagnetism. In $LiCoO_2$ and $LiNiO_2$, the trivalent transition metal ion appears to be in the low spin state. Lithium can be reversibly removed from some of these oxides by chemical or electrochemical means, rendering the oxides useful as electrode materials in solid state batteries. Lithium can be replaced by Ag by an ion exchange reaction to yield a delafossite.

EuO is an interesting oxide in the rock salt family. It is a ferromagnetic (T_C = 69 K) insulator when pure and stoichiometric. Unlike the case of 3*d* oxides, superexchange interaction in EuO is ferromagnetic because the charge transfer is from the 4*f* orbital to an empty 5*d* orbital. Electronic structure of EuO can be described as consisting of localized Eu $4f^7$ states separated from empty conduction bands of Eu $5d(t_{2g})$ and Eu 6*s* by about 1.4 eV. Below the localized $4f^7$

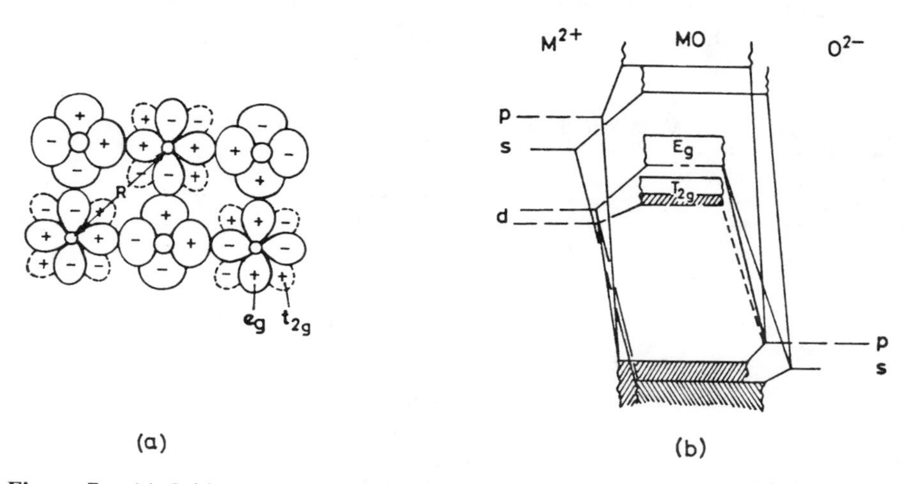

Figure 7. (a) Orbitals in the (100) plane of rock salt structure showing cation–cation and cation–anion–cation interactions. (b) Schematic energy band diagram of TiO. (Following Goodenough).

states is the filled band of oxygen $2p$ states. Stoichiometric EuO shows a insulator–metal transition at 300 kbar due to the promotion of a $4f^7$ electron into the $5d$ band. EuO shows nonstoichiometry on both sides of the stoichiometric composition ($Eu_{1-x}O$ and EuO_{1-x}). Transport properties differ for these two defect oxides. In $Eu_{1-x}O$, holes are introduced into the $4f^7$ level and conduction is due to small polaron hopping; there is no anomaly in the resistivity at T_c. In EuO_{1-x}, electrons are introduced into the $5d$ band. The samples show a sharp semiconductor–metal transition at 50 K $< T_C$ and a pronounced maximum in the resistivity above T_c. At this transition, the electrical resistivity changes by about 13 orders in some samples and the metallic state occurs on the low temperature side.

4.2 Dioxides

Dioxides of many of the transition metals crystallize in the rutile structure, and their electronic properties are shown in Table 4. The rutile structure provides the possibility of 135° cation–anion–cation interaction between corner-shared octahedra and 90° cation–anion–cation interaction between edge-shared octahedra. A direct cation–cation interaction is possible along the direction c. The octahedral crystal field of the cations and the tetragonal structure provide an axial field, which splits the triply degenerate metal t_{2g} orbitals into a nondegenerate $t_{\parallel}$ orbital and a doubly degenerate $t_{\perp}$ orbital. While $t_{\perp}$ can form π^* bands with anion p_{π} orbitals, $t_{\parallel}$ forms cation sublattice d bands if the cation–cation distance along the c-axis is less than R_c. The rutile structure therefore enables the oxide to become metallic through cation–cation interaction and/or through cation–anion–cation interaction.

Some of the transition metal dioxides exhibit a distorted (monoclinic) rutile structure in which the metal–metal distance changes alternately along the chain

Table 4 Transition Metal Dioxides Possessing Rutile-Type Structure

Oxide	R (Å)	R_c (Å)	Properties
TiO_2	2.96	3.0	Diamagnetic semiconductor, $E_g \sim 3$ eV
VO_2	2.88 (2.65, 3.12)[a]	2.94	Semiconductor–metal transition at 340 K
CrO_2	2.92	2.86	Ferromagnetic ($T_c = 398$ K) metal
β-MnO_2	2.87	2.76	Antiferromagnetic ($T_N = 94$ K) semiconductor; anomaly in resistivity at T_N
MoO_2	2.52, 3.10		Diagmagnetic metal; structure distorted to monoclinic symmetry as a result of metal–metal bonding
WO_2	2.49, 3.08		Diamagnetic metal; structure distorted to monoclinic symmetry as a result of metal–metal bonding
RuO_2[b]	3.14		Pauli paramagnetic metal

[a] Monoclinic structure.
[b] OsO_2, RhO_2, IrO_2, and PtO_2 show similar properties.

because of metal–metal bonding. There is a direct relation between the c/a ratio and the formation of metal–metal bonds in the rutile family of oxides. The c/a ratio is minimum for $3d^1$, $4d^2$, $5d^2$, $4d^3$, and $5d^3$ configurations, and the corresponding dioxides exhibit monoclinic distortion as a result of metal–metal bonding. Among the first-row transition metal dioxides, TiO_2 and VO_2 are the only rutile oxides in which the intercationic distance is less than R_c. Ti^{4+} has no d electrons and TiO_2 is therefore an insulator. Reduction of TiO_2 gives rise to the homologous Ti_nO_{2n-1} family, which has interesting electronic properties.

Tetragonal CrO_2, having the largest c/a ratio among the $3d$ metal dioxides, is ferromagnetic and metallic, while MoO_2 (and also WO_2), with a distorted structure, is nonmagnetic and metallic. With two d electrons per transition metal atom, the properties critically depend on whether R is greater or less than R_c. In CrO_2, one of the d electrons is in $t_{\parallel}$ and the other in $t_{\perp}$. While the $t_{\parallel}$ electron is localized ($R > R_c$ due to large c/a ratio), the $t_{\perp}$ electron is in the π^* band formed through cation–anion–cation interaction. Intra-atomic exchange splits the states into α and β spin components (Fig. 8a). The α spin states are singly occupied, and a ferromagnetic cation–anion–cation interaction between localized $t_{\parallel}$ electrons and collective π^* electrons makes this oxide simultaneously ferromagnetic and metallic.

In MoO_2 (also in WO_2), $R < R_c$ and one d electron per metal atom is involved in cation–cation homopolar bonding through $t_{\parallel}$ orbitals. The extra d electron partially fills the π^* band, rendering the oxides metallic (Fig. 8b). This picture does not account for the fact that the metal–metal bond order is greater than one in MoO_2. The true situation is however more complex, involving simultaneous participation of $t_{\perp}$ electrons in M—M bonding as well as M—O π-bonding. In

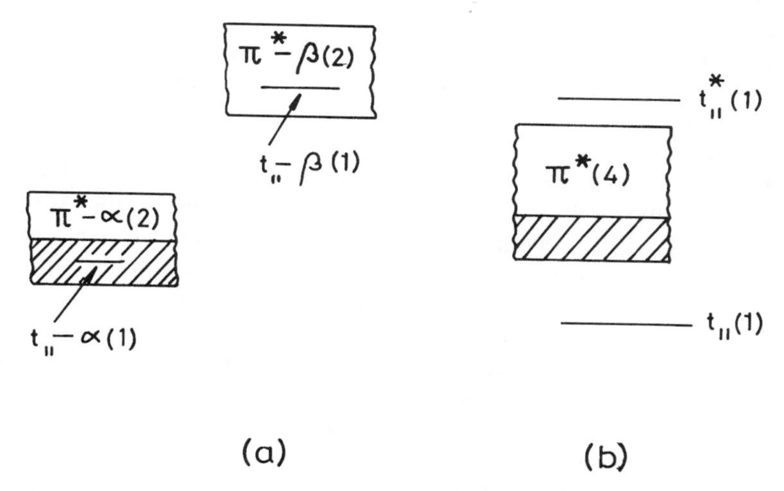

Figure 8. Schematic energy level diagrams of (a) ferromagnetic CrO_2, showing the presence of both localized ($t_{\parallel}$) and itinerant (π^*) electrons and (b) MoO_2 (or WO_2) showing the presence of metal–metal bonding and itinerant π^* electrons. (Following Goodenough).

MnO_2, with three d electrons per cation, both $t_{\|}$ and π^* electrons are localized ($R > R_c$ and $b_\pi < b_c$), and the oxide exhibits spin-only magnetism and is semiconducting. There is an anomaly in the electrical resistivity and specific heat at the magnetic ordering temperature ($T_N = 94$ K) in this oxide.

VO_2 is monoclinic at room temperature and transforms to the tetragonal structure at 340 K. The structural transition is accompanied by a semiconductor–metal transition. In the tetragonal form, since $R < R_c$, $t_{\|}$ orbitals overlap to produce a cation sublattice d band; this band overlaps the π^* band formed by $t_\perp$ orbitals. Since the highest band is partially filled, tetragonal VO_2 is metallic. In the monoclinic form, the cations are displaced to produce cation–cation pairs along the c-axis, the interaction distances being alternately 2.65 and 3.12 Å, instead of the uniform 2.88 Å distance in the tetragonal phase. There is a ferroelectric component to the monoclinic distortion. The distorted structure traps vanadium d electrons in homopolar metal–metal bonding, and the material is nonmagnetic and semiconducting. There is no magnetic ordering in the low temperature phase, and a sharp drop in magnetic susceptibility occurs at the transition temperature because the Fermi surface disappears.

4.3 Sesquioxides

The important sesquioxides of the first-row transition metals crystallizing in the corundum structure are Ti_2O_3, V_2O_3, Cr_2O_3, and α-Fe_2O_3. Both Cr_2O_3 and α-Fe_2O_3 are antiferromagnetic insulators. Ti_2O_3 and V_2O_3 exhibit semiconductor–metal transitions at 410 and 150 K, respectively. The transition in Ti_2O_3 is broad, occurring over a large temperature interval. In V_2O_3, the transition is accompanied by an antiferromagnetic ordering. These properties can be understood in terms of the cation–cation and cation–anion–cation interactions in the corundum structure. Two different cation–anion–cation interactions (135° and 90°) and cation–cation interactions (in the basal plane and along the hexagonal c-axis) are relevant here. The structure provides a trigonal component to the octahedral crystal field, which splits the t_{2g} orbitals into a_{1g} (directed along c_H) and $e_g(\pi)$ directed in the basal plane. In Ti_2O_3, with one d electron per atom and a small hexagonal c/a ratio, both the cation–cation distances are smaller than the critical value ($R_{cc} < R_{bb} < R_c$). The a_{1g} band is therefore filled and separated from the empty $e_g(\pi)$ band by a finite gap, accounting for the semiconducting behavior at low temperatures. In V_2O_3, with a large c/a ratio and with two d electrons per vanadium, $R_{cc} < R_{bb} < R_c$; the $e_g(\pi)$ band is more stable than the a_{1g} band. The metallic nature of rhombohedral V_2O_3 indicates that these bands overlap to some extent.

In Cr_2O_3 and Fe_2O_3, the a_{1g} and $e_g(\pi)$ orbitals are half-filled. In Cr_2O_3, the e_g electrons are localized, but the a_{1g} electrons are likely to be intermediate, since the intercation distance in the c direction is less than R_c. In Cr_2O_3, with no $e_g(\sigma)$ electrons, the cation–anion–cation interaction is weaker than in Fe_2O_3, which has two $e_g(\sigma)$ electrons. Accordingly, T_N in Fe_2O_3 is higher (953 K) than in Cr_2O_3 (307 K). Fe_2O_3 exhibits weak, parasitic ferromagnetism in the range

$253 < T < 953$ K. In this temperature range, the atomic moments lie nearly in the basal plane. Antisymmetric spin coupling is parallel to the *c*-axis, and the anisotropic superexchange cants spins in the basal plane to produce a net moment. Upon application of a strong magnetic field in the *c* direction, a field-induced first-order spin–flip transition (Morin transition) is observed around 260 K.

4.4 Perovskites and Spinels

A number of isostructural ternary oxide families exhibit perovskite, spinel, garnet, pyrochlore, K_2NiF_4, and other structures whose electrical and magnetic properties have been extensively studied. We shall examine properties of perovskite oxides in some detail. The perovskite structure is ideally suited for the study of 180° cation–anion–cation interaction of octahedral site cations (Fig. 9). The possibility of cation–cation interaction is remote because of the large interaction distance along the face diagonal. The variety in the properties of perovskites is illustrated by the following examples: $BaTiO_3$ is ferroelectric, $SrRuO_3$

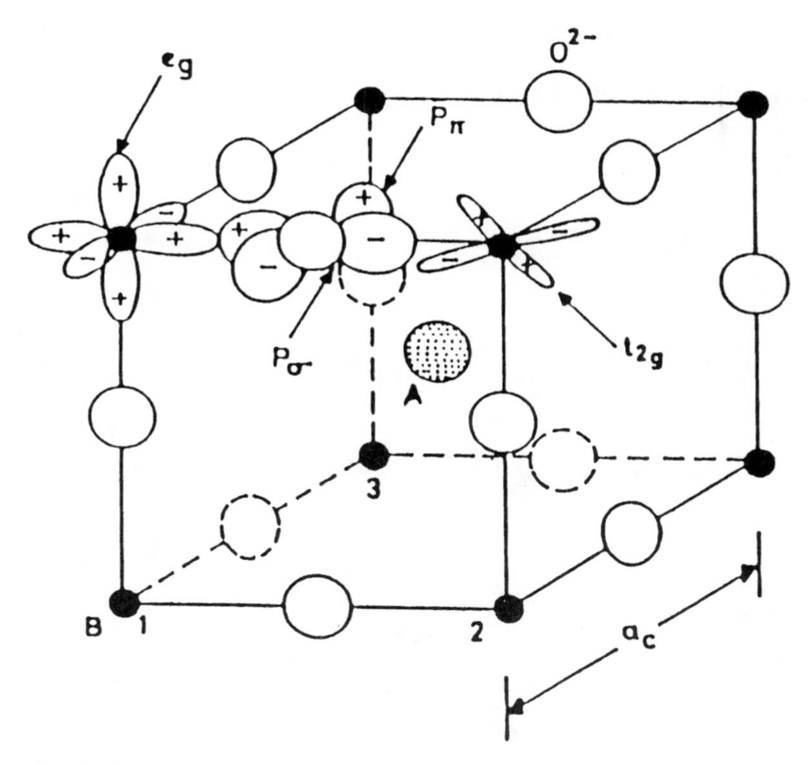

Figure 9. Perovskite structure showing the possibility of cation–anion–cation interaction along the cube edge.

is ferromagnetic, $LaFeO_3$ is weakly ferromagnetic, and $BaPb_{1-x}Bi_xO_3$ is superconducting, while $LaCoO_3$ shows a nonmetal-to-metal transition. Several perovskite oxides exhibit metallic conductivity; typical examples are ReO_3, A_xWO_3, $LaTiO_3$, $AMoO_3$ (A = Ca, Sr, Ba), $SrVO_3$, and $LaNiO_3$. Metallic conductivity in perovskites is entirely due to the strong cation–anion–cation interaction. Tables of perovskites have been published by Goodenough and Longo as well as by Nomura in Landolt–Börnstein.

The important perovskite oxides containing B-site transition metal ions are listed in Figure 10, which groups oxides with the same *d* electron configuration together in the columns. The entries in each column are arranged in the decreasing order of B cation–anion transfer energy *b* (B—O covalency) from top to bottom. Covalent mixing parameters λ_σ and λ_π (and therefore the transfer energies b_σ and b_π) increase with the increasing oxidation state of the B cation. For the same oxidation state, mixing varies as $5d > 4d > 3d$. The influence of A cations on B—O covalency is indirect; acidic A cations decrease B—O covalency. In all the perovskites, λ_σ exceeds λ_π.

The dashed lines in Figure 10, representing $b_\pi = b_m$ (b_m is the critical value for spontaneous magnetism), $b_\pi = b_c$ and $b_\sigma = b_c$ (where b_c is the critical transfer energy) separate oxides that exhibit localized electron behavior from those having itinerant electron properties. Oxides in column 1 are insulators because the B cations are of d^0 electron configuration. Most of the oxides in column 2 ($S = \frac{1}{2}$) are metallic and Pauli paramagnetic; the line $b_\pi = b_m$ separates $LaTiO_3$ from

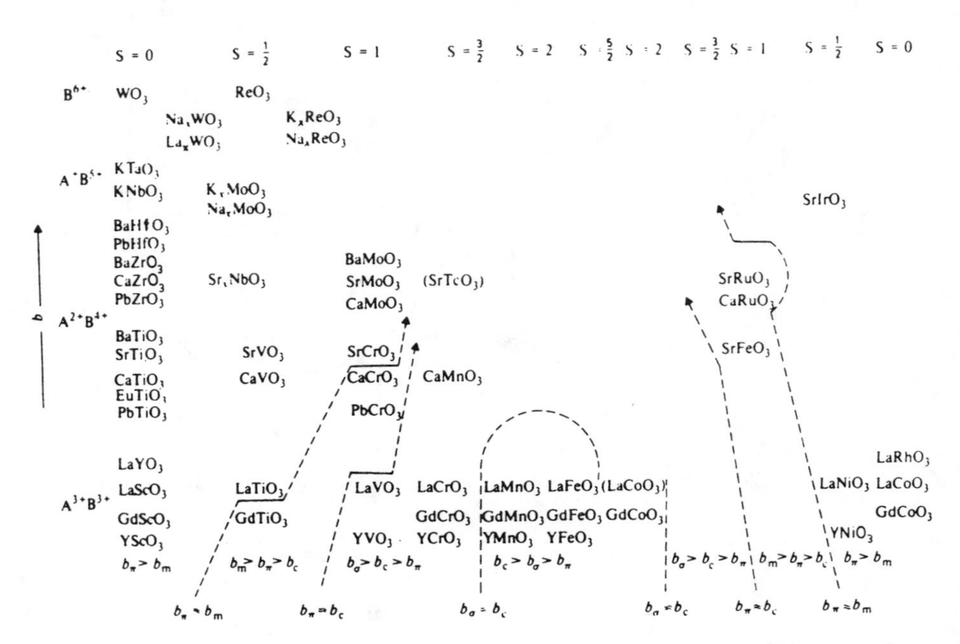

Figure 10. Oxide perovskites with transition metal ions in different electron configurations. (Following Goodenough.[4])

$GdTiO_3$ because $GdTiO_3$ is a semiconductor with a ferromagnetic $T_c = 21$ K. (Recent measurements however show that stoichiometric $LaTiO_3$ is not metallic.) In the third column ($S = 1$), $AMoO_3$ (A = Ca, Sr, Ba) and $SrCrO_3$ are metallic and Pauli paramagnetic. Other oxides in this column are semiconducting and antiferromagnetic. The line $b_\pi = b_m$ separates metallic and Pauli paramagnetic $SrCrO_3$ from the antiferromagnetic semimetal $CaCrO_3$. The line $b_\pi = b_c$ separates $PbCrO_3$ from $LaVO_3$ because the latter exhibits a crystallographic transition at $T < T_N$ characteristic of localized electrons. The region $b_m > b_\pi > b_c$ appears to be quite narrow, as revealed by electrical, magnetic, and associated properties. Variation of T_N with pressure is interesting in this region. Thus, $dT_N/dP < 0$ in $CaCrO_3$ while $dT_N/dP > 0$ in $YCrO_3$ and $CaMnO_3$. Increasing the pressure increases b_π (by decreasing the lattice dimensions), and therefore $dT_N/dP > 0$ for $b_\pi < b_c$ (localized electron behavior) and $dT_N/dP < 0$ for $b_m > b_\pi > b_c$ (itinerant electron behavior).

Oxides in columns 4, 5, and 6 of Figure 10 are antiferromagnetic insulators. Since the intra-atomic exchange, $S(S + 1)$, decreases the covalent mixing, maxima in the the curves $b_\pi = b_c$ and $b_\sigma = b_c$ corresponding to the smallest values of b_π and b_σ occur in the middle of the columns with $S = \frac{5}{2}$. Rare earth orthoferrites with $S = \frac{5}{2}$ are antiferromagnetic insulators and exhibit parasitic ferromagnetism. In these materials, Fe^{3+} spins are canted in a common direction either by cooperative buckling of oxygen octahedra or by anisotropic superexchange, and the antiferromagnetic rare earth sublattice is canted because of the interaction between two sublattices.

In Figure 10, $LaCoO_3$ is shown both in $S = 2$ ($t_{2g}^4e_g^2$) and $S = 0$ (t_{2g}^6) columns at the end, since Co^{3+} in this solid can have either the low spin or the high spin configuration. At very low temperatures, Co^{3+} is entirely in the low spin state. The low spin cobalt ions transform to the high spin state with increase in temperature, and the two states order themselves on unique sites in the lattice. Around 700 K, this oxide becomes metallic. The ninth column of Figure 10, ($S = 1$) lists perovskites containing d^4 cations. Of the three oxides in this column, $SrRuO_3$ is a ferromagnetic metal ($T_c = 160$ K) and $CaRuO_3$ is antiferromagnetic ($T_N = 110$ K) with a weak ferromagnetism. Since both the oxides have the same RuO_3 array, the change from ferromagnetic to antiferromagnetic coupling is significant. $SrFeO_3$ is placed in the same column on the assumption that Fe^{4+}: $3d^4$ is in the low spin state, but recent work shows that Fe^{4+} in this oxide is in the high spin state down to 4 K. $CaFeO_3$, however, shows disproportionation of Fe^{4+} to Fe^{3+} and Fe^{5+} below 290 K. In the penultimate column, containing $S = \frac{1}{2}$ cations, metallic and Pauli paramagnetic $LaNiO_3$ should be distinguished from antiferromagnetic $YNiO_3$, indicating that in $LaNiO_3$ $b_\sigma > b_m$ and in $YNiO_3$ $b_\sigma < b_m$. Similarly, in the last column, $LaCoO_3$ should be separated from $LaRhO_3$ because the latter is a narrow gap semiconductor with a filled $t_{2g}(\pi^*)$ band and an empty $e_g(\sigma^*)$ band. Metallic $BaPbO_3$ does not appear in Figure 10; it has a partially filled 6*s* band. $BaPb_{0.75}Bi_{0.25}O_3$ is a superconductor ($T_c \sim 13$ K), while $BaBiO_3$ (with Bi in 3+ and 5+ states in equal proportions) is an insulator.

Oxides of spinel structure, AB_2O_4, are well-known magnetic materials. Ferrimagnetic $CoCr_2O_4$ has a conical spiral configuration. The cooperative Jahn–Teller effect shown by some of the spinels (e.g., $FeCr_2O_4$) is interesting. Other oxides showing this effect are rare earth zircons (e.g., $TbVO_4$, $DyVO_4$) and $PrAlO_3$. In vandate spinels, $AV_2^{3+}O_4$, the *d* electrons are localized when 2.88 Å $< R_{V—V} <$ 2.97 Å. Fe_3O_4, which is an inverse spinel, has been of much interest in the past several decades. It is noteworthy that the spinels $Li_{1-x}M_x^{2+}Ti_2O_4$ (M = Mg, Mn) and $Li_{1+x}Ti_{2-x}O_4$ show superconductivity.

Hexagonal, cubic, and intergrowth bronzes formed by WO_3 with alkali, hydrogen, and other metals have been well-documented in the literature. Of these, the intergrowth bronzes, in which strips of the hexagonal bronze intergrow with strips of WO_3, sometimes recurrently, are especially interesting. Electrical transport and other properties of WO_3 and related bronzes have been extensively reviewed in the literature,[13,14] especially by Greenblatt.[15] MoO_3 forms different varieties of bronzes: blue bronzes of the type $A_{0.3}MoO_3$ (A = K, Tl, Rb), which are quasi-one-dimensional metals with charge-density wave (CDW) instability; purple bronzes, $A_{0.9}Mo_6O_{17}$ (A = Na, K), which are quasi-two-dimensional metals with CDW instability; $Li_{0.9}Mo_6O_{17}$, which is one-dimensional and superconducting ($T_c \approx 2$ K); red bronzes $A_{0.33}MoO_3$ (A = K, Tl, Rb), which are semiconducting, and $Li_{0.33}MoO_3$, which is violet and three-dimensional with low resistivity. Hydrogen molybdenum bronzes, H_xMoO_3, of different compositions ($0 < x \leq 2.0$) with structures related to MoO_3, have been characterized. Conductivity measurements have been made on some of these hydrogen bronzes. Di- and monophosphate tungsten bronzes of the type $A_x(P_2O_4)_2(WO_3)_{2m}$ and $A_x(PO_2)_4(WO_3)_{2m}$ (A = Na, K, Rb, or Ba) possess hexagonal tunnels. The tunnels may be empty, as in the monophosphate bronze $P_4W_8O_{32}$ ($m = 4$), or occupied, as in the diphosphate tungsten bronzes. Anisotropic electronic properties of $CsP_8W_8O_{40}$, which has a unique structure, have been measured.

5 Mixed Valence

Many transition metal oxides contain cations in more than one oxidation state. Properties of these oxides are generally determined by the rate of electron transfer between the different oxidation states.[6,16] A classification of such compounds has been made by Robin and Day[16] on the basis of the valence delocalization coefficient (Table 5). This type of mixed valence is different from that prevalent in rare earth and actinide materials, in which valence fluctuation, heavy fermion behavior, and superconductivity are found. Depending on the relative energies of the f^n configuration and the Fermi level due to non-*f* electrons, three electronic regimes are distinguished: the magnetic regime, the Kondo regime, and the fluctuating valence regime. EuO and SmS exhibit valence fluctuation under pressure as a result of the promotion of an *f* electron to the conduction band.

In rare earth oxides of the types Pr_6O_{11}, Pr_7O_{12}, and Tb_4O_7, related to the fluorite structure, the electronic conductivity is controlled by the hopping mech-

Table 5 Different Classes of Mixed Valent Solids[a]

Oxide	Classification	Remarks
Pb_3O_4	Class I	Red lead, insulator
$BaBiO_3$	Class I	Insulator
Sb_2O_4	Class I	Mineral cervanite
$Li_xNi_{1-x}O$	Class II	Hopping semiconductor
$La_{1-x}Sr_xMnO_3$	Class II	Ferromagnet
$BaBi_{1-x}Pb_xO_3$	Class III	Superconductivity
$LiTi_2O_4$	Class III	Superconductivity
Na_xWO_3	Class III	Bronze luster; metallic at high x

[a] Following Robin and Day[16], based on the magnitude of the valence delocalization coefficient α, which depends on the energy difference ΔE between the two states $M_A^{n+}M_B^{(n+1)+}$ and $M_A^{(n+1)+}M_B^{n+}$. When ΔE is large, α is small. In class I, α is small while in class III, α is large. In class III the two sites are difficult to distinguish, but in class II the differences are clear.

anism, hence is proportional to the product $[M^3]$ $[M^{4+}]$. The conductivity reaches a maximum when this product becomes a maximum, the point at which the sign of the charge carriers also changes from n- to p-type. Oxides of the type Ti_3O_5 and V_3O_5 undergo metal–insulator transitions, whereas oxides of the type Co_3O_4 are insulators. The presence of more than one oxidation state is readily recognizable from the formula in oxides such as Fe_3O_4, Pb_3O_4, V_nO_{2n-1}, Ti_nO_{2n-1}, and Pr_6O_{11}, but not in oxides such as $BaBiO_3$ and Sb_2O_4, which have Bi/Sb ions in +3 and +5 states. Metal ions in the lower oxidation state (+2 or +3) can be leached out by acid dissolution from Pb_3O_4, Pr_6O_{11}, and Tb_4O_7 leaving only the dioxides in the solid state.

Fe_3O_4, with the inverse spinel structure, undergoes a ferrimagnetic–paramagnetic transition around 850 K and a transition associated with charge ordering (Verwey transition) around 123 K. The latter transition and the electronic properties of the oxide through the transition have been a subject of much study. An entire issue of *Philosophical Magazine* (Vol. B 42, No. 10, 1980) was devoted to this topic. Many papers have appeared since then. Yet, there is considerable uncertainty about the transition and the mechanism of conduction. Honig and co-workers have shown that the transition is markedly dependent on stoichiometry, and becomes second-order at large stoichiometric deviations.[17,18] In Figure 11 we show how the Verwey transition varies with oxygen stoichiometry. Most of the data on transport properties seem to suggest a small polaron model. The observed entropy change [R ln 2/mol Fe_3O_4] suggests the presence of dimer units below the transition temperature. Randomization of the Fe^{3+} and Fe^{2+} ions from an ordered state seems to be too naive a description of this fascinating transition.

$Ln_{1-x}Sr_xCoO_3$ (Ln = La, Nd, etc.) becomes metallic when $x \geq 0.3$, and the itinerancy of the d electrons is associated with ferromagnetism. Mössbauer spectroscopic studies clearly show that cobalt has an average oxidation state between +3 and +4. The rate of electron transfer obviously determines the nature of

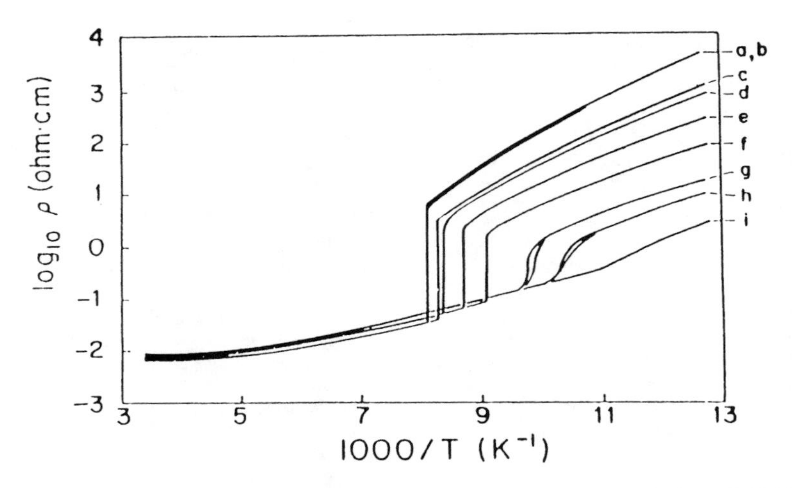

Figure 11. Resistivity data for $Fe_{3(1-\delta)}O_4$; δ values: a, −0.00053; b, −0.00017; c, 0.00021; d, 0.00018; e, 0.00069; f, 0.0017; g, 0.0050; h, 0.0068, and i, 0.0097. (From Honig.[17])

mixed valency in such oxides.[19] $Fe_{3-x}Mo_xO_4$ is another oxide with fast electron transfer with an average oxidation state of +2.5 for iron; Mo occurs in both +3 and +4 states in this system.[20] Many other oxides show such behavior (e.g., $La_{1-x}Sr_xRuO_3$, $LaMn^{3+}_{1-x}Mn^{4+}_xO_{3+\delta}$). In the $LaMn^{3+}_{1-x}Mn^{4+}_xO_3$ and $La_{1-x}Sr_xMnO_3$ systems, the Mn^{4+} content directly determines the ferromagnetic Curie temperature and the metal–insulator transition temperature; when x is large ($\gtrsim 0.5$), the material becomes antiferromagnetic.

6 Metal–Nonmetal Transitions

The band structure of a crystalline solid made up of an even number of electrons can be made to change over to a structure in which the empty and filled bands cross or overlap in response to a change in pressure or temperature or by suitable doping. Such band overlap or crossover transitions are accompanied by a change in the crystal structure and, in certain instances, magnetic ordering as well. The Mott transition from a metallic to a nonmetallic state can occur when the bandwidth decreases until it is smaller than the intrasite electron–electron energy (because of localization induced by electron correlation). Localization can also occur because of disorder, as in amorphous materials, thus giving rise to a metal–nonmetal (M–NM) transition *(Anderson transition);* in such a transition, the bandwidth becomes less than the width of the distribution of random site energies.

In spite of extensive studies, we do not yet fully understand this fascinating phenomenon, which occurs in many transition metal oxides.[21,22] The different types of metal–nonmetal transitions found in metal oxides are as follows:

1. Pressure-induced transitions, as in NiO, in which the pressure increases the orbital overlap between neighboring atoms to induce a change from localized to itinerant behavior of electrons.
2. Transitions as in Fe_3O_4 involving charge ordering.
3. Transitions as in $LaCoO_3$, which are induced initially because of the different spin configurations of the transition metal ion; electron transfer between the two spin states initiates a process that eventually renders the oxide metallic.
4. Transitions as in EuO, arising from the disappearance of spin polarization band-splitting effects when the ferromagnetic Curie temperature is reached.
5. Compositionally induced transitions, as in $La_{1-x}Sr_xCoO_3$ and $LaNi_{1-x}Mn_xO_3$, in which changes of band structure in the vicinity of the Fermi level are brought about by a change in composition or are due to disorder-induced localization.
6. Transitions in two-dimensional systems, such as La_2NiO_4, in which metal–oxygen–metal interaction can occur only in the *ab* plane (unlike the case of the three-dimensional $LaNiO_3$).
7. Temperature-induced transitions in a large class of oxides (e.g., Ti_2O_3, VO_2, V_2O_3), some of which were briefly discussed in Section 4 earlier.

Category 7, involving temperature-induced M–NM transitions, deserves some elaboration. In Ti_2O_3, a second-order transition occurs around 410 K, accompanied by a gradual change in the rhombohedral *c/a* ratio and a hundredfold jump in conductivity; the oxide remains paramagnetic throughout. A simple band-crossing mechanism accompanying the change in the *c/a* ratio explains this transition. Accordingly, substitution of Ti by V up to 10% in Ti_2O_3 makes the system metallic; the *c/a* ratio of this metallic solid solution and the high temperature phase of TiO_3 are similar. In VO_2, a first-order transition around 340 K is accompanied by a change in structure (monoclinic to tetragonal) and a 10^4-fold jump in conductivity; the material remains paramagnetic throughout. A crystal distortion model wherein a gap opens up in the low temperature, low symmetry structure adequately explains the transition. Substitution of trivalent ions such as Cr^{3+} and Al^{3+} for vanadium in VO_2 leads to a complex phase diagram with at least two insulating phases whose properties are significantly different from those of the insulating phase of pure VO_2. These phases are now fairly well understood.

The M–NM transition in V_2O_3 and its alloys has been a subject of a large number of publications. Pure V_2O_3 undergoes a first-order transition (monoclinic–rhombohedral) at 150 K accompanied by a 10^7-fold jump in conductivity and an antiferromagnetic–paramagnetic transition. Application of pressure makes V_2O_3 increasingly metallic, thus suggesting that it is near a critical region; accordingly, doping with Ti or Cr has a marked effect on the transition; the former has a positive pressure effect and the latter a negative pressure effect. V_2O_3 also shows a second-order transition around 400 K with a small conductivity anomaly. Mere crystal distortion or magnetic ordering cannot explain the large conductivity jump at 150 K.

The current status of the V_2O_3 transition is best presented in terms of a plot of log ρ versus reciprocal temperature, and the diagrams in Figure 12[23] show several interesting features. There are three transitions in the V_2O_3/Cr_2O_3 (98.5:1.5) system (Fig. 12a). In the 0–150 K range, the alloy is an antiferromagnetic insulator (AFI). There is a sharp transformation to the paramagnetic metallic state (PM), which prevails between 150 and 300 K. At that point, another first-order transition transforms the oxide to a paramagnetic insulating (PI) state. In the 300–1000 K range, the resistivity gradually diminishes with increasing temperature, and beyond 650 K, the alloy is in another paramagnetic metallic phase (PM′), which resembles the PM state. The three transitions are altered by minor changes through alloying. Electrical properties of the V_2O_3/Cr_2O_3 (97:3) system are given in Figure 12(b). The PM phase is eliminated altogether, and the alloy goes directly from the AFI to the PI state via a sharp

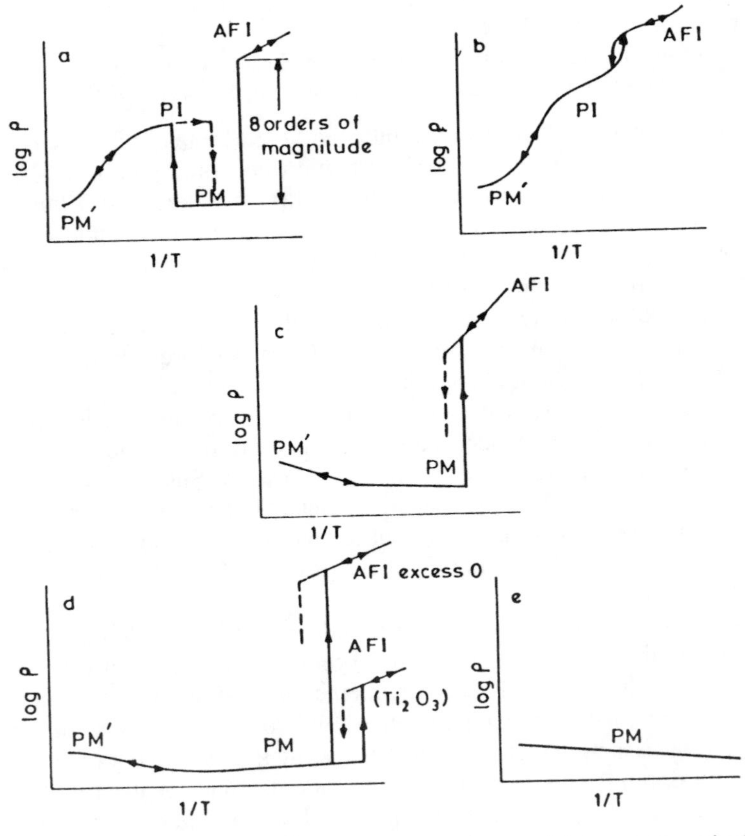

Figure 12. Changes in the logarithm of resistivity, log ρ, with temperature, in the V_2O_3 system; see text for descriptions of individual diagrams. (Following Honig and Spalek.[23])

transition; this is followed by a gradual transition to the PM′ state. In pure V_2O_3, the resistivity exhibits a different pattern (Fig. 12c): a first-order transition near 150 K links the AFI and PM phases. With increasing temperature in the 350–650 K range, an anomalous rise in resistivity occurs, and the undoped or lightly Cr_2O_3-doped V_2O_3 passes continuously from the PM to the PM′ state. Figure 12d gives the situation for the V_2O_3/Ti_2O_3 (99:1) system and for nonstoichiometric V_2O_3. The AFI–PM transition is shifted to lower temperature, and the resistivity discontinuity diminishes with increasing Ti/cation vacancy concentration. When the Ti content is increased beyond 5.5 or the vacancy concentration passes beyond 0.9, all transitions are eliminated and the PM phase is retained (Fig. 12e). Below approximately 10–15 K, however, the metallic phase transforms to an antiferromagnetic metal (AFM, not shown in Fig. 12).

A variety of theoretical models have been proposed to explain M–NM transition in metal oxides.[7,21,22,24] Let us briefly examine the mechanisms involving electron correlation and disorder. The simplest model for correlation effects in solids is that due to Hubbard discussed earlier. The Hubbard gap increases with U, and at large U the material is an AFI, whereas at small U it is a PM. The Hartree–Fock approximation is not satisfactory for M–NM transitions in which charge fluctuations, spin fluctuations, temperature, and mean field have comparable energy scales. Gutzwiller introduced a different kind of approach that emphasizes local correlation effects and the relevance of this model to M–NM transitions due to correlation was pointed out by Brinkman and Rice. This model is satisfactory near the transition, but because spatial correlations are completely ignored, properties at $T \neq 0$ and of the insulating state are not described. Features such as orbital effects, electron–lattice interaction, Coulomb interaction, and disorder have also been ignored in this treatment.

P. W. Anderson showed that when randomness exceeds a critical value, an electron locates itself around an appropriate potential fluctuation so that the state is spatially localized rather than extended, as in the case of weak disorder. Consequences of localization were clearly enunciated by Mott. An oxide becomes metallic or insulating depending on whether the states near the Fermi energy are extended or localized. With increasing disorder or decreasing Fermi energy, the mobility edge crosses the Fermi energy and the system becomes insulating. From this model, one obtains the *minimum metallic conductivity*. The conductivity of a metal drops from σ_{min} to zero at the transition. There is considerable evidence for the change of the transport regime in disordered systems at σ_{min}. At very low temperatures close to critical disorder, however, it has been found that conductivity goes continuously to zero at the localization transition. A scaling theory of localization has been proposed to circumvent this difficulty.

The well-known criterion for the M–NM transition is that due to Mott. It states

$$n^{1/3}a_{\mathrm{H}}^{*} \approx 0.25$$

where a_{H}^{*} is the shallow state radius or atomic orbital size and n is the carrier density. This criterion is spectacularly successful over a wide density range, as shown by Edwards and Sienko, although the transition is not discontinuous as predicted. It can be shown that the Anderson localization criterion $\xi t = W$ (where ξ is the localization length) and the Hubbard criterion $U \approx Zt$, are similar to Mott's criterion. A criterion due to Herzfeld states that for a metal, the ratio of molar refractivity to molar volume (R/V) is greater than or equal to unity. This criterion holds for all elemental metals. A thermodynamic criterion based on latent heat of evaporation is found to be equally satisfactory in separating metallic elements from nonmetallic ones.

In the absence of exact models, M–NM transitions in real systems have been explained qualitatively in terms of the available models. For example, for the V_2O_3 transition, Honig proposed a simple model wherein the density-of-states curve for the *d* band has a set of high peaks alternating with deep valleys. The Fermi level in close proximity to one of the minima can be replaced by a band gap that is opened by Cr or Al doping. Change in oxygen stoichiometry or Ti doping shifts the Fermi level to render the material metallic. Honig and Spalek[23] have worked out a thermodynamic model for V_2O_3 that uses different free energy expressions for electrons in the localized and itinerant regimes.

Disorder has been invoked to explain transitions in oxide systems such as $La_{1-x}Sr_xVO_3$ and $La_{1-x}Sr_xCoO_3$.[25] Studies of systematics of M—NM transitions across a related series of oxides have yielded valuable results, as discussed in a preceding section. It is noteworthy that in binary transition metal compounds, the transfer from insulating behavior to metallic behavior occurs in the vicinity of oxides.

Complex oxides exhibiting compositionally controlled M–NM transitions are especially appealing to chemists. $Ln_{1-x}Sr_xMO_3$ (Ln = La, Pr, Nd; M = V, Mn, Co) show M–NM transitions with increase in x. Thus, $La_{1-x}Sr_xCoO_3$ becomes metallic when $x = 0.3$–0.5. Metallicity in other systems (M = Mn, Co) is accom-

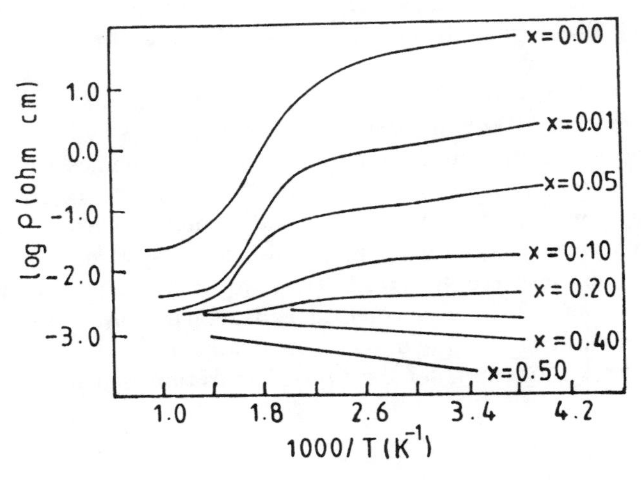

Figure 13. Resistivity behavior of $Nd_{1-x}Sr_xCoO_3$ showing compositionally controlled nonmetal-to-metal transition with increase in x. (From Rao et al.[19])

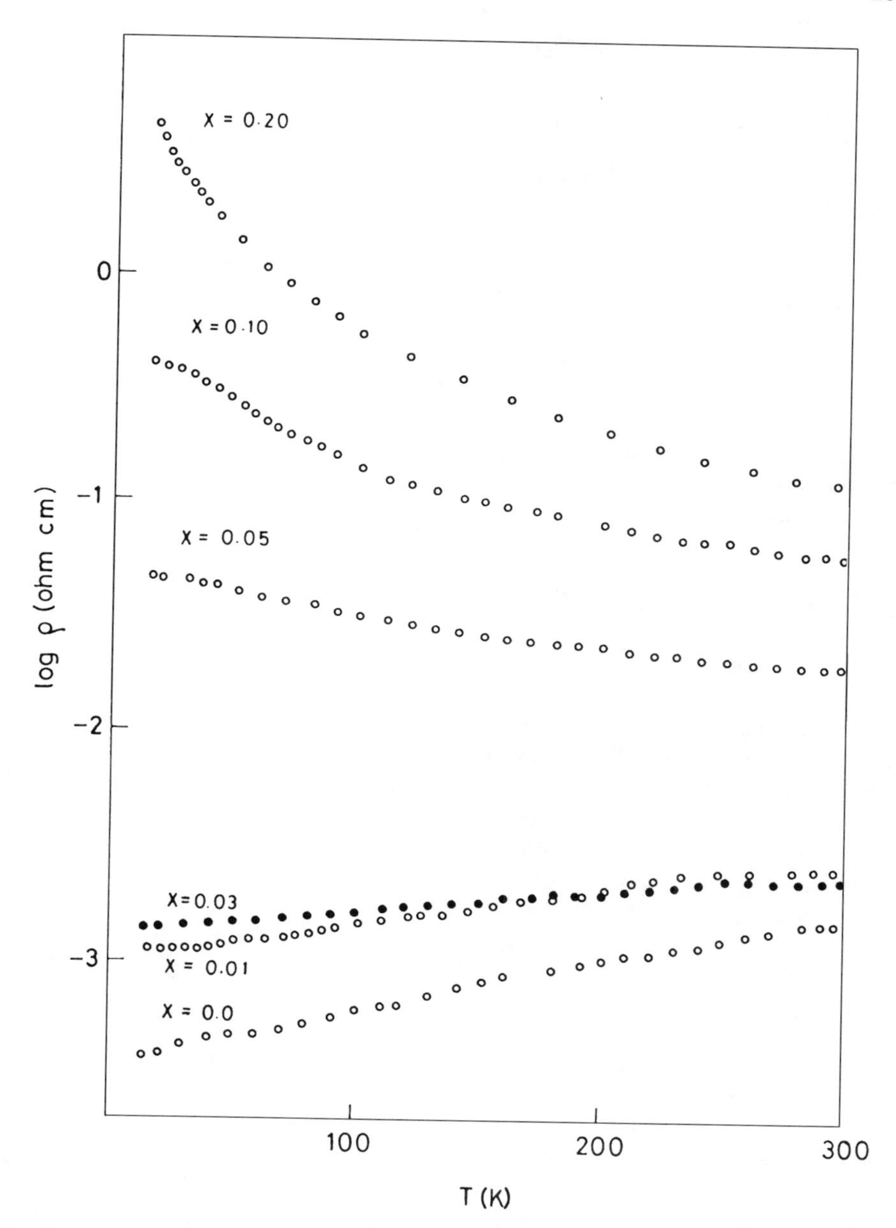

Figure 14. Resistivity behavior of $LaNi_{1-x}Mn_xO_3$ Å $LaNiO_3$ ($x = 0$) is a metal while $LaMnO_3$ ($x = 1.0$) is an insulator. (From Rao and Ganguly.[25])

panied by ferromagnetism. In $LaNi_{1-x}M_xO_3$ (M = Cr, Mn, Fe, or Co), the system goes from metallic to insulator behavior above a critical value of x; at the cross-over, the system has Mott's σ_{min} value. $La_{4-x}Ba_{1+x}Cu_5O_{13+\delta}$ is metallic when $x \approx 0$, but as x increases, it becomes insulating. The pyrochlore system $Bi_{2-x}Gd_x/Ru_2O_7$ exhibits an M–NM transition with increase in x. Figures 13 and 14[19,25] show typical electrical resistivity behavior of $Ln_{1-x}Sr_xCoO_3$ (Ln = La, Nd, . . .) and $LaNi_{1-x}Mn_xO_3$.

Insulator–metal transitions in $LaMnO_3$ and $La_{1-x}A_xMnO_3$ (A = Ca, Sr, or Ba) are interesting.[25] With the increase in the Mn^{4+} content, the oxide becomes rhombohedral or cubic (parent $LaMnO_3$ with ≤15% Mn^{4+} is orthorhombic) and the Mn^{3+}—O—Mn^{4+} interaction renders it ferromagnetic. Electron hopping between Mn^{3+} and Mn^{4+} makes the d electrons itinerant and accordingly, the oxide undergoes an insulator–metal transition just below the ferromagnetic Curie temperature. On application of a high magnetic field, these oxides show a large change in electrical resistivity (magnetoresistance), especially at the insulator–metal transition temperature.

7 Low-Dimensional Oxides

Chemists generally deal with three-dimensional structures. However, there is great interest in the lower dimensional solids (one-dimensional chain and two-dimensional layer compounds), which show spectacular anisotropy in their properties.[6,8] It may be recalled that graphite is metallic in two dimensions and a semiconductor in the third dimension. Striking differences in the properties of mica and asbestos (fiber) are routinely experienced. The platinum chain compound $K_2Pt(CN)_4Br_{0.30}{\cdot}3H_2O$, reflects visible light and conducts electricity like a metal, only in the chain direction.

In understanding the magnetic behavior of solids, it is necessary to take into account not only the dimensionality of the lattice (1–3), but also the dimensionality of the spin or of the order parameter (1–3), which together give rise to nine possible types of magnetic system.[26] In addition, the coupling parameter J can be positive (ferromagnetic) or negative (antiferromagnetic), and this makes 18 different types of system possible. Magnetism in various model systems has been discussed by de Jongh and Miedema.[26] One-dimensional magnetic systems in which there is magnetic interaction along only one direction do not show a magnetic phase transition (one-dimensional Ising systems) if there are only short-range interactions. If there are deviations from ideal one-dimensional behavior, it is possible to observe a phase transition at nonzero temperatures to a three-dimensionally ordered structure. If J is the intrachain interaction strength and J' is the interchain interaction strength, the ratio J'/J determines the temperature at which the phase transition takes place. The two-dimensional spin–half-Ising model on a square lattice shows a phase transition in the absence of an external magnetic field. This is an exact result and was first solved by Onsager. A two-dimensional Heisenberg system cannot show any long-range order, and the magnetic sublattice in such a system cannot have spontaneous magnetization. However, there is no thermodynamic argument that excludes the possibility of the susceptibility divergence, although the zero-field magnetization may vanish. The temperature at which the susceptibility diverges in the absence of deviation from ideal behavior, the so-called Stanley–Kaplan temperature T_{SK}, is the lower limit to the transition temperature that is attributable to deviations from ideal two-dimensional behavior. Only the three-dimensional magnetic system shows long-range order irrespective of the spin dimension.

Oxides of K_2NiF_4 structure are well-known two-dimensional magnetic systems.[27] Investigations of the electronic and magnetic properties of low-dimensional solids have been pursued for some time, and many interesting oxide systems have come to light in recent years. V_3O_5 exhibits a metal–nonmetal transition at 425 K and a magnetic transition at lower temperatures. A maximum in the magnetic susceptibility occurs at 12 K, but the actual transition found from heat capacity measurements is at 76 K. The 76 K transition reflects a changeover from three- to one-dimensional chain ordering in V_3O_5. Both La_2NiO_4 and La_2CuO_4, possessing the K_2NiF_4 structure, show antiferromagnetic ordering. Both show anisotropic electrical conduction, as well; the conductivity is much higher along the *ab* plane than along *c*. In the $(LaO)(LaNiO_3)_n$ and related families, conductivity increases with n as a result of the increase in dimensionality (note that $n = 0$ is metallic $LaNiO_3$ and $n = 1$ is La_2NiO_4). Ganguly and Rao[27] have reviewed antiferromagnetic (e.g., Ca_2MnO_4) and ferromagnetic (e.g., $La_{0.5}Sr_{1.5}CoO_4$) oxides of K_2NiF_4 structure.

Comparison of properties of two- and three-dimensional oxides is interesting. $LaNiO_3$ and $LaCuO_3$ are metallic, while La_2NiO_4 and La_2CuO_4 are not. $SrRuO_3$ is a metallic ferromagnet, but Sr_2RuO_4 is a paramagnetic insulator. $LaCoO_3$ is a paramagnetic insulator, while La_2CoO_4 is an antiferromagnetic insulator. A strict comparison of the properties of two- and three-dimensional oxides can be made only when the *d*-electron configuration of the transition metal ion, B, is the same in both cases. A comparative study of two such systems has been made with respect to their electrical and magnetic properties. For example, members of the $La_{1-x}Sr_{1+x}CoO_4$ system are all paramagnetic semiconductors with a high activation energy for conduction, unlike $La_{1-x}Sr_xCoO_3$ ($x \geq 0.3$), which is metallic and ferromagnetic. $La_{0.5}Sr_{1.5}CoO_4$ shows a magnetization of 0.5 Bohr magneton (μ_B) at 0 K (vs. 1.5 μ_B of $La_{0.5}Sr_{0.5}CoO_3$), but the high temperature susceptibilities of the two systems are comparable. In $SrO(La_{0.5}Sr_{0.5}MnO_3)_n$, both magnetization and electrical conductivity increase with increase in n, approaching the value of the perovskite $La_{0.5}Sr_{0.5}MnO_3$. The system $LaSrMn_{0.5}Ni_{0.5}/(Co_{0.5})O_4$ shows no evidence of long-range ferromagnetic ordering, unlike the perovskite $LaMn_{0.5}Ni_{0.5}/(Co_{0.5})O_3$; these two insulating systems are similar in their high temperature susceptibility behavior. $LaSr_{1-x}Ba_xNiO_4$ exhibits high electrical resistivity, with the resistivity increasing in proportion to the magnetic susceptibility. High temperature susceptibilities of $LaSrNiO_4$ and $LaNiO_3$ are comparable. Susceptibility measurements show no evidence for long-range ordering in $LaSrFe_{1-x}Ni_xO_4$, but not in $LaFe_{1-x}Ni_xO_3$ ($x \leq 0.35$), and the electrical resistivity of the former system is considerably higher.

8 Superconducting Oxides

With the discovery of superconductivity in the La/Ba/Cu/O system ($T_c \sim 30$ K) by Bednorz and Müller[28] late in 1986, transition metal oxides gained great respectability. Oxides themselves are not new to superconductivity. Transition temperatures of about 13 K had been obtained earlier in $Li_{1-x}Ti_{2-x}O_4$ and $BaBi_{1-x}Pb_xO_3$. Superconductivity in oxide systems has been the most studied subject in physical sciences in recent times, and the number of publications in this area is formidable (see refs. 29–33 for details). We shall briefly present some of the highlights in this section.

The superconducting phase in the La/Ba/Cu/O system has the quasi-two-dimensional K_2NiF_4 structure, the parent oxide being La_2CuO_4. While stoichiometric La_2CuO_4 is an antiferromagnetic insulator, doping it with holes by means of excess oxygen or by part substitution of La^{3+} by a divalent ion such as Ba^{2+} or Sr^{2+} renders it superconducting, with a maximum T_c of approximately 35 K. The structure of this oxide system appears in Figure 15a. Early in 1987, superconductivity was found above liquid nitrogen temperature in the Y/Ba/Cu/O system ($T_c \sim 90$ K), the composition of the cuprate being $YBa_2Cu_3O_{7-\delta}$. For values of δ upto approximately 0.6, this cuprate has an orthorhombic structure (Fig. 15b), becoming tetragonal and nonsuperconducting when δ reaches or exceeds 0.6. The variation of T_c with δ in $YBa_2Cu_3O_{7-\delta}$ is shown in Figure 16 along with the variation of the Cu valence in the CuO_2 sheet. When $\delta = 0.5$, T_c is 45 K. The 60 K plateau region is noteworthy, but compositions in this region appear to be metastable. All the rare earth cuprates having the general formula $LnBa_2Cu_3O_7$ (123) with the exception of Ce, Pr, and Tb, are 90 K superconductors. Cuprates of the type $LnBa_2Cu_3O_8$ (124) with two Cu—O chains show a T_c of ~80 K. In between 123 and 124, we have the intergrowth 247 cuprate, with alternating 123 and 124 units, which is also superconducting (Fig. 17). $YBa_2Cu_3O_{6.5}$ is itself an intergrowth of alternate units of $YBa_2Cu_3O_7$ (with Cu—O chains) and $YBa_2Cu_3O_6$ (with no Cu—O chains).

Many superconducting cuprate systems have been discovered since 1986 (Table 6), the notable ones including bismuth cuprates of the general formula $Bi_2(Ca,Sr)_{n+1}Cu_nO_{2n+4}$, $Tl_2Ca_{n-1}Ba_2Cu_nO_{2n+4}$, $TlCa_{n-1}Ba_2Cu_nO_{2n+3}$, and its Sr analogues, layered Pb cuprates of the types $Pb_2Sr_2(Ln, Ca)Cu_3O_8$ and $H_gCa_2Ba_2Cu_3O_y$. The bismuth and thallium cuprates contain Bi(Tl)—O layers and CuO_2 sheets (Fig. 18) with nominal mixed valence of Cu. $Pb_2Sr_2Ca_{1-x}Ln_xCu_3O_8$ contains CuO_2 sheets, with the PbO layers joined by O—Cu(I)—O sticks. Structure and properties of the different families of cuprates have been discussed extensively in the literature. The highest T_c as of today occurs in $HgBa_2Ca_2Cu_3O_y$ ($T_c \sim 133$ K); under pressure, the T_c increases to 155

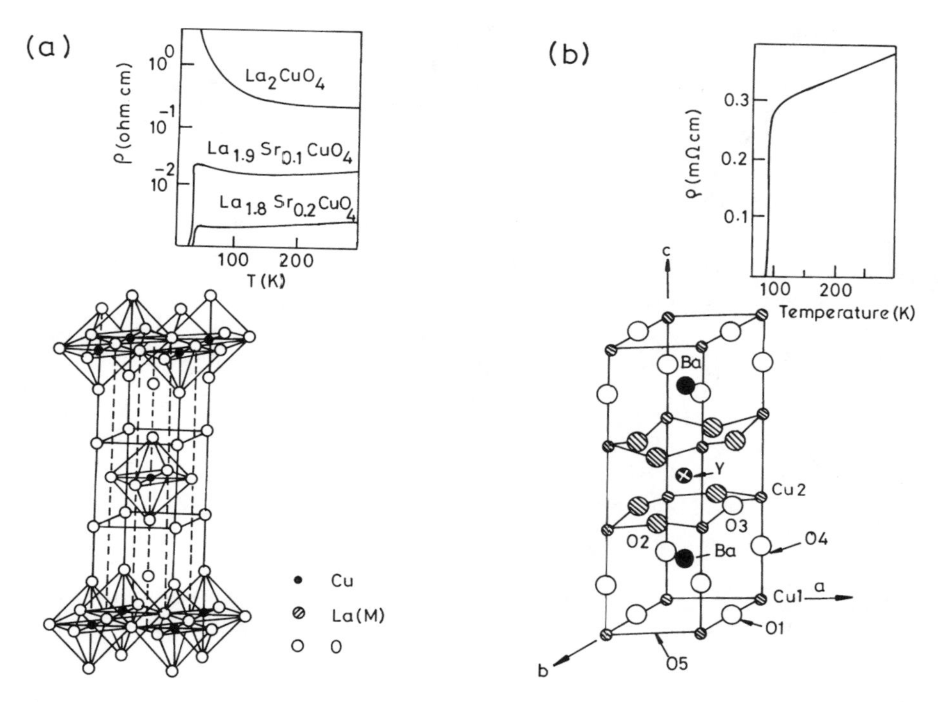

Figure 15. Structure and superconducting transitions of (a) $La_{2-x}Sr_xCuO_4$ and (b) $YBa_2Cu_3O_7$.

K.[34a] In the K_2NiF_4 family of cuprates, besides $La_{2-x}Ba_x(Sr_x)CuO_4$, the second member, $(La, Sr)_2CaCu_2O_6$ prepared under high oxygen pressures, is also superconducting ($T_c \sim 60$ K). Infinitely layered cuprates of the types $Ca_{1-x}Sr_xCuO_2$ and $Sr_{1-x}Nd_xCuO_2$ have been prepared under high pressures, and they show T_c values between 40 and 120 K. All the cuprates mentioned hitherto are hole superconductors. Electron superconductors of the types $Nd_{2-x}Ce_xCuO_4$,

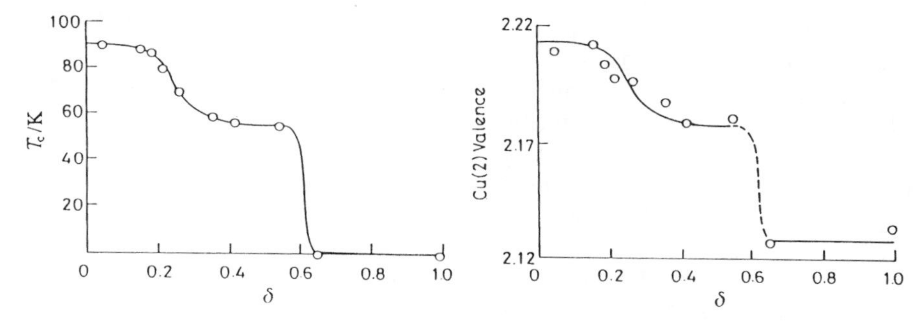

Figure 16. Variation of T_c and Cu(2) valence with δ in $YBa_2Cu_3O_{7-\delta}$. (From Cava et al.[34b])

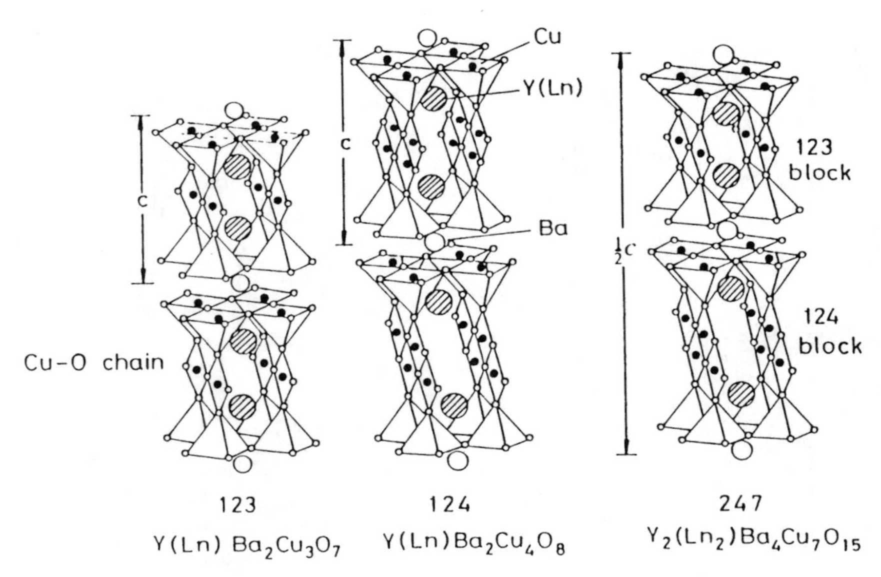

Figure 17. Schematic structures of $YBa_2Cu_3O_7$ (123), $YBa_2Cu_4O_8$ (124), and $Y_2Ba_4Cu_7O_{15}$ (247).

$Pr_{2-x}Th_xCuO_4$, and $Nd_2CuO_{4-x}F_x$ have been prepared. These cuprates have the T′ tetragonal structure (Fig. 19), unlike the T structure of $La_{2-x}M_xCuO_4$.

There are many striking commonalities in the structure and properties of the high T_c cuprates.[29–33] All the cuprates can be considered to be a result of the intergrowth of defect perovskite layers of $ACuO_{3-x}$ with AO-type rock salt layers (see Part I for details) leading to the general formula $[ACuO_{3-x}]_n[AO]_{n'}$. The 123 compounds, however, do not have rock salt layers and may be considered to be the $n' = 0$ member of this general family. (See Part I, Table 8). The more important common features are listed below.

1. All the cuprates possess CuO_2 layers sandwiched between M—O layers (e.g., TlO, BiO) acting as charge reservoirs or spacers. In 123 and 124 cuprates, the Cu—O chains are charge reservoirs. The seat of superconductivity is in the CuO_2 layers. The T_c in Bi and Tl cuprates increases up to $n = 3$ and then decreases. Interaction or spacing between the Cu—O layers is, however, crucial. This is demonstrated by experiments in which the introduction of a fluorite layer $[Ln_{1-x}Ce_x]_2O_2$ between two CuO_2 sheets in Bi or Tl cuprates lowers the T_c markedly. However, intercalation of iodine between the BiO layers in $Bi_2CaSr_2Cu_2O_8$ (causing a substantial increase in the c parameter) does not affect the T_c.

An examination of the interplanar Cu—Cu distances in the cuprates shows that the highest values of T_c are found in the 3.0–3.6 Å range. When the Cu—Cu distance is greater than ~6 Å, the T_c becomes low.

Infinitely layered cuprates ($Ca_{1-x}Sr_xCuO_2$ and related compounds) show values of T_c up to 120 K. The origin of superconductivity in these materials is

Table 6 Structural Parameters and Approximate T_c Values of High Temperature Superconductors

Cuprate	Structure	Unit Cell Dimensions (Å)	T_c (K), Maximum Value
$La_2CuO_{4+\delta}$	*Bmab*	$a = 5.355$ $b = 5.401$ $c = 13.15$	39
$La_{2-x}Sr_x(Ba_x)CuO_4$	I4/*mmm*	$a = 3.779$ $c = 13.23$	35
$La_2Ca_{1-x}Sr_xCu_2O_6$	I4/*mmm*	$a = 3.825$ $c = 19.42$	60
$YBa_2Cu_3O_7$	P*mmm*	$a = 3.82$ $b = 3.885$ $c = 11.676$	93
$Yba_2Cu_4O_8$	*Ammm*	$a = 3.84$ $b = 3.87$ $c = 27.24$	80
$Y_2Ba_4Cu_7O_{15}$	*Ammm*	$a = 3.851$ $b = 3.869$ $c = 50.29$	90
$Bi_2Sr_2CuO_6$	*Amaa*	$a = 5.362$ $b = 5.374$ $c = 24.622$	10
$Bi_2CaSr_2Cu_2O_8$	*A2aa*	$a = 5.409$ $b = 5.420$ $c = 30.93$	92
$Bi_2Ca_2Sr_2Cu_3O_{10}$	*A2aa*	$a = 5.39$ $b = 5.40$ $c = 37$	110
$Bi_2Sr_2(Ln_{1-x}Ce_x)_2Cu_2O_{10}$	*P4/mmm*	$a = 3.888$ $c = 17.28$	25
$Tl_2Ba_2CuO_6$	*A2aa*	$a = 5.468$ $b = 5.472$ $c = 23.238$	92
	I4/mmm	$a = 3.866$ $c = 23.239$	
$Tl_2CaBa_2Cu_2O_8$	*I4/mmm*	$a = 3.855$ $c = 29.318$	119
$Tl_2Ca_2Ba_2Cu_3O_{10}$	*I4/mmm*	$a = 3.85$ $c = 35.9$	128
$Tl(BaLa)CuO_5$	*P4/mmm*	$a = 3.83$ $c = 9.55$	40
$Tl(SrLa)CuO_5$	*P4/mmm*	$a = 3.7$ $c = 9$	40
$TlCaBa_2Cu_2O_7$	*P4/mmm*	$a = 3.856$ $c = 12.754$	103
$(Tl_{0.5}Pb_{0.5})CaSr_2Cu_2O_7$	*P4/mmm*	$a = 3.80$ $c = 12.05$	90
$TlSr_2Y_{0.5}Ca_{0.5}Cu_2O_7$	*P4/mmm*	$a = 3.80$ $c = 12.10$	90

(Continued)

Table 6 *(Continued)*

Cuprate	Structure	Unit Cell Dimensions (Å)	T_c (K), Maximum Value
$TlCa_2Ba_2Cu_3O_8$	*P4/mmm*	$a = 3.853$ $c = 15.913$	110
$(Tl_{0.5}Pb_{0.5})Sr_2Ca_2Cu_3O_9$	*P4/mmm*	$a = 3.81$ $c = 15.23$	120
$TlBa_2(Ln_{1-x}Ce_x)_2Cu_2O_9$	*I4/mmm*	$a = 3.8$ $c = 29.5$	40
$Pb_2Sr_2Ln_{0.5}Ca_{0.5}Cu_3O_8$	*Cmmm*	$a = 5.435$ $b = 5.463$ $c = 15.817$	70
$Pb_2(Sr,La)_2Cu_2O_6$	*P22₁2*	$a = 5.333$ $b = 5.421$ $c = 12.609$	32
$(Pb,Cu)Sr_2(Ln,Ca)Cu_2O_7$	*P4/mmm*	$a = 3.820$ $c = 11.826$	50
$(Pb,Cu)(Sr,Eu)(Eu,Ce)Cu_2O_x$	*I4/mmm*	$a = 3.837$ $c = 29.01$	25
$HgBa_2CuO_4$	*P4/mmm*	$a = 3.8797$ $c = 9.509$	90
$HgBa_2Ca_2Cu_3O_8$	*P4/mmm*	$a = 3.93$ $c = 16.1$ Å (×2)	133[a]
$Nd_{2-x}Ce_xCuO_4$	*I4/mmm*	$a = 3.95$ $c = 12.07$	30
$Ca_{1-x}Sr_xCuO_2$	*P4/mmm*	$a = 3.902$ $c = 3.35$	110
$Sr_{1-x}Nd_xCuO_2$	*P4/mmm*	$a = 3.942$ $c = 3.393$	40
$Ba_{0.6}K_{0.4}BiO_3$	*Pm3m*	$a = 4.287$	31

[a] 155 K under high pressure.

considered to be due to the presence of defect Sr—O layers corresponding to $Sr_3O_{2\pm x}$ blocks, which in turn introduce an apical oxygen for interaction with the Cu—O sheets.[35a]

2. The Cu—O bonds in the cuprates are highly covalent. That is, Cu($d_{x^2-y^2}$)—O($2p$) hybridization strength is large and the Cu—O charge-transfer energy is small. The antibonding nature of the Cu—O(π) interaction makes it sensitive to hole concentration, hole doping decreasing the in-plane Cu—O distance.

3. An interesting comparison can be made between the Cu—O sheets in the hole and the electron superconductors. Cuprates with the T′ structure, where Cu has a square-planar coordination, can be doped with electrons, while those with the T structure ($La_{2-x}Sr_xCuO_4$) can be doped with holes (Fig. 19). There is a certain symmetry between these two situations, as shown in the phase diagram in Figure 20. The Cu–O–Cu angle is less than 180° in the electron supercon-

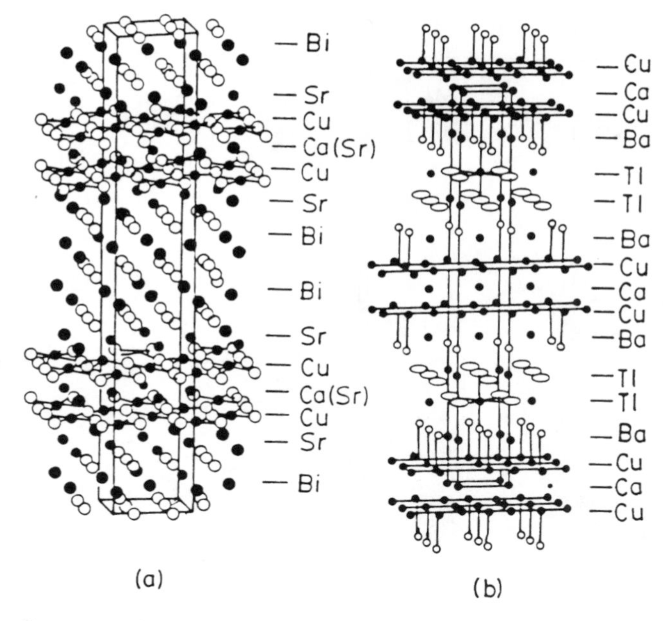

Figure 18. Structure of (a) $Bi_2(CaSr)_3Cu_2O_8$ and (b) $Tl_2CaBa_2Cu_2O_8$.

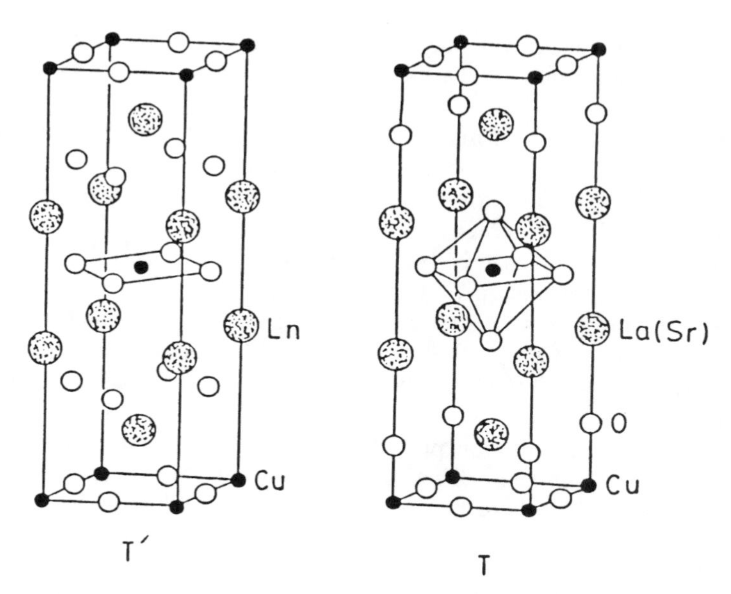

Figure 19. The T′ structure of $Nd_{2-x}Ce_xCuO_4$ and the T structure of $La_{2-x}Sr_xCuO_4$. Electrons are doped (Ce^{4+} in place of Nd^{3+}) into square-planar CuO_4, while holes are doped into square-pyramidal or octahedral Cu—O polyhedra.

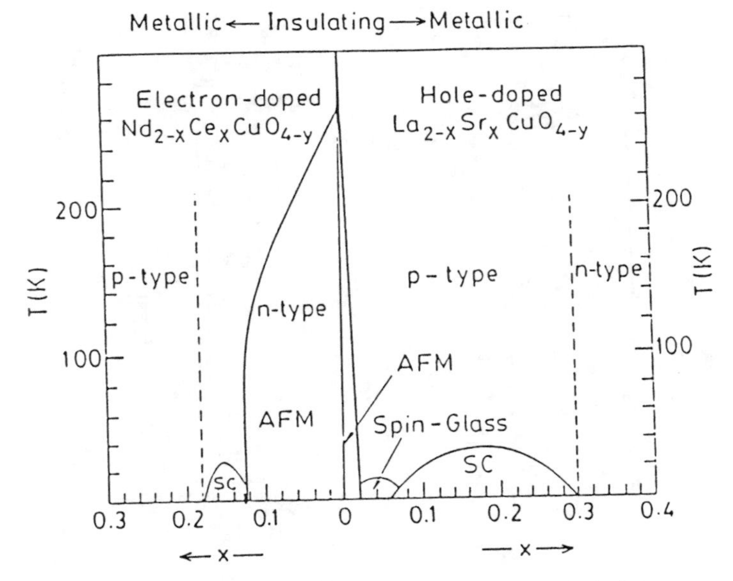

Figure 20. Symmetry in-phase diagrams of electron and hole superconductors $Nd_{2-x}Sr_xCuO_{4-y}$ and $La_{2-x}Sr_xCuO_{4-y}$.

ductors. The position of the apical oxygen in the Cu—O square pyramids (or octahedra) in the hole superconductors seems to modulate the width of the conduction band. The apical Cu—O distance in $YBa_2Cu_3O_{7-\delta}$ mirrors the change in T_c with δ.

4. The parent cuprates in many of the cuprate superconductors are antiferromagnetic insulators. For example, La_2CuO_4, $YBa_2Cu_3O_6$, $Bi_2Sr_2LnCu_2O_8$, and $Pb_2Sr_2LnCu_3O_8$ are the antiferromagnetic insulators corresponding to the superconductors $La_{2-x}Sr_xCuO_4$, $YBa_2Cu_3O_7$, $Bi_2Ca_{1-x}Ln_2Sr_2Cu_2O_8$, and $Pb_2Sr_2Ca_{1-x}Ln_xCu_3O_8$, respectively. In the case of $Nd_{2-x}Ce_xCuO_4$, the parent Nd_2CuO_4 is the antiferromagnetic insulator.

5. The cuprates nominally contain mixed valent copper, which can disproportionate as follows: $Cu^{II} \rightarrow Cu^{III} + Cu^{I}$. In other words, the phenomenon is associated with the stability of empty and completely filled bands compared to a half-filled band.

6. Local charge distribution provides a basis for understanding superconductivity in the cuprates. The phenomenon is well demonstrated in $Pb_2Sr_2Ln_{1-x}Ca_xCu_3O_8$, where excess oxygen oxidizes Pb^{2+} and Cu^{+} without affecting the CuO_2 sheets. It is necessary to replace the yttrium by calcium between the CuO_2 sheets to render this system superconducting.[35b]

7. Oxygen stoichiometry, homogeneity, and disorder play important roles in the superconductivity of the cuprates, as is well demonstrated in the 123 system. Oxygen excess La_2CuO_4 is biphasic: the stoichiometric composition is antiferromagnetic and the superconducting composition is one of oxygen excess.[35b]

8. The superconducting cuprates are marginally metallic in the normal state, sitting on a metal–insulator boundary. We would, therefore, expect abnormal properties in the normal state. One of the striking abnormal normal-state properties of these materials is the linearity of resistivity over a wide range of temperatures (Fig. 21).

9. In the hole superconductors, the T_c value reaches a maximum at an optimal value of the hole concentration, n_h (Fig. 22).[36] The value of n_h at maximum T_c appears to be around the same for all cuprates with the same number of CuO_2 sheets (e.g., $n_h = 0.2$ for cuprates with two CuO_2 sheets). The normalized plot in the inset shows breaks corresponding to insulating (antiferromagnetic) and metallic regimes. The in-plane Cu—O distance also reflects the hole concentration, and the variation of T_c with this distance is significant.

Table 7 lists some of the novel properties of cuprate superconductors; these include resistivity in the *ab* plane and along *c* (to show the anisotropy), Hall density, hole concentration, and the values of the energy gap, $\Delta/k_B T_c$.

All the high T_c superconductors discussed hitherto were cuprates. The only high T_c oxide without copper is $Ba_{1-x}K_xBiO_3$, which has a T_c of 30 K.[37] There is mixed valency of Bi in this oxide, and this is considered to be a negative

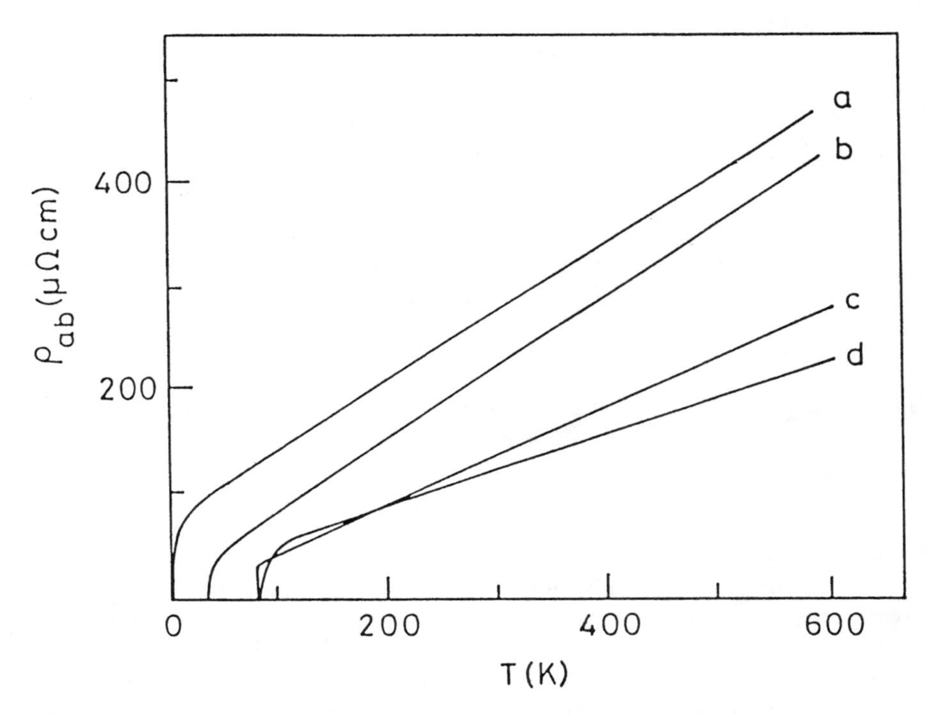

Figure 21. The *ab*-plane resistivity in a few cuprate superconductors as a function of temperature: a, Bi cuprate, 2201 ($T_c \sim 10$ K); b, $La_{1.85}Sr_{0.15}CuO_4$ ($T_c \sim 40$ K); c, Bi cuprate, 2212 ($T_c \sim 80$ K); d, $YBa_2Cu_3O_7$ ($T_c \sim 90$ K). Upon multiplication by a scale factor, resistivity data of polycrystalline samples exactly match this behavior.

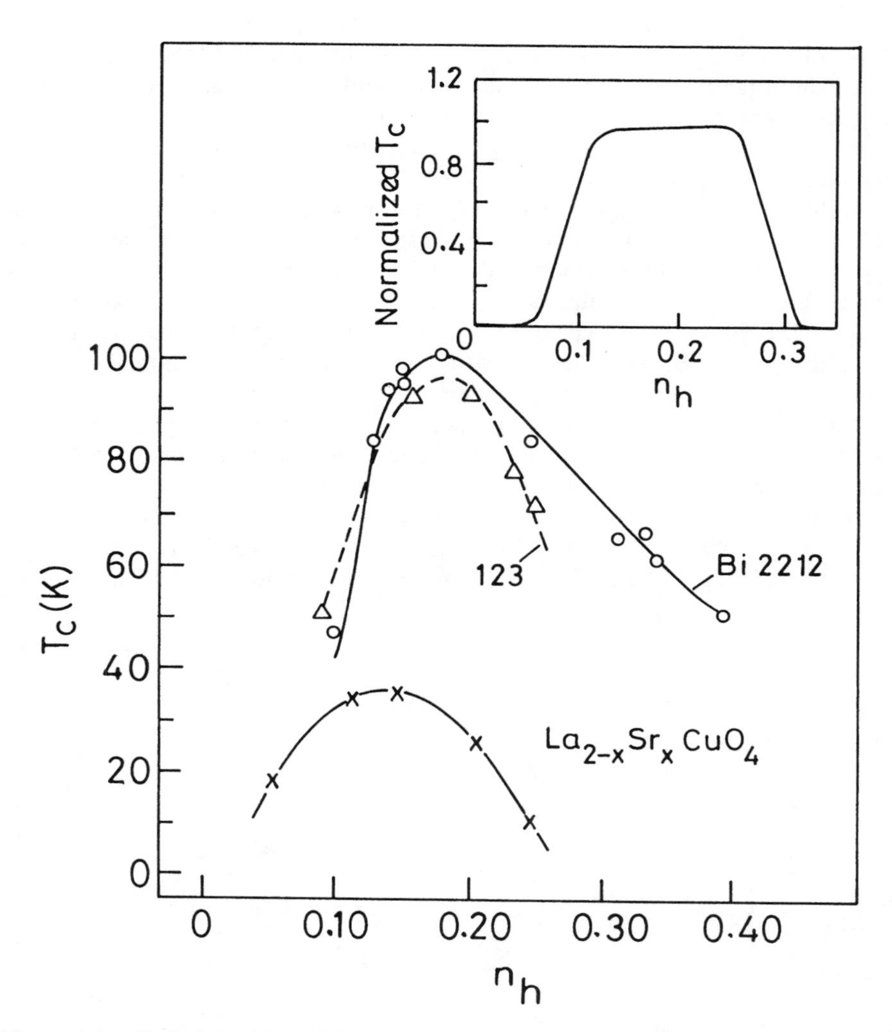

Figure 22. Variation of T_c with hole concentration in cuprates (From Rao et al [36].) Inset shows the normalized T_c plot of Zhang and Sato.[36]

Hubbard U system. Other perovskites with various combinations of metals have been generally found to exhibit at best low values of T_c. $La_{2-x}Sr_xNiO_4$ and other oxide systems that initially showed some promise are not superconducting.

Figure 23 shows the single site energy levels of the $[CuO_2]$ unit cell of cuprates. The one d hole of Cu^{2+} (or the d^9 level) lies lowest at energy ϵ_d^h with respect to the no-hole or d^{10} configuration. To add another d hole (Cu^{3+} or d^8) costs extra energy ($\epsilon_d^h + U$), where U ($\simeq$8 eV) is the single site or Mott–Hubbard repulsion. The large value of U would invalidate noninteracting electron band

Table 7 Normal State and Superconducting Properties of Cuprates

Material		Form	$\rho_{ab}(\mu\Omega \cdot cm)$		$\rho_c(m\Omega \cdot cm)$	$d\rho_c/dT$	Hall Density, η_H 10^{21} cm^{-3}	
			300 K	100 K	300 K		300 K	100 K
$YBa_2Cu_3O_7$[a]		Single crystal	110	35	5	+	11–16	4–6
		film	200–300	60–100			5–9	2–3
$YBa_2Cu_4O_8$		Single crystal	75	20	10	–	14	17
		film	100–200	20–50			22	
$Bi_2Sr_2CuO_6$		Single crystal	300	150	5000	–	6	5
$Bi_2Sr_2CaCu_2O_6$[b]		Single crystal	150	50	>1000	–	4	3
$Tl_2Ba_2CuO_6$		Single crystal	300–400	50–75	200–300	+	3.1	2.5
$Tl_2Ba_2Ca_2Cu_3O_{10}$[c]		Ceramic	(120 K)	$\rho \approx 0.4\ \Omega\cdot$cm	(300 K)	$\rho \approx 1.5\ m\Omega \cdot cm$	(200 K) ≈ 2	
$La_{2-x}Sr_xCuO_4$	$x = 0.12$	Single crystal	900	350	200	+ for $T > 225$ K	2.5	
$La_{2-x}Sr_xCuO_4$	$x = 0.20$	Single crystal	400	200	80	+ for $T > 150$ K	10	
		film	400	160			8.4	6.3
$Nd_2Ce_xCuO_4$	$x = 0.17$[d]	Single crystal	500	275			53	17
	$x = 0.15$	film	140–180	35			32	11

[a] Energy gap, $2\Delta/k_BT_c$, is 5–6; J_c of films ~ 30 = $\times 10^6$ A/cm^2 has been attained; λ_{ab} ~ 1400 Å (penetration depth).
[b] Energy gap, $2\Delta/k_BT_c$, is 8–9; J_c ~ 2×10^6 A/cm^2; λab ~ 2700 Å.
[c] With $Tl_2Ba_2CaCu_2O_8$, $2\Delta/k_BT_c$ ~ 6–7 (ceramic sample); J_c ~ 10×10^6 A/cm^2 (film); λ_{ab} ~ 2000 Å.
[d] $2\Delta/k_BT_c$ ~ 7 (ceramic sample); J_c ~ 0.2×10^6 A/cm^2 (film).

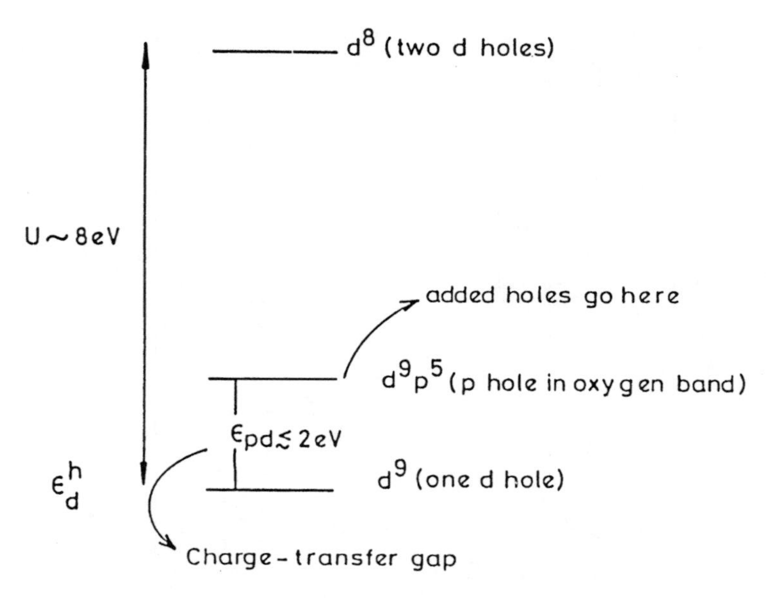

Figure 23. Relevant energy levels of Cu and oxide ions in the CuO_2 sheets of cuprates.

theory, leading to the insulating behavior of the stoichiometric material having one d hole per site (e.g., La_2CuO_4). The feature that distinguishes the cuprates (as well as many transition metal oxides) from one-band Mott insulators is the presence of oxygen p-hole levels, ϵ_p^h, between d^8 and d^9 energies. The minimum energy required to transfer a d hole to an unoccupied site is $(\epsilon_p^h - \epsilon_p^d) \simeq 2$ eV, the charge transfer being to the same linear combination of p_x, p_y oxygen hole orbitals with local $d_{x^2-y^2}$ symmetry. When holes are added to La_2CuO_4 by the substitution of trivalent La with divalent Sr, they go into the oxygen p-hole band and hop from site to site, both directly (with hopping amplitude $t_{pp} \simeq 0.7$ eV) and through admixture with the d states ($t_{pd} \simeq 1$ eV). All the superconducting cuprates have a fraction x (0.05–0.30) of mobile holes in the plane per [CuO_2] unit cell and may be considered to be doped charge-transfer insulators in the strong local correlation or Mott limit.

From the phase diagram in Figure 20 we see that the antiferromagnetic order of La_2CuO_4 is destroyed rapidly by doping, with T_N falling from over 210 K to 0 K as x changes from 0 to about 2%. Beyond this, there is an insulating or a barely metallic spin-glass phase at zero temperature up to $x \simeq 0.07$. The ground state is either insulating or superconducting. If the holes move, the insulating antiferromagnetic state is destroyed and a superconducting state obtains. The normal metal region above T_c is rather unusual in the range $0.05 < x \simeq 0.30$. Overdoped $La_{2-x}M_xCuO_4$ ($x > 0.3$) seems to be a conventional metal.

The ab-plane dc electrical resistivity of cuprate superconductors above T_c is approximately linear in temperature over a wide range of several hundred degrees (Fig. 21). The slopes are nearly the same ($\simeq 1$ μΩ·cm/K) to within a factor of 2 for a wide range of single crystal specimens. The extrapolated zero-temperature intercept can be small. A linear temperature dependence of resistivity in the range 10–400 K is unusual indeed. Thus, the resistivity of clean metals due to electron–phonon interaction rises initially as T^n with n ranging from 3 to 5. A number of properties are consistent with the linear temperature dependence of resistivity; these include the low frequency background in Raman scattering, the ac conductivity, and the weakly temperature-dependent nuclear spin relaxation rate. This behavior suggests that low energy electronic excitations are not well defined, unlike the case in conventional metals. The Hall effect in the normal state of the superconducting cuprates is also unusual, being strongly temperature dependent. For example, the ratio (σ/σ_H) is proportional to T^2, σ being the electrical conductivity and σ_H, the Hall conductivity. Unusual disorder effects have been seen at low temperatures in $Bi_2Sr_2CuO_6$, which has a linear resistivity but no superconducting transition. At temperatures well below 5 K, inelastic processes are a small perturbation on the elastic random scattering of carriers, and weak localization effects become noticeable (e.g., in the upturn of the resistivity). A negative orbital magnetoresistance exists that can be fitted to a weak localization form, from which one can extract an inelastic decay time for carriers. All these results indicate that the lowest energy ($\epsilon < 0.01$ eV) electronic excitations in cuprates are unlike those of any known metal. It is generally believed that the key to high temperature superconductivity is in understanding the unusual normal-state properties.

Superconductivity in cuprates is due to coherence between electron pairs, as confirmed by experiments showing that the magnetic flux is quantized in units of (hc/2e). There is disagreement as to what causes pairing. In conventional superconductors, pairing is due to retarded phonon-induced attraction. There is no evidence for large electron–phonon coupling in cuprates giving rise to high T_c. Theories for cuprate superconductors fall into two categories. In the conventional approaches, one looks for bosonic excitations that have the right energy scale, and coupling to electrons and the metallic state is conventional. Bosonic excitations include excitons, plasmons, magnons, and spin fluctuations. Such a system generally has a normal metallic state above T_c unless radical assumptions are made about the boson spectrum and the electron–boson coupling. In other theories, it is assumed that because of strong local correlations as well as quantum effects, novel ground states arise.

P. W. Anderson has proposed several new ideas to account for the superconductivity in cuprates. According to Anderson, high T_c superconductivity in cuprates is a strong correlation effect connected with the formation of a coherent assembly of singlet spin pairs. The quantum mechanical overlap of a large number of such singlet configurations with different pair lengths and pair members renders such a resonating valence bond (RVB) state highly stable. Doping the RVB system with holes leads to neutral spinlike excitations (spinons) and charged holelike excitations (holons). The question to be answered relates to the degree to which these mutually interacting excitations are approximated by relatively weakly coupled spinons and holons. In the one-band Hubbard model, if the Hubbard correlation energy U is large, electron configurations with doubly occupied orbitals are disfavored. In the cuprates, this corresponds to the two-hole configuration d^8, which has an energy $2\epsilon + U$. For large U, the properties of the Hubbard system depend on the filling. At one electron per site, the system is a Mott insulator with the nearest-neighbor spins coupled antiferromagnetically. (This situation corresponds to the parent La_2CuO_4.) Doping with holes introduces holes in the oxygen p band. Each such spin ($\frac{1}{2}$) is strongly and antiferromagnetically coupled to a copper d^9 spin, and the two form a singlet. This composite singlet can be thought of as a spinless hole in the Hubbard band. Several types of spin–liquid ground states, depending on the spin-pairing amplitudes and relative phases, have been proposed by many workers. Some of these are the uniform RVB phase, and the spiral, chiral, and flux phases. One strategy has been to explore possible spin–liquid solutions in the undoped Mott insulator (the actual ground state solution is a Néel antiferromagnet), hoping that they describe the actual spin–liquid ground states in the presence of holes.

9 Ferroics

As discussed earlier, ferromagnetic materials show the characteristic hysteresis loop in the relationship between magnetization and magnetic field. The electrical analogue of a ferromagnet ic a *ferroelectric,*which shows a hysteresis loop in the relationship between polarization and electric field and exhibits spontaneous polarization in the absence of an external electric field.[38] Ferroelectric materials possess permanent dipole moments, the dipoles arising from the absence of a center of symmetry. If energy considerations favor an antiparallel arrangement of the permanent dipoles on adjacent planes instead of a parallel arrangement, such a crystal is called an *antiferroelectric.* Both ferroelectric and antiferroelectric materials show a dielectric constant anomaly at a critical temperature, but only the former exhibit the polarization–electric field hysteresis loop.

Following Newnham and Cross,[39] a ferroic may be defined as a material possessing two or more orientation states or domains that can be switched from one to another through the application of one or more appropriate forces. In a *ferromagnet,* the orientation state of magnetization (M) in domains can be switched by the application of a magnetic field (M). In a *ferroelectric,* the orientation state of spontaneous electric polarization (P) can be altered by the application of an electrical field (F). In a *ferroelastic,* the direction of spontaneous strain (ϵ) in a domain can be switched by the application of mechanical stress (σ). Such transitions are described as ferroic transitions. In all these cases, the boundaries of domains are moved by the application of force that accomplishes change in orientation. $BaTiO_3$ is a typical ferroelectric, CrO_2 is a ferromagnet, and $CaAl_2Si_2O_8$ is a ferroelastic. The properties for which directionality changes in the foregoing examples—namely, electric polarization, magnetic polarization, and elastic strain—are primary quantities in the sense that their magnitudes directly determine the free energy of the system. Ferroics governed by the switchability of these properties are therefore called *primary ferroics.* In *secondary ferroics,* these properties occur as induced quantities. The orientation states differ in derivative quantities, which characterize the induced effects. Thus, the induced electric polarization is characterized by dielectric susceptibility k_{ij}, induced magnetic polarization by magnetic susceptibility χ_{ij}, and the strain induced by the elastic compliance, C_{ijkl}. The orientation states in secondary ferroics, therefore, differ in k_{ij}, χ_{ij}, and C_{ijkl}; these are tensor quantities, and the rank of the tensor is equal to the number of subscript(s). The induced effects such as polarization or magnetization can also result from cross-coupled effects, such as stress-induced polarization *(piezoelectric)* and stress-induced magnetization *(piezomagnetic),* or as a combined effect of two types of field such as in *elastoelectric* or *elastomagnetic* effects. The directional change can then be visualized to occur in the corresponding derivative quantities. Table 8 lists different types of ferroic effects. The classification scheme of ferroics into primary, secondary, and so on is of thermodynamic origin.

Remembering that both spontaneous and induced effects must be taken into account, we write the difference in the Gibbs free energies of two orientation states as follows:

$$\Delta G = \Delta P_i^{s} E_i + \Delta M_i^{s} H_i + \Delta \epsilon_{ij}^{s} \sigma_{ij}$$
$$+\tfrac{1}{2}[\Delta k_{ij} E_i E_j + \Delta \chi_{ij} H_i H_j + \Delta C_{ijkl} \sigma_{ij} \sigma_{kl}]$$
$$+2[\Delta \alpha_{ij} \mathrm{E}_i \mathrm{H}_j + \Delta d_{ijk} E_i \sigma_{jk} + \Delta Q_{ijk} H_i \sigma_{jk}]$$

where Δ represents the difference in the relevant quantity between orientation states II and I and the superscript s stands for spontaneous quantities. That is, $\Delta \epsilon_{ij}^{s} = \epsilon_{ij}^{s}(\mathrm{II}) - \epsilon_{ij}^{s}(\mathrm{I})$, and so on, are the differences in the ij components of $\epsilon^{s}, \ldots$ in states II and I. When no external force is acting on the system, $\Delta G =$

Table 8 Typical Primary and Secondary Ferroics

Ferroic Type	Ferroic Property	Switching Field	Example
Primary			
Ferroelectric	Spontaneous polarization	Electric field	$BaTiO_3$
Ferromagnetic	Spontaneous magnetization	Magnetic field	CrO_2, γ-Fe_2O_3
Ferroelastic	Spontaneous strain	Mechanical stress	$CaAl_2Si_2O_8$
Secondary			
Ferrobielectric	Dielectric susceptibility	Electric field	$SrTiO_3$, $NaNbO_3$
Ferrobimagnetic	Magnetic susceptibility	Magnetic field	NiO
Ferrobielastic	Elastic compliance	Mechanical stress	α-Quartz
Ferromagnetoelectric	Magnetoelectric coefficients	Magnetic field and electric field	Cr_2O_3

0 and the orientation states are energetically degenerate. Several possible ferroic phenomena become evident from the preceding equation, depending on the dominance of particular terms. In a material that has a large value of spontaneous polarization, other terms become unimportant and the free energy in an electric field is governed by $\Delta p^s E$. Similar terms can be written for primary ferroics involving spontaneous magnetization and spontaneous strain when they are the dominant quantities and interact with the corresponding external fields. When $\Delta P_i^s = \Delta M_i^s = \Delta\epsilon_{ij}^s = 0$, the quantities that determine the ΔG values arise from terms in the two sets of square brackets in the equation.

A ferroelectric transition from one orientation state to another (observed by means of the hysteresis loop) is electrically a first-order transition. The order of the transition between domain states in ferroics is the sum of the exponents of the field terms in the free energy expression. Ferrobielastic, ferrobimagnetic, ferroelastoelectric, ferromagnetoelectric, and other such transitions are all second-order. The various coefficients are themselves interdependent. For example, spontaneous polarization is associated with a field, which in turn produces a strain through strong coupling to the lattice, thus activating electrostrictive or piezoelectric coefficients. Thus, wherever a spontaneous polarization exists, spontaneous strain also is present, and vice versa. Whenever spontaneous polarization and spontaneous strain produce orientation states fully independent of each other, both ferroelastic and ferroelectric properties can be fully realized and the materials can be referred to as fully ferroelectric and fully ferroelastic materials. Similar relations can be visualized from the interdependence of the other quantities.

Ferroelectric materials transform to the paraelectric state (where dipoles are randomly oriented) at some temperature, just as ferromagnetic materials transform to the paramagnetic state and ferroelastic materials to the twin-free state. The transitions are characterized through order parameters—that is, characteristic properties parametrized in such a way that the resulting quantity is unity for the ferroic state at a temperature sufficiently below the transition temperature and zero in the nonferroic phase beyond the transition temperature. Polarization, magnetization, and strain are the *proper order parameters* for the ferroelectric, ferromagnetic, and ferroelastic transitions, respectively. Whenever transitions are governed by the expected variations of these order parameters, they are called *proper ferroics.* Ferroics for which the order parameter does not represent a "proper" property are called *improper ferroics.* Accordingly, in terbium molybdate, the order parameter is a condensed optic mode. The optic mode causes a spontaneous strain, which in turn causes a spontaneous polarization through piezoelectric coupling. An improper primary ferroic is distinguished from a true secondary ferroic by the appearance of the prefix "ferro" with the primary ferroic quantity only, not for both the coupled quantities. Thus, the term "magnetoferroelectric" implies that the material (e.g., Cr_2BeO_4) is an improper ferroic, whereas a "ferromagnetoelectric" material (e.g., Cr_2O_3) is a secondary ferroic.

Among the 32 crystal classes, 11 possess a center of symmetry and are centrosymmetric, hence do not possess polar properties. Of the 21 noncentrosymmetric classes, 20 exhibit electric polarity when subjected to a stress and are called *piezoelectric;* one of the noncentrosymmetric classes (cubic 432) has other symmetry elements that combine to exclude piezoelectric character. Piezoelectric crystals obey a linear relationship between polarization P and force F, $P_i = g_{ij}F_j$ where g_{ij} is the piezoelectric coefficient. An inverse piezoelectric effect leads to mechanical deformation or strain under the influence of an electric field. Ten of the 20 piezoelectric classes possess a unique polar axis. In nonconducting *pyroelectric,* crystals, a change in polarization can be observed by a change in temperature. If the polarity of a pyroelectric crystal can be reversed by the application of an electric field, we call such a crystal a *ferroelectric.* Knowledge of the crystal class is therefore sufficient to establish the nature of a solid (i.e., piezoelectric or pyroelectric), but reversible polarization is a necessary condition for ferroelectricity. All ferroelectric materials are piezoelectric, but the converse is not true. For example, quartz is piezoelectric, but not ferroelectric.

An important characteristic of ferroelectric materials is that the dielectric constant obeys the *Curie–Weiss law,* similar to the equation relating magnetic susceptibility to temperature in ferromagnetic materials. The temperature variation of dielectric constant of $BaTiO_3$ is shown in Figure 24a. Above 393 K, $BaTiO_3$ becomes paraelectric (dipoles are randomized). Polycrystalline samples show less marked changes at the transition temperature.

$BaTiO_3$, which crystallizes in the perovskite structure, has cubic symmetry above 393 K with Ba^{2+} in the body center and TiO_6 octahedra in the corners. It undergoes the following transformations: to a tetragonal structure at 393 K, to an orthorhombic structure at 273 K, and to a rhombohedral structure at 183 K (Fig. 24, top). Relative to the cubic phase, elongation occurs along one of the edges ($\langle 100 \rangle$ direction) in the tetragonal phase, along one of the face diagonals ($\langle 110 \rangle$ direction) in the orthorhombic phase, and along one of the body diagonals ($\langle 111 \rangle$ direction) in the rhombohedral phase. The Ti^{4+} ion moves in these three directions successively as the crystal is cooled from the cubic phase (which has Ti^{4+} in the center of the octahedron; also see Fig. 21 in Part I). Besides the dielectric constant and polarization, heat capacity and other properties show anomalous changes at the three phase transitions of $BaTiO_3$.

Polarization of a ferroelectric material varies nonlinearly with the applied electric field, the *P–E* behavior being characterized by a *hysteresis loop.* The hysteresis loop has its origin in the rearrangement of domains under the influence of an applied electric field. Generally, the domains are randomly distributed, giving a net zero polarization. Under an applied field or mechanical stress, favorably oriented domains grow at the expense of the less favorably oriented domains until a single domain configuration is obtained. The domain structure itself is related to the crystallography of the ferroelectric phase with respect to the paraelectric phase. For example, in the tetragonal phase of $BaTiO_3$, adjacent domains may have their polar axes making angles of 90° or 180°.

A large number of materials exhibit other interesting ferroic properties besides ferroelectricity. Some examples of typical paired properties are as follows:

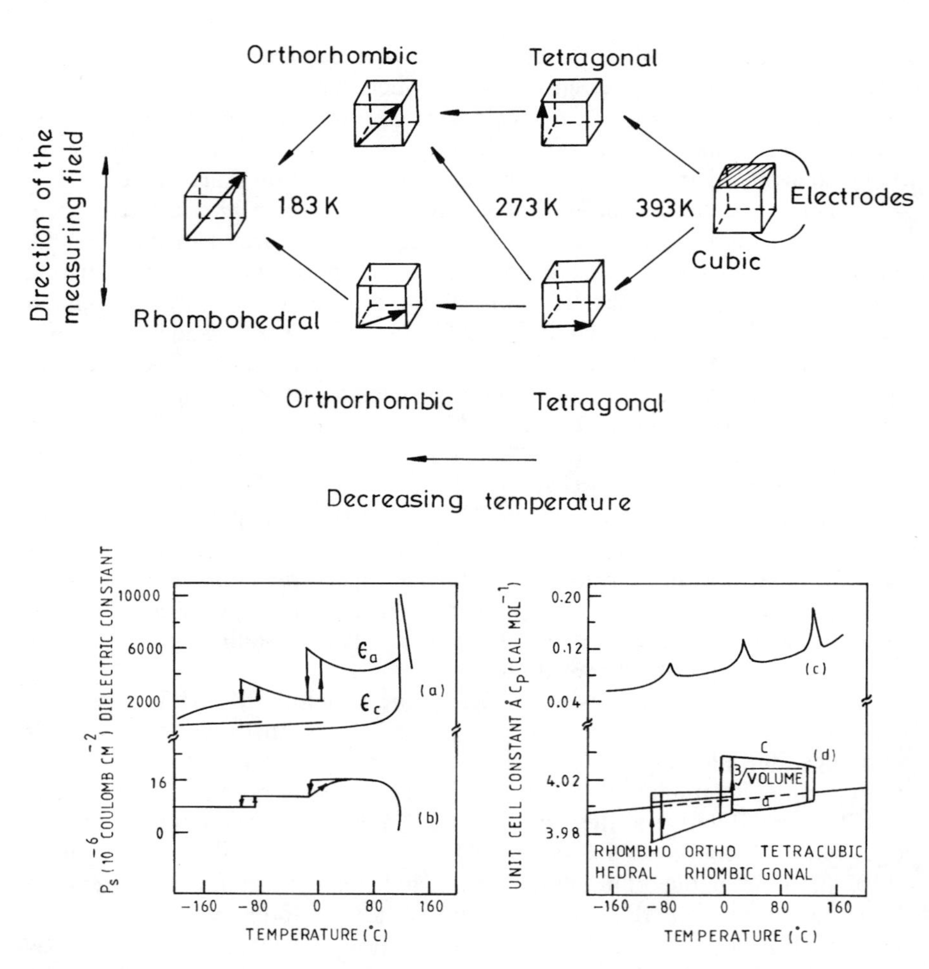

Figure 24. Phase transitions in $BaTiO_3$. *Top:* Changes in the orientation of the polar axis when an electric field is applied along the pseudocubic (001) direction; arrows indicate polar axes. *Bottom:* Curves indicating changes in dielectric constant (a), spontaneous polarization (b), heat capacity (c), and lattice dimensions (d).

Ferroelectric–ferroelastic, $Gd_2(MoO_4)_3$, $KNbO_3$
Ferroelectric–antiferromagnetic, $YMnO_3$, $HoMnO_3$
Ferroelectric–ferromagnetic, $Bi_9Ti_3Fe_5O_{27}$
Antiferroelectric–antiferromagnetic, $BiFeO_3$, $Cu(HCOO)_2{\cdot}4H_2O$
Ferroelectric–semiconducting, reduced $SrTiO_3$, $YMnO_3$
Ferroelectric–superconducting, $SrTiO_3$

The class of materials known as *relaxor ferroelectrics* shows diffuse transitions (extending over a large temperature range) and is associated with large changes in dielectric constant. The magnitudes of the peak dielectric constants decrease with increasing frequency and the peaks shift to higher temperatures, suggesting features of a relaxational mechanism. Optical and electrical measurements show that microdomains arising from compositional fluctuations are responsible for the phenomenon. Several relaxor ferroelectrics are known to possess low thermal expansivities, some of the good examples being $Pb(Mg_{0.33}Nb_{0.67})O_3$, 0.9 $Pb(Mg_{0.33}Nb_{0.67})O_3$, 0.1 $PbTiO_3$ and $Pb(Zn_{0.33}Nb_{0.67})O_3$.

The dependence of refractive index on the electric field or the lattice polarization is commonly called the *electro-optic effect.* Large values of the electro-optic coefficients are associated with large linear susceptibilities, ferroelectrics being the obvious candidates. Ferroelectric–electro-optic crystals can be centric, as in $LiNbO_3$, $LiTaO_3$, and $Ba_2NaNb_5O_{15}$, or acentric, as in KH_2PO_4(KDP), $NH_4H_2PO_4$(ADP), and CsH_2AsO_4. Among the advantages of electro-optic composites over single crystal materials are ease of fabrication, low cost, and better control of optical axis orientation. An electro-optic ceramic is a nearly transparent ferroelectric material, $Pb(Zr_{1-x}Ti_x)O_3$ (PZT) being the best-known example. The composition (Zr/Ti ratio) can be varied and Pb partly substituted by other ions (La, Bi, etc.). Hot pressed PZT ceramics show electrically controlled light scattering (e.g., PBiZT) and birefringence (e.g., PLZT and PBiZT).

Ferroelectric tungsten bronzes are an interesting family of oxides exhibiting a wide range of ferroelectric, piezoelectric, and electro-optic properties. The properties achieved are equal to or better than those of $BaTiO_3$ and PLZT. The following are the four important classes of bronzes:

1. Crystals exhibiting strong transverse effects (4*mm* symmetry), including $Sr_{1-x}Ba_xNb_2O_6$, $Sr_2KNb_5O_{15}$, and $Ba_6Ti_2Nb_8O_{30}$.
2. Crystals exhibiting strong longitudinal effects (4*mm* symmetry), including $K_3Li_2Nb_5O_{15}$ and $Ba_{2-x}Sr_xK_{1-y}Na_yNb_5O_{15}$.
3. Crystals exhibiting strong transverse and longitudinal effects (*mm*2 symmetry), including $Sr_{1.9}Ca_{0.1}NaNb_5O_{15}$ and $Sr_4Ca_2Ti_2Nb_8O_{30}$.
4. Crystals exhibiting strong longitudinal effects (*mm*2 symmetry), including $Pb_2KNb_5O_{15}$ and $Pb_{0.6}Ba_{0.4}Nb_2O_6$.

10 Results from Empirical Theory

A better understanding of the properties of certain complex transition metal oxides has become possible in the light of the theoretically calculated electronic band structure.[40–42] This applies particularly to such low-dimensional transition metal oxide systems as the molybdates and tungstates. Structures of these oxides involve MO_6 octahedra present in corner- and edge-sharing geometries, giving them a low-dimensional character. Electronic properties of these materials derive from the interactions of the transition metal *d* orbitals and the oxygen *p* orbitals. Since the transition metal ions are present in their highest formal oxidation state, the metal levels influencing the electronic properties are the low-energy t_{2g} states. Recognition of this feature has led to the development of simple and reliable methods for investigating energy band structures. These methods are generally based on the extended Hückel tight binding (EHTB) approximation and have been developed by Hoffmann, Whangbo, and others.[40–42] The importance of these methods lies in their ability to bridge results from the ligand field treatment of molecular complexes with the results of rigorous band structure calculations of crystalline solids. EHTB approximation cannot provide optimized geometries, but it gives reliable energy dispersions in the valence band region in known structures and is insensitive to small inaccuracies in atomic parameters. If properly parametrized, the EHTB technique provides a reasonable density of states. In addition, the method gives a reliable Fermi surface and is convenient for investigating band structures and visualizing the dispositions of metal *d* orbitals. Several important experimental results have been examined in the light of these calculations, as illustrated by the following examples.

Red molybdenum bronzes of the formula $A_{0.33}MoO_3$ (A = K, Rb, Cs, or Tl) possess two-dimensional structures, wherein the MoO_3 layers consist of edge-sharing MoO_6 octahedra. They are present in eclipsed and staggered double chains, which are humped, and the cations are located between the layers. It was not clear whether these bronzes were semiconductors with a band gap, since semiconducting behavior can result from partially filled bands when they are Mott insulators. The repeat unit in these bronzes is an Mo_6O_{18} slab, and the t_{2g} bands of this system are shown in Figure 25.[43] Since there are ($0.33 \times 6 \approx 2$) electrons, they fill the **a** band completely, and there would be an indirect gap of

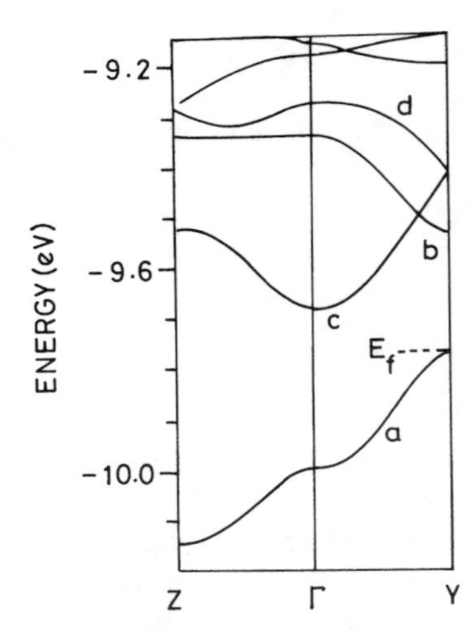

Figure 25. Energy dispersion of the t_{2g} block bands in Mo_6O_{18} slabs as in $Tl_{0.3}MoO_3$. (After Ganne et al.[43])

0.12 eV to band **c**. The bands are quite dispersive unlike the situation in a Mott insulator. The red bronzes are therefore considered to be semiconductors based on this analysis. Hump octahedra present in these chains do not contribute to the lower t_{2g} bands of the bronzes, but they influence the structures through strong intrachain distortions.

Blue bronzes of the formula $A_{0.3}MoO_3$ (A = K, Rb, or Tl) are metallic at room temperature. They contain $Mo_{10}O_{30}$ layers (10 MoO_6 octahedra in a repeat unit), but the hump octahedra do not contribute to the low energy bands. The t_{2g} bands of $Mo_{10}O_{30}$ block are shown in Figure 26a, where the dashed line indicates electron filling (three electrons). In Figure 26b, Fermi surfaces are shown for the two bands [separately in (i) and (ii) and together in (iii)].[44] The nested portions of the Fermi surfaces are clearly evident in (iii). The Fermi surfaces suggest that blue bronzes are one-dimensional metals, in agreement with experiment. An incommensurate CDW transition occurs, leading to the formation of a semiconductor around 180 K.

The Magnéli phase, Mo_8O_{23}, is metallic above 360 K; it forms an incommensurate phase between 360 and 285 K and transforms to a commensurate phase at 285 K. The MoO_6 octahedra in Mo_8O_{23} are all distorted, and the structure can be considered to be built up of Mo_4O_{15} slabs and Mo_4O_{14} chains on the one hand and as three-dimensional Mo_8O_{23} on the other. In the first case, the band structure would be a combination of the band structures of the slabs and chains, provided there is no appreciable interaction between them. The unit cell

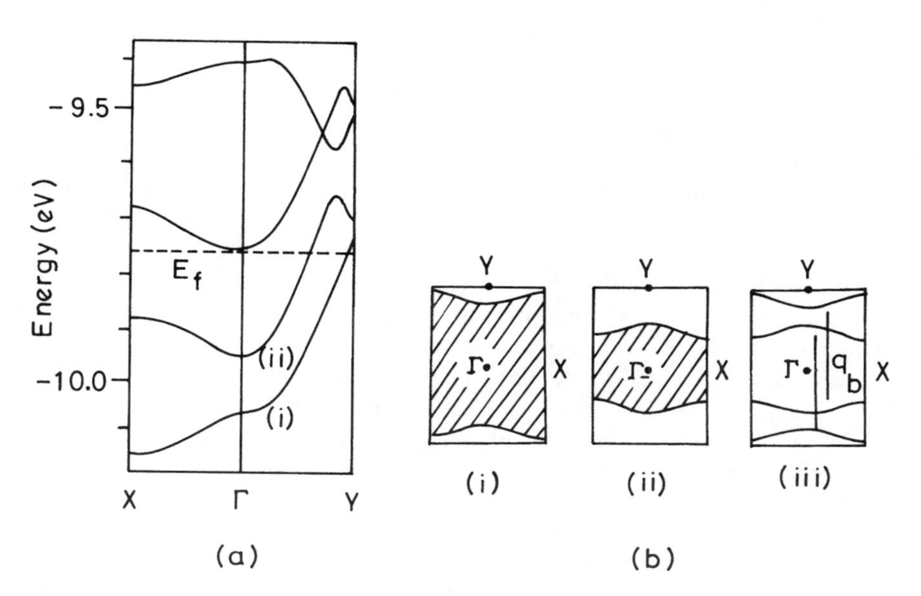

Figure 26. (a) Energy dispersion of the t_{2g} bands in $Mo_{10}O_{30}$ blocks. (b) Fermi surfaces of the two lowest bands (i) and (ii) and the combined Fermi surface. (Following Whangbo and Schneemeyer.[44])

corresponds to $(Mo_8O_{23})_2$ with four electrons, which fill the two bottom-most bands arising from the Mo_4O_{15} slabs. Calculations have revealed the presence of two rather closely spaced bands in addition to a band of even lower energy. Thus, Mo_8O_{23} turns out to be a semimetal. As shown by Sato et al.,[45] electronic instability in Mo_8O_{23} is not caused by the nesting of Fermi surfaces but by a classical mechanism of concerted pairwise rotation of MoO_6 octahedra within the layers.

Structures of disphosphate tungsten bronzes (DPTB) with the general formula $A_x(P_2O_4)_4\,(WO_3)_{4m}$ (x varying between 1 and 2 and m generally equal to 4) are related to the structure of Mo_8O_{23}. DPTBs are quasi-two-dimensional metallic materials, and their quasi-two-dimensional behavior is due not to two-dimensional bands but to one-dimensional bands present in orthogonal directions. The band structure of $Li_{0.9}Mo_6O_{17}$ (the lithium purple bronze) proposed by Whangbo et al[44] is interesting. It exhibits a pseudo-one-dimensional metallic character while the structure is three-dimensional. The low energy bands originate from Mo_4O_{18} chains possessing two essentially flat bands. They are partially filled in one direction, thereby exhibiting instabilities at temperatures lower than 25 K.

Analysis of the two-dimensional metals Γ- and η-Mo_4O_{11} is complicated by the presence of step layers. In these structures the Mo_6O_{22} layers are linked via MoO_4 tetrahedra. Monophosphate tungsten bronzes (MPTB) of the general formula $A_x(PO_2)(WO_3)_4(WO_3)_6$ also possess step layers. Band structure calculations of the Magnéli phases as well as MPTBs by Whangbo and others[44] reveal the presence of complex Fermi surfaces that are nested in pieces. Thus, formation of charge-density waves leads to pockets in the Fermi surface, giving rise to anomalies in low temperature conductivity. For example, Mo_4O_{11} exhibits a conductivity anomaly at 30 K.

Purple molybdenum bronzes, AMo_6O_{17} (A = Na, K), are two-dimensional metals. Unlike the case of the blue bronzes, two-dimensional metallic character in these bronzes persists even after CDW formation, and the nature of carriers changes from electrons in the normal state (above the CDW transition at 120 K) to holes in the CDW state. About 50% of the carriers are removed below 120 K. The crystal structure of KMo_6O_{17} can be considered to be made up of hexagonal Mo_6O_{17} layers built from Mo_4O_{21} units, the latter consisting of an array of four MoO_6 octahedra sharing three axial corner oxygens. Band structures of KMo_6O_{17} and Mo_2O_9 layers are very similar.[46]

The three-dimensional oxide $Na_{0.25}TiO_2$ has a structure consisting of Ti_2O_8 double octahedral chains from which layers and double chains are built. The pseudo-one-dimensional character of $Na_{0.25}TiO_2$ originates from such a structure, giving rise to an interesting metal–insulator transition at 630 K, followed by a transition to an incommensurate phase below 430 K. On the other hand, $Li_{0.3}MoO_3$ is a violet-blue bronze and, unlike the Na and K bronzes, has a triclinic structure and is semiconducting. Its structure consists of Mo_6O_{24} chains, which determine the band structure. The band structure of $Li_{0.3}MoO_3$ suggests that it has to be semiconducting in all directions.

Hückel calculations have been employed extensively in other approaches such as the angular overlap model developed by Burdett and co-workers.[42] Several problems—in particular, stabilities of crystal structures, pressure- and temperature-induced transitions, and dynamical pathways in reactions—have been analyzed using angular overlap models. Burdett and co-workers[42] have examined the electronic control of rutile structures and the stability of the defect structure of NbO by molecular orbital methods. In the case of NbO, the structure is stable at d^3, involving the formation of square-planar Nb in which the nonbonding Nb($4d_{z^2}$) orbital mixes with the Nb($5s$) orbital and enhances the Nb—Nb bonding. On the basis of studies using pair potentials, the defect structure is attributed to local repulsive forces. The method of moments has been employed to probe the cause of defect ordering in perovskites (e.g., $Ca_2Mn_2O_5$). Distortions in the NiO_6 octahedra in $BaNiLnO_5$ have been explained on the basis of a model that combines results from MO theory, bond structure, and atom–atom potential arguments.

11 Understanding Electronic Structure from Electron Spectroscopy Combined with Empirical Theory

Electronic structures of transition metal and rare earth compounds have been extensively studied using techniques of electron spectroscopy, which have the advantage of directly probing the electronic structure. Such investigations have provided estimates of various interaction strengths that determine the electronic structure of oxides with strongly interacting electrons. Let us consider the classic case of NiO. Several efforts have been made to explain the insulating behavior of NiO, particularly from band structure calculations within the single particle approximation. These studies assume that the origin of the insulating property of NiO is not related to intra-atomic correlation effects. An insulating state of NiO has been obtained within such an approach, but the band gap was considerably lower than that found experimentally. Mott had pointed out many years ago the need to go beyond the single particle model and to take the Coulomb correlations into account in explaining the properties of transition metal compounds. The Hubbard Hamiltonian gives ground states that are metallic or insulating, depending on the relative magnitude of the hybridization and the Coulomb interaction strengths. Such a model, wherein only the transition metal d^n energy levels are taken into account and the ligand $2p$ levels are neglected altogether is of very limited use. While this picture works to some extent for compounds of the early transition metals, in the case of cobalt, nickel, and copper compounds, the band gap seems to be directly related to the electronegativity of the anion, indicating the role played by the ligand levels.

A highly successful approach to understanding the electronic structures of transition metal compounds is due to Zaanen, Sawatzky, and Allen (ZSA).[47] These investigators proposed a phase diagram for the insulating/metallic properties in particular, and the electronic structure in general, of transition metal compounds in terms of the bare charge-transfer energy Δ, between the ligand $2p$ and metal $3d$ levels (i.e., the energy associated with $d^n \rightarrow d^{n+1}L^1$ transition with L^1 representing a hole in the ligand levels) and the on-site Coulomb correlation energy U_{dd}, establishing thereby that the cationic and anionic energy levels have equally important roles. These workers calculated the gap arising in the charge excitation spectrum of a single transition metal impurity hybridizing with a ligand-derived band within the charge-transfer impurity model. The interaction parameters characterizing the system in this case are (1) the charge-transfer energy Δ, (2) the intra-atomic Coulomb interaction strength U_{dd}, within the $3d$ manifold $[U_{dd} = E(d^{n+1}) + E(d^{n-1}) - 2E(d^n)]$, and (3) the hybridization strength t between the $3d$ orbital and the ligand level. With this model, ZSA calculated the phase line separating metals from insulators in terms of U_{dd}/t and Δ/t. The ZSA diagram could explain differences between compounds of the early transition metals and of the late transition metals, as well as the dependence of the conductivity gap, within a related series of compounds of any transition metal, on the electronegativity of the anion. The limitation of the ZSA approach was the neglect of transition metal translational symmetry implicit in the impurity model. While such a model is expected to describe systems with small hopping interactions reasonably well, the large hybridization strengths between the transition metal $3d$ and the oxygen $2p$ states in oxides indicate the need to go beyond the impurity model. Furthermore, such an impurity model cannot describe the magnetic properties of these systems.

Figure 27 shows the ZSA-type phase diagram obtained by Sarma et al.,[48] who took into account the translational symmetry as well. This phase diagram is very similar to the original ZSA phase diagram and suggests the existence of different regimes of insulating and metallic properties giving rise to charge-transfer insulators, Mott–Hubbard insulators, p-d type metals, and d-band metals (A–D in Fig. 27). The phase diagram also identifies a new insulating region (E in Fig. 27) in addition to the two insulating regions (A and B) obtained by ZSA.

The features of the different regions of the phase diagram are as follows. Region A, corresponding to the charge-transfer insulators, is characterized by $U_{dd} \gg \Delta > \Delta_c(U)$, where $\Delta_c(U)$ is a critical Δ value of the bare charge-transfer energy depending on U_{dd}. The electronic structure of this region can be schematically represented by the one-particle excitation spectrum shown in Figure 28a.[49] Thus, the essentially oxygen-$2p$-derived band appears between the lower and upper Hubbard bands. This region has the lowest energy charge excitation primarily involving a configuration $d^n p^m$ going to a $d^{n+1}p^{m-1}$ configuration, and the band gap in this region is mainly dictated by Δ. Late transition metal compounds are expected to belong to this regime. In contrast, the Mott–Hubbard region (B in Fig. 27) is characterized by $\Delta \gg U_{dd} > \Delta_c(U)$, where $U_c(\Delta)$ is the critical U_{dd} value depending on Δ. In this limit, one obtains the schematic electronic structure shown in Figure 28*b*. Here the lowest energy charge excitation involves a $d_i^n d_j^n$ state being excited to $d_i^{n+1} d_j^{n-1}$ state; the band gap is seen to be primarily controlled by the on-site Coulomb correlation energy U_{dd} (see Fig.

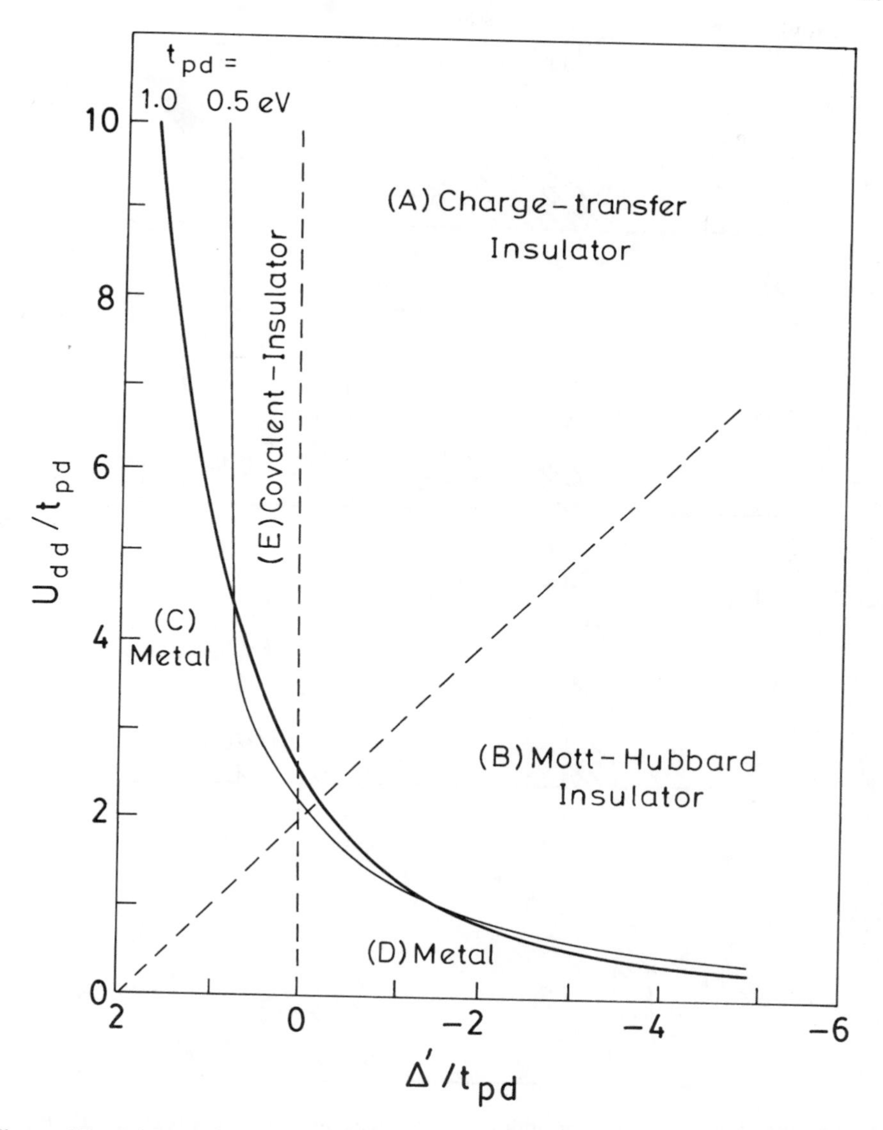

Figure 27. Phase diagram for transition metal compounds in the space of U/t and Δ/t showing the parameter ranges for the various metallic and insulating states; for description of regions A–E, see text. (From Sarma et al.[48])

28b). This is the expected situation for the early transition metal compounds. From Figure 28a, it seems obvious that for a small Δ in the $U_{dd} \gg \Delta$ regime, the upper Hubbard band overlaps the top of the oxygen-2p-derived band, thereby giving rise to a metallic ground state. This is the origin of the metallic region C in Figure 27. In the other limit of $\Delta \gg U_{dd}$ (see Fig. 28b), a sufficiently small

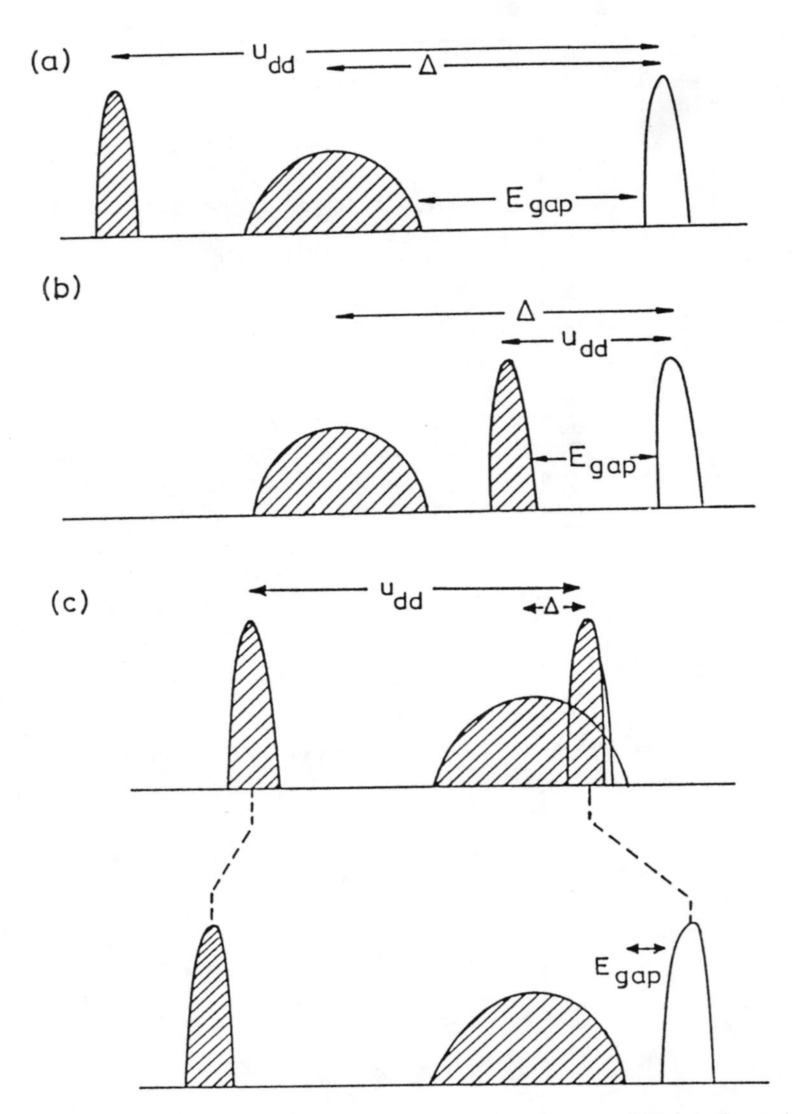

Figure 28. Schematic electron addition (open areas) and removal (hatched areas) spectra representing the electronic structure of transition metal compounds for different regimes of parameter values: (a) charge-transfer insulator regime with $U > \Delta$, (b) Mott–Hubbard insulator with $\Delta > U$, and (c) covalent insulator where the band gap is driven by the hybridization strength. (From Rao et al.[49])

U_{dd} leads to a metallic ground state (D in Fig. 27) that is due to the overlap of the lower and upper Hubbard bands.

The foregoing descriptions are somewhat modified in the presence of finite hybridization strength, mixing the transition metal d and oxygen p states. It is to be noted however that both charge-transfer and Mott–Hubbard insulators retain a finite gap in the charge excitation spectrum in the limit of vanishing hybridization strength. Region E is the covalent insulator regime, characterized as a state that would have been a metal in the absence of any hybridization mixing, but is driven to the insulating state because large hybridization matrix elements are present. Figure 28c shows the schematic electronic structure for the covalent insulator, which in the absence of any hybridization strength would have been a metal because of the presence of small Δ, but is driven to an insulating state by the sizable hybridization strength.

The ZSA phase diagram and its variants provide a consistent description of the overall electronic structure of stoichiometric and ordered transition metal compounds. Within the preceding description, the electronic properties of transition metal oxides are primarily determined by the values of Δ, U_{dd}, and t. Several electron spectroscopic (photoemission) investigations have been undertaken to estimate the interaction strengths. Valence band as well as core level spectra have been analyzed for a large number of transition metal and rare earth compounds. Calculations of the spectra have been performed at different levels of complexity, but generally within an Anderson impurity Hamiltonian. In the case of metallic systems, the situation is complicated by the presence of a continuum of low energy electron–hole excitations across the Fermi level. In the case of the rare earths and their intermetallics, these excitations play an important role. This effect is particularly important for the valence band spectra. The spectroscopy of rare earth compounds and their electronic properties have been reviewed in detail in the literature.

Analysis of the valence band spectrum of NiO has helped to understand the electronic structure of transition metal compounds. It is to be noted that the crystal field theory cannot explain the features over the entire valence band region of NiO. To explain the spectral features satisfactorily, therefore, it becomes necessary to take explicit account of the ligand (i.e., O $2p$)–metal (Ni $3d$) hybridization and the intra-atomic Coulomb interaction U_{dd}. This has been done by approximating bulk NiO by a cluster $(NiO_6)^{10-}$. The ground state wave function ψ_g of this cluster is given by,

$$\Psi_g = a \mid d^2L^0 > +\mathrm{b} \mid d^1L^1 > + c \mid d^0L^2>,$$

where $|d^2\mathrm{L}^0>$ (in the hole configuration) denotes the ionic $Ni^{2+}O^{2-}$ (or the $3d^82p^6$ electron configuration) state, $|d^1L^1>$ represents the $Ni^{1+}O^{1-}$ (or the $3d^92p^5$) state, and the $|d^0L^2>$ is the $3d^{10}2p^4$ state. Hybridization between the Ni $3d$ and 0 $2p$ states mixes the three states, to achieve a substantial lowering of the energy of the ground state compared to the ionic state $|d^2L^0>$ or the $Ni^{2+}O^{2-}$ state. (Note the importance of covalence.) Assuming the contribution of $|d^0L^2>$ in Ψ_g to be small, ignoring its effects, the ground state is found to have the symmetry $^3A_{2g}$. This implies that the $3d^8$ configuration ($|d^2L^0>$) with $|t^3_{2g\uparrow}t^3_{2g\downarrow}e^2_{g\uparrow}, {}^3A_{2g}>$ hybridizes with various $|d^1L^1>$ configurations having the same symmetry to form the ground state wave function. Dipole selection rules permit the final state following the photoemission process to have one less electron and ^2E_g, $^2T_{1g}$, and $^4T_{1g}$ symmetries. The corresponding $3d^7$ (or $|d^3L^0>$) configurations are given by $|t^6_{2g}e^1_g, {}^2E_g>$, $|t^5_{2g}e^2_g, {}^2T_{1g}>$, and $|t^5_{2g}e^2_g, {}^4T_{1g}>$; the $|d^2L^1>$ and $|d^1L^2>$ states also are chosen with the proper symmetries. Estimating the transfer integrals from band structure calculations and using the various energies (e.g., U_{dd} and Δ) as fitting parameters, the photoemission spectrum for the cluster has been calculated. From the best fit to the experimental spectrum, the charge-transfer energy Δ and the Coulomb energy U_{dd} are estimated to be about -3.5 and 9 eV, respectively. These estimates suggest that NiO is a charge-transfer insulator in terms of the phase diagram shown in Figure 27 (regime A).

Use of core level spectra to understand the electronic structure of oxides is illustrated by the Cu $2p$ core level spectrum of insulating cuprates (e.g., La_2CuO_4), which are the parent compounds for the high T_c superconducting cuprates. These cuprates are modeled by the CuO_2 square-planar sheet. At the simplest level, one Cu $3d_{x^2-y^2}$ orbital and four oxygens (with only p_x and p_y orbitals) around it are taken into account. One then includes within the Cu $3d$ levels the hybridization matrix elements t and the bare site energies ϵ, as well as the Coulomb interaction strength U_{dd}. Insulating cuprates have one hole per Cu site as in stoichiometric La_2CuO_4. This property implies that U_{dd} plays no role in this case. Since the Cu $3d_{x^2-y^2}$ orbitals belong to CuO_4 cluster, only the linear combination, $0.5(p^1_x - p^2_y - p^3_x + p^4_y)$, belongs to the b_{1g} representation. The electronic structure of the CuO_4 cluster within the foregoing approximation will be controlled entirely by the interaction between the Cu $3d_{x^2-y^2}$ and the oxygen-derived b_{1g} orbitals. The relevant parameters in the model are therefore the energy difference and the hybridization strength between these two levels and the core hole–valence hole repulsion strength U_{dx}. The value of Δ would be equal to $\epsilon_p - \epsilon_d$ in the absence of oxygen–oxygen interaction; inclusion of oxygen–oxygen near-neighbor interactions shifts Δ without making any qualitative change in the results. The effective hybridization strength, t, between the two states is $2t_{pd}$, where t_{pd} is the interaction strength between the Cu $3d_{x^2-y^2}$ and the oxygen p^1_x orbitals.

Experimental Cu $2p$ core level spectra of cuprates exhibit a two-peak structure with one peak at about 933 eV (main peak) and the other at 942 eV (satellite). Although the model described in the preceding paragraph is simplistic, it provides an analytical expression for the Cu $2p$ core level spectrum. The spectrum is calculated as a function of Δ, t, and U_{dx} to obtain the energy difference between the main Cu $2p$ peak and the satellite ΔE, and the ratio of their intensities, I_s/I_m. In the specific case of La_2CuO_4, I_s/I_m is 0.33 and ΔE is 8.5 eV. To obtain a range of parameters suitable to describe the core level spectrum of La_2CuO_4 within the cluster approximation, one constrains t, the hybridization strength, within the reasonably wide bounds of 1.5 and 3.0 eV, while freely varying Δ and U_{dx}. To fit the experimental spectrum within the experimental uncertainties, Sarma finds that the value of U_{dx} must be between 7.0 and 8.0 eV. The experimental core level spectrum thus provides a reasonably narrow range of estimate of U_{dx}, although the Δ and t values can have a wide range.

The calculation of the core levels of La_2CuO_4 just discussed was performed within a cluster approximation that neglects the dependence of the spectral shape on the oxygen bandwidth. To include the effect of the oxygen-2p-derived band on the spectral shape, the core level photoemission calculation was performed within the impurity model, which includes the O 2p bandwidth (instead of a single O 2p level). The result of such a calculation is shown in Figure 29.[50] Just as in the cluster approximation, several sets of values of the parameters Δ and t can explain the observed photoemission spectrum. For reasonable values of hybridization strength t, the range of values of Δ is such that the Cu $3d_{x^2-y^2}$ appears close to the bottom of the oxygen-2p-derived bandwidth or lies within it. This is the regime of the covalent insulator in terms of region E in Figure 27. The ground state of La_2CuO_4 is associated with a strongly mixed valent character from photoemission evidence. Furthermore, the band gap as well as magnetic properties of this cuprate are described within the preceding model for the same set of parameter values used in the photoemission calculations.

Analysis of the kind just described has been carried out for the core level spectra of many cuprates. From such analyses, it would appear that the cuprates can be described by a narrow range of parameter values characterizing them. The average occupancy of the Cu 3d level is generally fractional, suggesting a strongly mixed valent ground state in all the cuprates. The effect of hole-doping on I_s/I_m in $La_{2-x}Sr_xCuO_4$, $Bi_2Ca_{1-x}Ln_xSr_2Cu_2O_8$, and other cuprates has been

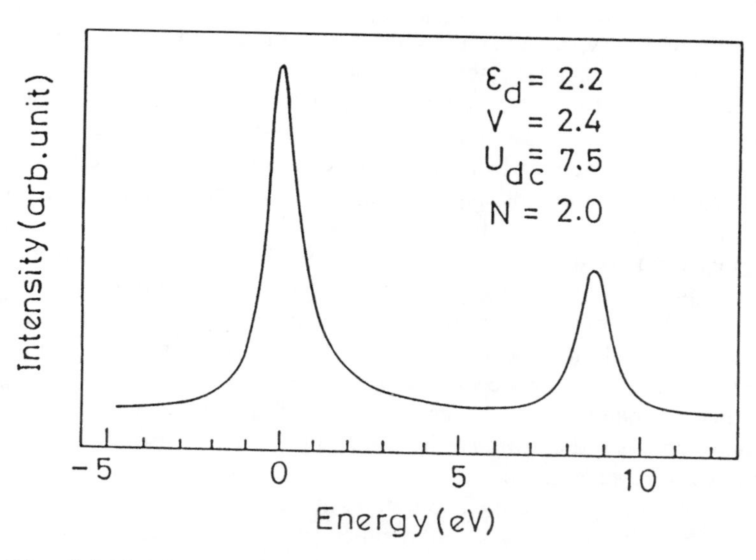

Figure 29. Calculated core level spectrum of La_2CuO_4 within an impurity model for the given parameter values. (From Sarma and Taraphder.[50])

examined in detail; I_s/I_m decreases smoothly with increasing hole concentration, thereby indicating the crucial role of Δ and t in determining the superconducting properties.[51]

We know from the analysis of core level spectra of cuprates that contain only a single hole in the d band per Cu ion that U_{dd} is an irrelevant parameter within an impurity model. However analysis can also be carried out for systems in which U_{dd} plays a significant role. This is illustrated by the analysis of the core level spectrum of $LaCoO_3$ carried out within the impurity model.[52] This oxide is modeled by the $(CoO_6)^{9-}$ cluster, with the transition metal ion being formally in the 3+ oxidation state and in an octahedral crystal field. One must therefore take into account interactions between various configurations such as $|d^6>$, $|d^7L^1>$, and $|d^8L^2>$. Moreover, the $3d$ manifold is split into t_{2g} and e_g subbands as a result of crystal field splitting, and exchange interaction causes these to be further split into up-spin and down-spin levels. The competition between the crystal field splitting, which stabilizes the low spin configuration, and the exchange interaction, which favors the high spin configuration, gives rise to the spin state transition in $LaCoO_3$. The low spin configuration is the ground state of this oxide. Further complication arises because in these cases two hybridization interactions characterize the system, namely, the symmetry-adopted O $2p$ states within the $3d$-derived t_{2g} and e_g states. Taking into account all these interaction strengths, calculations have been carried out for the low as well as the high spin states of $LaCoO_3$. The analysis shows that the results of the calculation within the high spin state are incompatible with the experimentally determined spectrum, whereas the calculation based on the low spin configuration provides a good description. The values of the parameters are as follows: $\Delta = 4.0$, $U_{dd} = 3.4$, and $t = 3.8$ eV. The ground state with these parameters turns out to be a very mixed state, with approximately 38.5% d^6, 45.4% d^7L^1, and 14.5% d^8L^2 characters. Such a ground state with U/t and $\Delta/t \simeq 1$ places $LaCoO_3$ in the mixed character region between the Mott–Hubbard, charge-transfer, and covalent insulators in terms of the phase diagram shown in Figure 27. The electron spectroscopic features have been calculated in such studies for different parameter strengths explicitly appearing in the Hamiltonian, but the various parameters giving the best fit to experiment are by no means unique.

The preceding discussion of the method of extracting various electronic interaction parameters from the analysis of electron spectroscopic results serves to describe the electronic structure of correlated systems. Electron spectroscopies and related methods have been used in different ways to provide important information concerning a wide variety of systems. Thus, extensive investigations have helped in understanding different aspects of mixed valent as well as heavy fermion systems based on rare earths and actinides. High resolution photoemission studies have revealed the opening of the superconducting gap in high T_c cuprates. The nature of the doped hole states at the Fermi energy in these cuprates has been established by carrying out detailed site- and symmetry-projected electron spectroscopic measurements on single crystals. The studies have shown that these states have strongly admixed $Cu(3d_{x^2-y^2})$ and $O(2p_{xy})$ character. Electron spectroscopy has been employed extensively by Sarma and co-workers to investigate the electronic structures of a variety of oxides including TiO_x, $LaNiO_3$, and $La_{1-x}A_xMO_3$ (A = Ca or Sr, M = V, Mn, or Co).

12 Nanomaterials

Particles of materials with diameters in the range of 1–50 nm constitute nanoscale clusters. Physical properties of such clusters correspond neither to those of the free atoms or molecules making up the particle nor to those of bulk solids of the same chemical composition. The clusters are characterized by a large ratio of surface area to volume, which implies that a large fraction of the atoms reside at the grain boundary. Metastable structures can therefore be generated in nanoclusters, and these are different from those of the bulk. Employing such clusters as precursors, it is possible to generate new metastable phases of a given substance. Nanoscale clusters have attracted considerable attention in recent years.[9,53–55] A variety of techniques have been developed to prepare clusters of different systems, and various tools of characterization have been employed to investigate their properties. Most investigations have been carried out on clusters prepared by gas phase condensation, which provides a substrate-free configuration for the nanoscale particles. Clusters grown in liquid and solid media have also been investigated.

The methods of preparation employed to prepare nanoscale clusters include evaporation in an inert gas atmosphere, laser pyrolysis, sputtering techniques, plasma techniques, and chemical methods. Microemulsions have been employed to prepare some oxides in nanomaterial form. Table 9 lists oxides that typically are prepared by chemical methods.

Nanophase materials are prepared by compacting the nanosized clusters, generally under high vacuum. Synthesis of such nanomaterials has been accomplished in the last few years. The average grain size in these materials ranges from 5 to 25 nm. Properties exhibited by the materials synthesized thus far are quite different, and often markedly improved, from coarser grained counterparts of the same chemical composition. The atomic arrangements at the interfaces of

Table 9 Oxide Nanomaterials Prepared by Chemical Methods

Material	Method	Cluster Size (nm)	Characterization Method
$Mn_{0.8}Zn_{0.2}Fe_2O_4$ and $MnFe_2O_4$	Coprecipitation	8	Magnetic, Mössbauer, and EXAFS spectroscopy
ZrO_2	Gel precipitation	5–20	Structure
Al_2O_3/ZrO_2	Chemical polymerization and precipitation	20–80	Structure
$PbTiO_3$	Sol–gel	12	Ferroelectricity

the nanosized grains permit different metastable configurations to be induced in them by changing the initial composition. Thus a wide range of materials with novel properties can be prepared.

Nanocomposites, as distinct from nanophase materials, lead to monophasic or multiphasic ceramics, glasses, or porous materials having tailored and improved properties.[56] Nanocomposites can be derived from sol–gel, intercalation, or entrapment modes of preparation. A large family of microcomposite electroceramic materials, superior in performance to single phase materials, have been prepared. Nanocomposites are likely to find applications in a variety of fields.

13 Catalysts and Gas Sensors

Heterogeneous catalysis is a vital aspect of modern chemical industry. In terms of tonnage, catalysts constitute a high proportion of materials synthesized today. Among the variety of solids employed as catalysts, transition metal oxides form an important group.[57] Oxides of nontransition metals are good acid–base catalysts. Oxides of transition metals can act as acid–base or redox catalysts, heteropolyacids being good examples of such bifunctional role. While the acid–base sites on the surfaces of such oxides are similar to those on oxides of non-transition metals, the redox sites are comparable to those in metal complex catalysts. Polynuclear metal complex catalysts provide good models to promote an understanding of transition metal ion catalysis. The comparison is closer to reality in the case of catalysts where the transition metals are dispersed on the surfaces of oxide support (Al_2O_3, SiO_2, TiO_2, etc.). In such cases, the interaction between the transition metal ion and the oxide support becomes important. In transition metal ion–support interactions, as well as other properties of oxide catalysts, there is considerable interplay between solid state chemistry and catalysis.

Adsorption of organic and inorganic molecules on the surfaces of certain transition metal oxides leads to the transfer of surface oxygen atoms to the molecules. It is this property of oxide surfaces that makes them good catalysts for oxidation reactions. The initial state of the oxide catalyst is restored by the molecular oxygen in the reaction mixture. The process involves adsorption, dissociation, and incorporation of the oxygen as ''oxide'' on the surface. For such oxygen transfer to occur, the surface oxygens have to be fairly mobile (or, not strongly bound to the surface). A good example is provided by solid solutions between Bi_2O_3 and oxides such as Nb_2O_5, V_2O_5, and MoO_3 possessing fluorite, pyrochlore, or perovskite structures. Oxide materials such as $LaCoO_3$, modified manganites of perovskite structure, and related oxide materials oxidize CO to CO_2. Transition metal oxide catalysts clearly have a major role in many oxidation reactions and offer candidates for auto exhaust catalysts.[6,57]

Bismuth molybdates that catalyze the selective oxidation and ammodixation of propylene to yield acrolein and acrylonitrile have received considerable attention:

$$CH_3CH{=}CH_2 + O_2 \rightarrow CH_2{=}CHCHO + H_2O$$

$$CH_3CH{=}CH_2 + NH_3 + \tfrac{3}{2}O_2 \rightarrow CH_2{=}CHCN + 3H_2O$$

Both the reactions involve formation of an allylic ($CH_2{\cdots}CH{\cdots}CH_2$) intermediate. Catalytic activity toward this reaction requires sequential reduction and oxidation of the catalyst. With bismuth molybdate catalysts, the chemisorbed hydrocarbon is oxidized by the lattice oxygen. The active sites are replenished by the rapid diffusion of lattice oxygens from the bulk to the catalyst surface. The catalysis cycle is completed by reducing gaseous oxygen and incorporating it in the form of oxide ions in the lattice. Experiments with ^{18}O have shown that the oxygen in the product originates primarily from the catalyst. Since the lattice oxygen is readily available, the rate of propylene oxidation is independent of the partial pressure of oxygen. Sleight[58] has suggested that reduction occurs at molybdenum sites and reoxidation at bismuth sites. The degeneracy (mixing) of Mo (4*d*) and Bi (6*p*) levels in the catalyst allows electrons added to Mo(4*d*) during reduction to be available at bismuth sites for reoxidation. Another requirement for selectivity is that the structure be able to permit easy removal and addition of oxygen. Coordinately unsaturated molybdenum atoms at the surface, which can undergo easy interconversion between octahedral and tetrahedral coordination, seem to make bismuth molybdate catalysts selective. Among the bismuth molybdates, $Bi_2Mo_3O_{12}$ possesses a defect scheelite structure (tetrahedrally coordinated molybdenum) and γ-Bi_2MoO_6, a layered structure with octahedrally coordinated molybdenum. The structure of $Bi_2Mo_2O_9$ seems to be related to that of scheelite. Selectivity and activity of the catalysts toward alkene oxidation are nearly constant, with composition from $Bi_2Mo_3O_{12}$ to Bi_2MoO_6. It seems that the coordination of molybdenum at the catalyst surface is different from that of the bulk solid.

Although it is generally believed that interaction between the metal and the oxide support in supported-metal catalysts is negligible, considerable evidence suggests that the support can influence catalytic reactions in many instances. One example is that of spillover of adsorbed species from the metal to the support. For example, hydrogen adsorbed on metals (e.g., Pt) undergoes dissociation and spills over to the support (e.g., Al_2O_3 or SiO_2), where it reacts with the adsorbed organic molecules. Catalytic properties of metals can be altered by interaction with certain supports. *Strong metal–support interaction (SMSI)* was first noticed with group 8 metals supported on TiO_2. The main features of SMSI are the suppression of H_2 and CO adsorption and radical changes in the activity and selectivity of reactions such as hydrogenation of CO. Thus, Pt/TiO_2 reduced above 750 K is 10 times as active as Pt/Al_2O_3 and 100 times as active as Pt/SiO_2 toward methanation. SMSI catalysts exhibit superior activity and selectivity in Fischer–Tropsch synthesis.

There has been considerable effort to understand the mechanism of SMSI. Experimental characterization of SMSI catalysts has shown considerable morphological changes accompanied by reduction of the support oxide. In Pt/TiO_2, formation of Ti_4O_7 is revealed by electron diffraction. The interaction seems to arise from a charge transfer from the reduced cation (Ti^{3+}) to the adjacent metal atom. There is also evidence for the diffusion of the titania to the metal surface. Hydrogen reduction at 700 K of a nickel film deposited on TiO_2 appears to result in the segregation of $TiO_x (x \sim 1.0)$ onto the nickel surface; reduced titania appears to diffuse rapidly through nickel. It seems that both the physical coverage of the nickel surface by TiO_x and chemical interaction between nickel and reduced titania on the surface contribute to SMSI behavior. SMSI is not special to TiO_2 alone. Other transition metal oxide supports such as Nb_2O_5 also exhibit this phenomenon. There appears to be a relation between SMSI and the reducibility of support oxides.

An important class of oxide catalysts makes use of transition metal oxides supported on other oxides of high surface area (e.g., Al_2O_3, SiO_2). A good example is the catalyst used in the oxidation of SO_2 to SO_3 in the manufacture of sulfuric acid. This catalyst contains V_2O_5 (with K_2O as promoter) supported on SiO_2. At the reaction temperature, the active component melts to form a thin liquid film on the surface of SiO_2. The oxidation of SO_2 to SO_3 proceeds through the diffusion of SO_2 and O_2 molecules into the film.

Catalytic properties of transition metal oxides also make them good candidates for service as *gas sensors*.[59] Thus, the oxidizing power of the oxide surfaces can be used to make gas sensors. For an oxide to be a good sensor, the reaction should occur fast and should produce a change in the resistivity, capacitance, or some such property of the oxide. Percentage conversion and yield are not important (as in catalysis), but sensor selectivity is an important criterion. The following oxides are able to sense the gases paired with them, in parentheses: ZnO (ethanol), Fe_2O_3 (CO or hydrocarbons, depending on whether Fe is in the γ or δ form), bismuth molybdates (NH_3), and V_2O_5 (hydrocarbons). Many of the conducting (metallic) transition metal oxides are good electrocatalysts. Typical of these are $La_{1-x}Sr_xCoO_3$, $Pb_2(Ru_{2-x}Pb_x)O_{7-y}$, and RuO_2.

References

1. J. B. Goodenough, *Magnetism and the Chemical Bond,* Wiley, New York, 1963.
2. C.N.R. Rao and G. V. Subba Rao, *Phys. Stat. Solidi, A1,* 597 (1970).
3. J. B. Goodenough, *Prog. Solid State Chem., 5,* 149 (1971).
4. J. B. Goodenough, in *Solid State Chemistry* (C.N.R. Rao, ed.), Dekker, New York, 1974.
5. C.N.R. Rao and G. V. Subba Rao, *Transition Metal Oxides, NSRDS-NBS Monograph* 49, U.S. National Bureau of Standards, Washington, DC, 1974.
6. C.N.R. Rao and J. Gopalakrishnan, *New Directions in Solid State Chemistry,* Cambridge University Press, Cambridge, 1986.

7. C.N.R. Rao, *Annu. Rev. Phys. Chem., 40,* 291 (1989).

8. A. K. Cheetham and P. Day (eds.), *Solid State Chemistry,* Clarendon Press, Oxford, 1987, 1992.

9. C.N.R. Rao (ed.), *Chemistry of Advanced Materials,* Blackwell, Oxford, 1992.

10. P. A. Cox, *Transition Metal Oxides: An Introduction to Their Electronic Structure and Properties,* Clarendon Press, Oxford, 1992.

11. C. Kittel, *Introduction to Solid State Physics,* Wiley, New York, 1977.

12. A. R. West, *Solid State Chemistry and Its Applications,* Wiley, Chichester, 1985.

13. P. Hagenmuller, *Prog. Solid State Chem., 5,* 71 (1971).

14. A. M. Chippindale, P. G. Dickens, and A. V. Powell, *Prog. Solid State Chem., 21,* 133 (1991)

15. M. Greenblatt, *Chem. Rev., 88,* 31 (1988); also see *Int. J. Mod. Phys. B, 7,* 3937–4145 (1993).

16. M. B. Robin and P. Day, *Adv. Inorg. Chem. Radiochem., 10,* 247 (1967); also see P. Day, *Int. Rev. Phys. Chem., 1,* 149 (1981).

17. J. M. Honig, *Proc. Indian Acad. Sci., 96,* 391 (1986).

18. R. Aragon, R. J. Rasmussen, J. P. Shepherd, J. W. Koenitzer, and J. M. Honig, *J. Magnets Magn. Mater., 54–57,* 1335 (1968); also see *Phys. Rev. B, 31,* 1818 (1985).

19. C.N.R. Rao, O. Parkash, D. Bahadur, P. Ganguly, and S. Nagabhushana, *J. Solid State Chem., 27,* 353 (1977).

20. L. Bouet, P. Tailhades, A. Rousset, K. R. Kannan, M. Verelst, G. U. Kulkarni, and C.N.R. Rao, *J. Solid State Chem., 102,* 414 (1993).

21. N. F. Mott, *The Metal–Insulator Transition,* 2nd ed, Taylor and Francis, London, 1993.

22. P. P. Edwards and C.N.R. Rao (eds.), *The Metallic and the Nonmetallic States of Matter,* Taylor and Francis, London, 1985.

23. J. M. Honig and J. Spalek, *Proc. Indian Natl. Sci. Acad., A52,* 232 (1986).

24. T. V. Ramakrishnan, in ref. 22; also see *Proc. Indian Natl. Sci. Acad., A52,* 217 (1986).

25. C.N.R. Rao and P. Ganguly, in ref. 22; also see R. Mahesh, R. Mahendiran, A. K. Raychandhuri, and C.N.R. Rao, *J. Solid State Chem., 114,* 297 (1995) and the references listed therein.

26. L. J. de Jongh and A. R. Miedema, *Adv. Phys., 23,* 1 (1974).

27. P. Ganguly and C.N.R. Rao, *J. Solid State Chem., 53,* 193 (1984).

28. J. G. Bednorz and K. A. Müller, *Z. Phys, B64,* 187 (1986).

29. A. W. Sleight, *Science, 242,* 1519 (1988).

30. R. J. Cava, *Science, 247,* 656 (1990).

31. C.N.R. Rao (ed.), *Chemistry of High-Temperature Superconductors,* World Scientific, Singapore, 1991; C.N.R. Rao, *Phil. Trans. R. Soc., London, A336,* 595 (1991); C.N.R. Rao and A. K. Ganguli, *Chem. Soc. Rev.* February 1995.

32. A. R. Armstrong and P. P. Edwards, *Chemistry of High-Temperature Superconducting Oxides,* Annual Reports C, The Royal Society of Chemistry, London, 1991, p. 259.

33. B. Raveau, C. Michel, M. Hervieu, and D. Groult, *Crystal Chemistry of High T_c Superconducting Copper Oxides,* Springer-Verlag, Berlin, 1991.

34. (a) M. N. Regueiro, J. L. Tholence, E. V. Antipov, J. J. Capponi, and M. Marezio, *Science, 262,* 97 (1993); (b) R. J. Cava et al., *Physica C, 165,* 419 (1990).

35. (a) H. Zhang et al., *Nature, 370,* 352 (1994); (b) J. D. Jorgensen, *Physics Today, 44,* 34 (1991).

36. C.N.R. Rao, J. Gopalakrishnan, V. Manivannan, and A. K. Santra, *Physica C, 174,* 11 (1991); H. Zhang and H. Sato, *Phys. Rev. Lett., 70,* 1697 (1993).

37. R. J. Cava et al., *Nature, 335,* 814 (1988).

38. F. Jona and G. Shirane, *Ferroelectric Crystals,* Pergamon Press, Oxford, 1962.

39. R. E. Newnham and L. E. Cross, in *Preparation and Characterization of Materials* (J. M. Honig and C.N.R. Rao, eds.), Academic Press, New York, 1981.

40. E. Canadell and M. H. Whangbo, *Chem. Rev., 91,* 965 (1991).

41. R. Hoffmann, *Solids and Surfaces: A Chemist's View of Bonding in Extended Structures,* VCH Publishers, New York, 1988.

42. J. K. Burdett, *Chem. Rev., 88,* 3 (1988). The following are representative references to the fine papers of Burdett and co-workers: *Inorg. Chem., 24,* 2244 (1985); *J. Am. Chem. Soc., 110,* 536 (1988); *J. Am. Chem. Soc., 112,* 6571 (1990); *Inorg. Chem., 32,* 5004 (1993). Also see S. Lee, *Acc. Chem. Res., 24,* 249 (1991).

43. M. Ganne, M. Dion, and A.C.R. Boumaza, *C.R. Acad. Sci. Paris,* Ser. 2, *302,* 635 (1986).

44. M. H. Whangbo et al, *Inorg. Chem., 25,* 2424 (1986); *J. Am. Chem. Soc., 110,* 358 (1988), *Inorg. Chem., 28,* 1466 (1989).

45. M. Sato, H. Fujishita, S. Sato, and S. Hoshino, *J. Phys. C, 19,* 3059 (1986).

46. M. H. Whangbo, E. Canadell, P. Foury, and J. P. Pouget, *Science, 252,* 96 (1991).

47. J. Zaanen, G. A. Sawatzky, and J. W. Allen, *Phys. Rev., 55,* 418 (1985).

48. D. D. Sarma et al., *Pramana-J. Phys., 38,* L531 (1992).

49. C.N.R. Rao, K. J. Rao, S. Ramasesha, D. D. Sarma, and S. Yashonath, *Annual Reports, C,* The Royal Society of Chemistry, London, Section 7, 1992.

50. D. D. Sarma and A. Taraphder, *Phys. Rev. B, 39,* 1570 (1989).

51. A. K. Santra, D. D. Sarma, and C.N.R. Rao, *Phys. Rev. B, 43,* 5612 (1991).

52. A. Chainani, M. Mathew, and D. D. Sarma, *Phys. Rev. B, 46,* 9976 (1992).

53. R. P. Andres, R. S. Averback, and W. L. Brown, *J. Mater. Res., 4,* 704 (1989).

54. P. Jena, B. K. Rao, and S. N. Khanna (eds.), *Physics and Chemistry of Finite Systems: From Clusters to Crystals,* Kluwer Publishers, Dordrecht, 1992.

55. R. W. Siegel, in *Advances in Materials and Their Applications* (P. Rama Rao, ed.), Wiley-Eastern, New Delhi, 1993.

56. S. Komarneni, *J. Mater. Chem., 2,* 1219 (1992).

57. J. M. Thomas and K. Zamaraev (eds.), *Perspectives in Catalysis,* Blackwell, Oxford, 1991.

58. A. W. Sleight, *Science, 208,* 895 (1980).

59. C.N.R. Rao, A. R. Raju, and K. Vijayamohanan, in *New Materials* (S. K. Joshi, C.N.R. Rao, T. Tsuruta, and S. Nagakura, eds.), Narosa, New Delhi, 1992.

PART

III

Preparation of Materials

PART

III

Preparation of Materials

1 Introduction

Synthesis of oxide materials involves considerable synthetic ingenuity. Although one can evolve a rational approach to the synthesis of oxides,[1] there is always an element of serendipity. A good example of an oxide discovered serendipitously is $NaMo_4O_6$ containing condensed Mo_6 octahedral metal clusters. This material was discovered by Torardi and McCarley[2] in their effort to prepare the lithium analogue of $NaZn_2Mo_3O_8$. Another serendipitous discovery is that of the phosphorus tungsten bronze $Rb_xP_8W_{32}O_{112}$, formed during the preparation of the Rb–W bronze by the reaction of phosphorus present in the silica of the ampoule. Since the material could not be prepared in a platinum crucible, Raveau and co-workers[3] suspected that a constituent of the silica ampoule must have been incorporated. This discovery led to the synthesis of the family of phosphorus tungsten bronzes of the type $A_xP_4O_8(WO_3)_{2m}$.

Rational synthesis of materials requires an understanding of the principles of crystal chemistry in addition to phase equilibria, thermodynamics, and reaction kinetics. There are many examples of rational synthesis. A good example is the fast Na^+ ion conductor nasicon, $Na_3Zr_2PSi_2O_{12}$, which was synthesized by Goodenough and co-workers[4] based on the understanding of coordination preferences of cations and the nature of oxide networks formed by them. The zero-expansion ceramic $Ca_{0.5}Ti_2P_3O_{12}$ and its derivatives possessing the nasicon framework were synthesized by Roy and co-workers[1] based on the idea that the property of zero expansion would be exhibited by two or three coordination polyhedra linked in a manner that left substantial empty space in the network. Synthesis of silicate-based porous materials, making use of organic templates to predetermine the pore or cage geometries, is well known. A microporous phosphate of the formula $(Me_4N)_{1.3}(H_3O)_{0.7}Mo_4O_8(PO_4)_2{\cdot}2H_2O$, where the tetramethylammonium ions fill the voids in the three-dimensional structure made up of Mo_4O_8 cubes and PO_4 tetrahedra, has been prepared in this manner by Haushalter and co-workers.[5]

A variety of oxides have been prepared in the last several years by the traditional ceramic method, which involves mixing and grinding powders of the constituent oxides, carbonates, and other compounds and heating them at high temperatures with intermediate grinding when necessary. A wide range of conditions, often bordering on the extreme, have been employed in materials synthesis; these include high temperatures or pressures, very low oxygen fugacities, and rapid quenching. The low temperature chemical routes, however, are of greater interest. The current trend is to avoid brute force methods in order to achieve better control of structure, stoichiometry, and phasic purity. The so-called *soft chemistry (chimie douce)* routes are indeed desirable, since they lead to novel products, often metastable, which otherwise are difficult to prepare.

To understand what soft chemistry implies, it is instructive to consider a few examples. Thus MoO_3, which ordinarily has a layered structure, can be prepared in a WO_3- or ReO_3-like structure by the dehydration of $MoO_3{\cdot}H_2O$ or by the oxidation of Mo_4O_{11}. A new form of TiO_2 was prepared by Tournoux et al.[6] by the dehydration of $H_2Ti_4O_9{\cdot}xH_2O$, obtained from $K_2Ti_4O_9$ by the exchange of K^+ by H^+. Raveau and co-workers[7] similarly prepared the empty tunnel structure of $Ti_2Nb_2O_9$ by the dehydration of $HTiNbO_5$, which in turn was obtained by cation exchange with $ATiNbO_5$ (A = K, Rb). Delmas and Borthomieu[8] prepared $Ni(OH)_2xH_2O$ with a large intersheet distance of 7.8 Å by the hydrolysis of $NaNiO_2$, followed by the reduction of NiOOH. Just as VS_2 was obtained by Murphy et al by the oxidative deintercalation of $LiVS_2$, several metastable oxides have been prepared by deintercalation of alkali metal ions.[9] We shall be examining a few cases of oxides obtained by soft chemistry routes[10] in subsequent sections.

Some of the important chemical methods of synthesis of oxides are coprecipitation and precursor methods, ion exchange and alkali flux methods, the sol–gel method, topochemical methods, electrochemical methods, high pressure methods (including hydrothermal synthesis), and the combustion method.[9–15] In this Part, we discuss the synthesis of transition metal oxides by chemical methods using several examples, including superconducting cuprates. Oxygen stoichiometry is an important factor in the synthesis of transition metal oxides. Indeed, to obtain the desired stoichiometry it often becomes necessary to heat the material in a suitable atmosphere (under high oxygen pressures or in an atmosphere of the appropriate oxygen fugacity provided by using CO/CO_2 mixtures or vacuum). We shall draw attention to this aspect wherever necessary.

2 Typical Reactions

Some of the reaction types employed in oxide synthesis are as follows:

1. $CaCO_3(s) \rightarrow CaO(s) + CO_2(g)$

   $M_mO_n(s) \rightarrow M_mO_{n-\delta}(s) + \frac{\delta}{2}O_2(g)$ (M = metal)
2. $YBa_2Cu_3O_6(s) + O_2(g) \rightarrow YBa_2Cu_3O_7(s)$
3. $Pr_6O_{11}(s) + 2H_2(g) \rightarrow 3Pr_2O_3(s) + 2H_2O(g)$
4. $ZnO(s) + Fe_2O_3(s) \rightarrow ZnFe_2O_4(s)$

   $BaO(s) + TiO_2(s) \rightarrow BaTiO_3(s)$
5. $ZnS(s) + CdO(s) \rightarrow CdS(s) + ZnO(s)$

   $LiFeO_2(s) + CuCl(l) \rightarrow CuFeO_2(s) + LiCl(s)$

In the preparation of complex oxides it is common to carry out thermal decomposition of a compound followed by oxidation (in air or O_2) essentially in one step:

$$2Ca_{0.5}Mn_{0.5}CO_3(s) + \tfrac{1}{2}O_2(g) \rightarrow CaMnO_3(s) + 2CO_2$$

Vapor phase reactions yield solid products in many instances. In chemical vapor transport reactions, a gaseous reagent acts as a carrier to transport a solid by transforming it into the vapor state. Thus, $MgCr_2O_4$ cannot be formed readily by the reaction of MgO and Cr_2O_3. However, Cr_2O_3 reacts with O_2, giving $CrO_3(g)$, which then reacts with MgO to give the chromate:

$$MgO(s) + Cr_2O_3(s) \xrightarrow{O_2} MgCr_2O_4(s)$$

Some of the typical transport reaction equilibria are:

$$TaOCl_2(s) + TaCl_5 \rightleftharpoons TaOCl_3 + TaCl_4$$

$$Nb_2O_5(s) + 3NbCl_5 \rightleftharpoons 5NbOCl_3$$

Table 1 lists a few examples of chemical transport system.

Specific reagents and reaction conditions are employed to carry out reduction of oxides. For example, oxides can be reduced in an atmosphere of (flowing) pure or dilute hydrogen (e.g., N_2/H_2 mixtures) or in an atmosphere of CO or CO/CO_2 mixtures. Reduction of oxides for the purpose of lowering the oxygen content is also achieved by heating oxides in argon or nitrogen; application of vacuum at an appropriate temperature (vacuum annealing or decomposition at low pressures) is also used. Exact control of oxygen stoichiometry in oxides such as Fe_3O_4 or V_2O_3 is accomplished by annealing the oxide in CO/CO_2 mixtures of known oxygen fugacity at an appropriate temperature.

$$M_2O_3(s) + H_2(g) \rightarrow 2MO(s) + H_2O(g) \qquad \text{(e.g., M = Fe)}$$

$$ABO_3(s) + H_2(g) \rightarrow ABO_{2.5}(s) + \tfrac{1}{2}H_2O(g)$$

Table 1 Examples of Chemical Transport

Solid	Transporting Agent
Nb_2O_5	Cl_2, $NbCl_5$
TiO_2	$I_2 + S_2$
IrO_2	O_2
WO_3	H_2O
$MnGeO_3$	HCl
$MgTiO_3$	Cl_2
CrOCl	Cl_2
$FeWO_4$	Cl_2
$MgFe_2O_4$	HCl
$CaNb_2O_6$	Cl_2, HCl
V_nO_{2n-1}	$TeCl_4$
ZrOS	I_2

Reduction of oxides is also accomplished by reacting with elemental carbon or with a metal:

$$M_2O_5 + 3M \rightarrow 5MO \qquad \text{(e.g., M = Nb)}$$

Metal oxychlorides are obtained by heating oxides with Cl_2 (LaOCl from La_2O_3). Examples exist of oxides reacted with a fluoride such as BaF_2 to attain partial fluorination.

Substitution of metal ions in oxide materials is carried out commonly to yield new structures and properties. For example, partial substitution of vanadium by Ti in V_2O_3 renders the oxide metallic. By substituting La^{3+} by Sr^{2+} in $LaCoO_3$, one brings about drastic changes in electrical and magnetic properties; thus, $La_{0.5}Sr_{0.5}CoO_3$ is a ferromagnetic metal.[16,17] Similar changes are brought about by substituting La^{3+} by Sr^{2+} or Ca^{2+} in $LaMnO_3$.[16] The main criteria in such substitutions are relative ionic size and charge neutrality. For example, Y in superconducting $YBa_2Cu_3O_7$ can be replaced by Ca while Ba can be replaced by La. In the nonlinear optical material $KTiOPO_4$, Gopalakrishnan et al.[18] replaced tetravalent Ti partly by pentavalent Nb and replaced P proportionally by Si as in $KTi_{0.5}Nb_{0.5}OP_{0.5}Si_{0.5}O_4$.

3 Ceramic Preparations

The ceramic method is most commonly employed to prepare solid materials. The procedure involves thoroughly grinding the powders of oxides, carbonates, oxalates, or other compounds containing the relevant metals and heating the mixture at the desired temperature, often after the material has been pelletized. A large number of oxides have been prepared by this method. A knowledge of the phase diagram is generally helpful in fixing the desired composition and conditions in this method of synthesis. Platinum, silica, and alumina containers are generally used for the synthesis of metal oxides. If one of the constituents is volatile or sensitive to the atmosphere, the reaction is carried out in sealed, evacuated capsules. Most ceramic preparations require relatively high temperatures, which generally are attained by resistance heating. Electric arc and skull techniques give temperatures up to 3300 K, while high power CO_2 lasers give temperatures up to 4300 K.

The ceramic method suffers from several disadvantages. When no melt is formed during the reaction, the entire reaction must occur in the solid state, initially by a phase boundary reaction at the points of contact between the components and later by diffusion of the constituents through the product phase. As the reaction progresses, diffusion paths keep becoming longer and the reaction rate slower. The product interface between the reacting particles acts as a barrier. The reaction can be speeded up to some extent by intermittent grinding between heating cycles. There is no simple way of monitoring the progress of the reaction in the ceramic method. Appropriate conditions that lead to the completion of the reaction can be decided only by trial and error (by carrying out X-ray diffraction and other measurements periodically). Because of this difficulty, one frequently ends up with mixtures of reactants and products. Separation of the desired oxide from such mixtures is always difficult, if not impossible. It is sometimes difficult to obtain a compositionally homogeneous product by the ceramic technique even when the reaction proceeds almost to completion. In spite of such limitations, the ceramic method has been widely used for the synthesis of solid materials, with considerable success. The ceramic method has been used routinely in cation substitutions, referred to earlier, in many oxide systems (e.g., $La_{1-x}M_xBO_3$, M = Ca, Sr, B = Mn, Co, V, or $LaMM'O_3$, M, M′ = Fe, Mn, Co, Ni).

Several modifications of the ceramic method have been employed to circumvent some of the limitations. One of these relates to decreasing the diffusion path lengths. In a polycrystalline mixture of reactants, the individual particles are approximately 10 μm in size, which represents diffusion distances of roughly 10,000 unit cells. By using freeze-drying, spray-draying, and coprecipitation, as well as sol–gel and other techniques, it is possible to bring down the particle size to a few hundred angstroms and thus effect a more intimate mixing of the reactants. In spray-drying, suitable constituents dissolved in a solvent are sprayed in the form of fine droplets into a hot chamber. The solvent evaporates instantaneously, leaving behind an intimate mixture of reactants, which gives the product upon heating at elevated temperatures. In freeze-drying, the reactants in a common solvent are frozen by immersing in liquid nitrogen and the solvent removed at low pressures. In coprecipitation, the required metal cations, taken as soluble salts (e.g., nitrates), are coprecipitated from a common medium, usually as hydroxides, carbonates, oxalates, formates, or citrates. In actual practice, oxides or carbonates of the relevant metals are digested with an acid (usually HNO_3) and the precipitating reagent added to the solution so obtained. After drying, the precipitate is heated to the required temperature in a desired atmosphere, to get the final product. For example, tetraethylammonium oxalate has been used to prepare superconducting $YBa_2Cu_4O_8$. The decomposition temperature of such precipitates is generally lower. Homogeneous coprecipitation can yield crystalline or amorphous solids. If all the relevant metal ions do not form really insoluble precipitates, it becomes difficult to control the stoichiometry. In some instances, coprecipitation yields monophasic precursor solid solutions.

An interesting variant of the ceramic method is *combustion synthesis* or *self-propagating high temperature synthesis*.[19] This versatile method for the synthesis of oxide materials makes use of a highly exothermic reaction between the reactants to produce a flame due to spontaneous combustion; the desired product or its precursor is yielded in finely divided form. For combustion to occur, one must ensure that the initial mixture of reactants is highly dispersed and contains high chemical energy. For example, one may add a fuel and an oxidizer in preparing oxides by the combustion method, both the additives being removed during combustion, to yield only the product or its precursor. Thus, one can take a mixture of nitrates (oxidizer) of the desired metals along with a fuel (e.g., hydrazine, glycine, urea) in solution, evaporate the solution to dryness, and heat the resulting solid to around 423 K to obtain spontaneous combustion. The yield will consist of an oxidic product in fine particulate form. Even if the desired product is not formed just after combustion, the fine particulate nature of the product facilitates its formation on further heating.

Sekar and Patil[20] used hydrazinate precursors to prepare powders of a variety of oxide materials by the combustion method. The powders from the combustion process must sometimes be further heated in pellet form to get the desired oxide. It seems that almost any ternary or quaternary oxide (e.g., $BaTiO_3$, $PbMoO_4$, $LaCrO_3$, $YBa_2Cu_3O_7$, $Bi_4Ti_3O_{12}$, $BaFe_{12}O_{19}$) can be prepared by this method. Thus, the combustion method has been used to prepare superconducting cuprates, although the resulting products, in fine particulate form, had to be heated to an appropriate high temperature in a desired atmosphere to obtain the final cuprate.[21]

4 Use of Precursors

It was mentioned earlier that diffusion distances for the reacting cations are rather large in the ceramic method. Diffusion distances are reduced markedly (to a few angstroms) by incorporating the cations in the same solid precursor.[9,15,22,23] Synthesis of complex oxides by the decomposition of compound precursors has been known for some time.[15] For example, thermal decomposition of $LaCo(CN)_6{\cdot}5H_2O$ and $LaFe(CN)_6{\cdot}6H_2O$ in air readily yields $LaCoO_3$ and $LaFeO_3$, respectively. $BaTiO_3$ is prepared by the thermal decomposition of $Ba[TiO(C_2O_4)_2]$, while $LiCrO_2$ is prepared from $Li[Cr(C_2O_4)_2(H_2O)_2]$. Ferrite spinels of the general formula MFe_2O_4 (M = Mg, Mn, Ni, Co) are prepared by the thermal decomposition of acetate precursors of the type $M_3Fe_6(CH_3COO)_{17}O_3OH{\cdot}12C_5H_5N$. Chromites of the type MCr_2O_4 are obtained by the decomposition of $(NH_4)_2M(CrO_4)_2{\cdot}6H_2O$. Metal–ceramic composites such as Fe/Al_2O_3 (or $Fe\text{-}Cr/Al_2O_3$) have been prepared by the thermal decomposition of the complex ammonium oxalate containing both iron and aluminum (or Fe, Cr, and Al) in the required proportions. Organometallic precursors, especially carboxylates and alkoxides, used for synthesis of some of the perovskite oxides have been reviewed by Chandler et al.[24]

Carbonates of certain metals (e.g., Ca, Mg, Mn, Fe, Co, Zn, Cd) are all isostructural, possessing the calcite structure. One can, therefore, prepare a large number of carbonate solid solutions containing two or more cations in different proportions.[23] The rhombohedral a_R or the carbonate solid solution varies systematically with the weighted mean cation radius (Fig. 1). Carbonate solid solutions are ideal precursors for the synthesis of monoxide solid solutions of rock salt structure, as demonstrated by Rao and co-workers.[15,23] For example, the carbonates are decomposed in vacuum or in flowing dry nitrogen, to obtain monoxides of the type $Mn_{1-x}M_xO$ (M = Mg, Ca, Co, Cd), having rock salt structure. Oxide solid solutions in which M is magnesium, calcium, or cobalt would require 770–970 K for their formation, while those containing cadmium are formed at still lower temperatures. The facile formation of oxides of rock salt structure by the decomposition of carbonates of calcite structure is due to the close (possibly topotactic) relationship between the structures of calcite and rock salt. The monoxide solid solutions can be used as precursors for preparing spinels and other complex oxides.

Besides monoxide solid solutions, a number of ternary and quaternary oxides

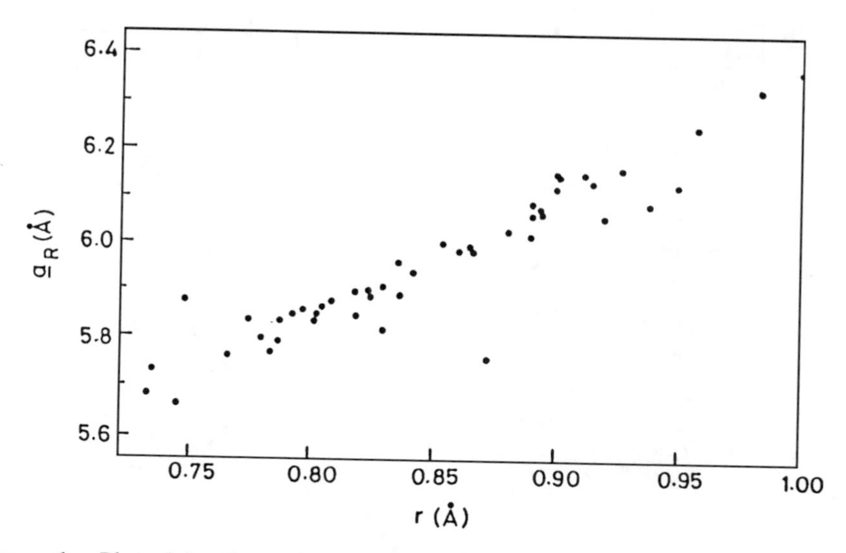

Figure 1. Plot of the rhombohedral lattice parameters a_R of a variety of binary and ternary carbonates of calcite structure (e.g., Ca—M, Ca—M—M′, Mg—M, and M—M′ where M, M′ = Mn, Fe, Co, Cd, etc.) against the mean cation radius. (Courtesy of C.N.R. Rao.)

of novel structures can be prepared by decomposing carbonate precursors containing the different cations in the required proportions. Thus, one can prepare $Ca_2Fe_2O_5$ and $CaFe_2O_4$ by heating the corresponding carbonate solid solutions in air at 1070 and 1270 K, respectively, for about an hour. $CaFe_2O_5$ is a defect perovskite with ordered oxide ion vacancies and has the well-known brownmillerite structure (Fig. 2a) with the Fe^{3+} ions in alternate octahedral (O) and tetrahedral (T) sites. Two new oxides of similar compositions, $Ca_2Co_2O_5$ and $CaCo_2O_4$, have been prepared by decomposing the appropriate carbonate precursors in oxygen atmosphere around 940 K. Unlike $Ca_2Fe_2O_5$, in $Ca_2Mn_2O_5$ (Fig. 2b), anion vacancy ordering in the perovskite structure gives a square-pyramidal coordination (SP) around the transition metal ion. One can also use the carbonate precursor route to synthesize quaternary oxides (Ca_2FeCoO_5, $Ca_2Fe_{1.6}Mn_{0.4}O_5$, $Ca_3Fe_2MnO_5$, etc.) belonging to the $A_nB_nO_{3n-1}$ family. In the Ca/Fe/O system, several other oxides, such as $CaFe_4O_7$, $CaFe_{12}O_{19}$, and $CaFe_2O_4(FeO)_n$ (n = 1, 2, 3), can, in principle, be synthesized by starting from the appropriate carbonate solid solutions and decomposing them in a proper atmosphere.

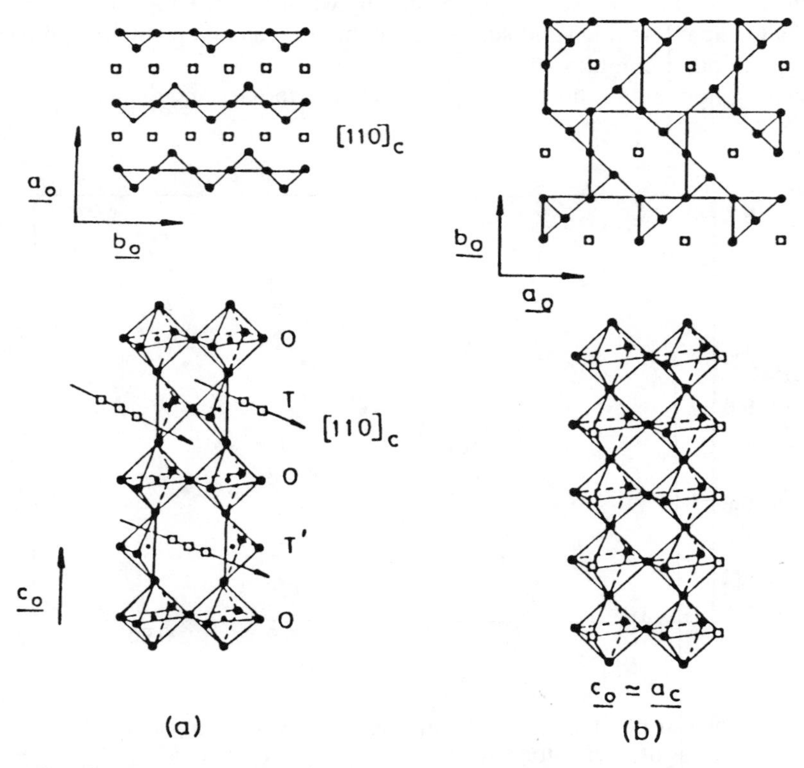

Figure 2. Structures of (a) $Ca_2Fe_2O_5$ (brownmillerite) and (b) $Ca_2Mn_2O_5$. Oxygen vacancy ordering in the *ab* plane is also shown.

An example of a multistep solid state synthesis achieved starting from carbonate solid solution precursors is provided by the $Ca_2Fe_{2-x}Mn_xO_5$ system.[23,25] Structures of the two end members, $Ca_2Fe_2O_5$ and $Ca_2Mn_2O_5$, are derived from the perovskite structure (Fig. 2). Solid solutions between the two oxides would be expected to show oxygen-vacancy-ordered superstructures with Fe^{3+} in octahedral and tetrahedral coordinations and Mn^{3+} in square-pyramidal coordination, but they cannot be prepared by the ceramic method. These solid solutions have indeed been prepared, starting from the carbonate solid solutions of the type $Ca_2Fe_{2-x}Mn_x(CO_3)_4$. The carbonates decompose in air around 1200–1350 K to give perovskitelike oxides $Ca_2Fe_{2-x}Mn_xO_{6-y}$ ($y < 1$). The compositions of the perovskites obtained with $x = 0.67$ and 1 are $Ca_3Fe_2MnO_8$ and $Ca_3Fe_{1.5}Mn_{1.5}O_{8.25}$. X-Ray and electron diffraction patterns show that they are members of the $A_nB_nO_{3n-1}$ homologous series with anion-vacancy-ordered superstructures with $n = 3$ ($A_3B_3O_{8+x}$). Careful reduction of $Ca_3Fe_2MnO_8$ in dilute hydrogen gives $Ca_3Fe_{1.33}Mn_{0.67}O_5 = Ca_3Fe_2MnO_{7.5}$ (Fig. 3). During this step, only Mn^{4+} in the parent oxides is topochemicaly reduced to Mn^{3+}; Fe^{3+} remains unreduced. The most probable structure of $Ca_3Fe_2MnO_{7.5}$ involves SP, O, and T polyhedra along the *b* direction. Upon heating in vacuum at 1140 K, however, the oxide transforms to the more stable brownmillerite structure with only O and T coordinations. Figure 4 shows typical oxides prepared from precursor solid solutions to illustrate the usefulness of the method.

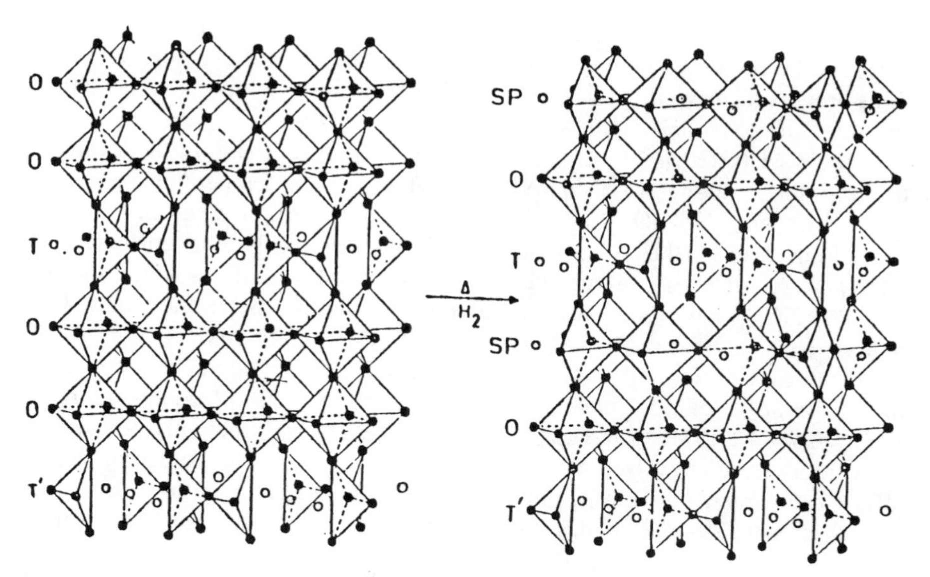

Figure 3. $Ca_3Fe_2MnO_{7.5}$ (with SP, O, T coordinations for Mn/Fe) obtained by the topotactic reduction of $Ca_3Fe_2MnO_8$ (O, T coordination). The latter is prepared by the decomposition of the precursor carbonate, $Ca_2Fe_{4/3}Mn_{2/3}(CO_3)_4$. (From Rao and Gopalakrishnan.[23])

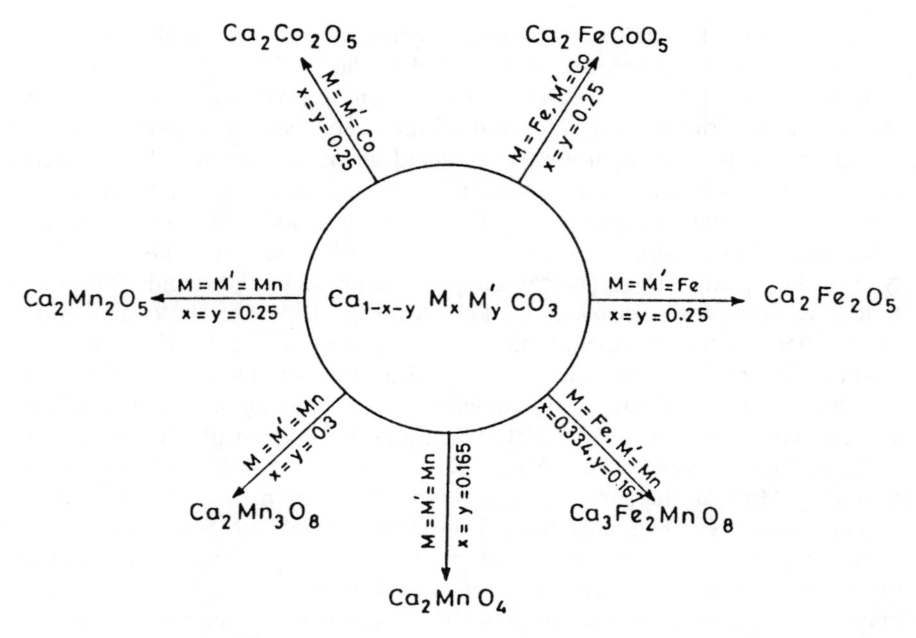

Figure 4. Some of the complex oxides prepared by the decomposition of carbonate precursors. (From Rao and Gopalakrishnan.[23])

Several ternary and quaternary metal oxides of perovskite and related structures can be prepared by employing hydroxide, nitrate, and cyanide solid solutions precursors as well.[23] For example, hydroxide solid solutions of the general formula $Ln_{1-x}M_x(OH)_3$ (Ln = La or Nd; M = Al, Cr, Fe, Co, or Ni) and $Ln_{1-x-y}M'_xM''_y(OH_3)$ (M′ = Ni; M″ = Co or Cu) crystallizing in the rare earth trihydroxide structure are decomposed at relatively low temperature (~870 K) to yield $LaNiO_3$, $NdNiO_3$, $LaNi_{1-x}Co_xO_3$, $LaNi_{1-x}Cu_xO_3$, and so on.

Making use of the isostructurality of anhydrous alkaline earth nitrates $A(NO_3)_2$ (A = Ca, Sr, Ba) and $Pb(NO_3)_2$, nitrate solid solutions of the formula $A_{1-x}Pb_x(NO_3)_2$ have been used as precursors for the preparation of ternary oxides such as $BaPbO_3$, Ba_2PbO_4, and Sr_2PbO_4.[23] Quaternary oxides of the types $LaFe_{0.5}Co_{0.5}O_3$ and $La_{0.5}Nd_{0.5}CoO_3$, which cannot be readily prepared by the ceramic method, have been obtained by the decomposition of $LaFe_{0.5}Co_{0.5}(CN)_6{\cdot}5H_2O$ and $La_{0.5}Nd_{0.5}Co(CN)_6{\cdot}5H_2O$, respectively. Sleight and co-workers have employed a hyponitrite precursor to prepare superconducting $YBa_2Cu_3O_7$ free from $BaCO_3$ impurity. Metal alloys have been used as precursors to prepare desired oxides after treatment with oxygen under appropriate conditions. For example, an alloy containing europium, barium, and copper has been oxidized at 1170 K to obtain superconducting $EuBa_2Cu_3O_7$. Organometallic precursors (e.g., β-diketonates) are used in the synthesis of a variety of oxide materials, including superconducting cuprates.[15,22,23]

5 Topochemical and Intercalation Reactions

A solid state reaction is said to be topochemically controlled when the reactivity is controlled by the crystal structure rather than by the chemical nature of the constituents.[25–28] The products obtained from many solid state decompositions are determined by topochemical factors, especially when the reaction occurs inside the solid without the separation of a new phase. Except for changes in dimension in one or more directions, the atomic arrangement in the reactant crystal remains largely unaffected during the course of a topotactic solid state reaction. The parent and product phases also bear orientational relationships. Günter showed some time ago that one such reaction is the dehydration of $WO_3{\cdot}H_2O$ (or $MoO_3{\cdot}H_2O$) to give WO_3 (or MoO_3). There are other topochemical transformations of the $WO_3/MoO_3/H_2O$ system yielding novel oxide structures. Dehydration of many other hydrates, such as $VOPO_4{\cdot}2H_2O$ and $HMoO_2PO_4{\cdot}H_2O$, also occurs topochemically. In addition, γ-FeOOH topochemically transforms to γ-Fe_2O_3 upon treatment with an organic base. Dehydration of β-$Ni(OH)_2$ to NiO and oxidation of $Ni(OH)_2$ to NiOOH are both topochemical reactions. Reduction of NiO to Ni metal also proceeds topochemically. Decomposition of WO_3, MoO_3, or TiO_2 to give *shear structures (Magnéli phases)* is accommodated by the collapse of the structure in specific crystallographic directions. The decomposition of V_2O_5 to form V_6O_{13} is such a reaction.

WO_3 crystallizes in a monoclinically distorted structure, but MoO_3 possesses a layered structure (Fig. 5). MoO_3 can be stabilized in the WO_3 structure by partly substituting tungsten for molybdenum. Solid solutions of $Mo_{1-x}W_xO_3$ can be prepared by the ceramic method (by heating MoO_3 and WO_3 in sealed tubes around 870 K) or by the thermal decomposition of mixed ammonium metallates. Because of the difference in volatilities of MoO_3 and WO_3, however, these methods do not always yield monophasic products. The $Mo_{1-x}W_xO_3$ solid solutions are conveniently prepared by the topochemical dehydration of the hydrates, the process being very gentle. $MoO_3{\cdot}H_2O$ and $WO_3{\cdot}H_2O$ are isostructural, and the solid solutions between the two hydrates are prepared readily by adding a solution of MoO_3 and WO_3 in ammonia to hot HNO_3(6M). The hydrates $Mo_{1-x}W_xO_3{\cdot}H_2O$ crystallize in the same structure as $MoO_3{\cdot}H_2O$ and $WO_3{\cdot}H_2O$ with a monoclinic unit cell. The hydrate solid solutions undergo dehydration under mild conditions (around 500 K) yielding the oxides $Mo_{1-x}W_xO_3$, which crystallize in the WO_3 structure. The nature of the dehydration of these hydrates has been studied by an in situ electron diffraction study in which the decomposition is brought about by beam heating. Electron diffraction patterns clearly show how $WO_3{\cdot}H_2O$ transforms to WO_3 topotactically with the required orientational relationships. The mixed hydrates $Mo_{1-x}W_xO_3{\cdot}H_2O$ also undergo topotactic dehydration with similar orientational relations. It is more interesting to observe that when the dehydration of $MoO_3{\cdot}H_2O$ is carried out under electron beam heating, MoO_3 seems to appear in the WO_3-type structure, instead of the expected layered structure. This structure of MoO_3 is metastable and is produced

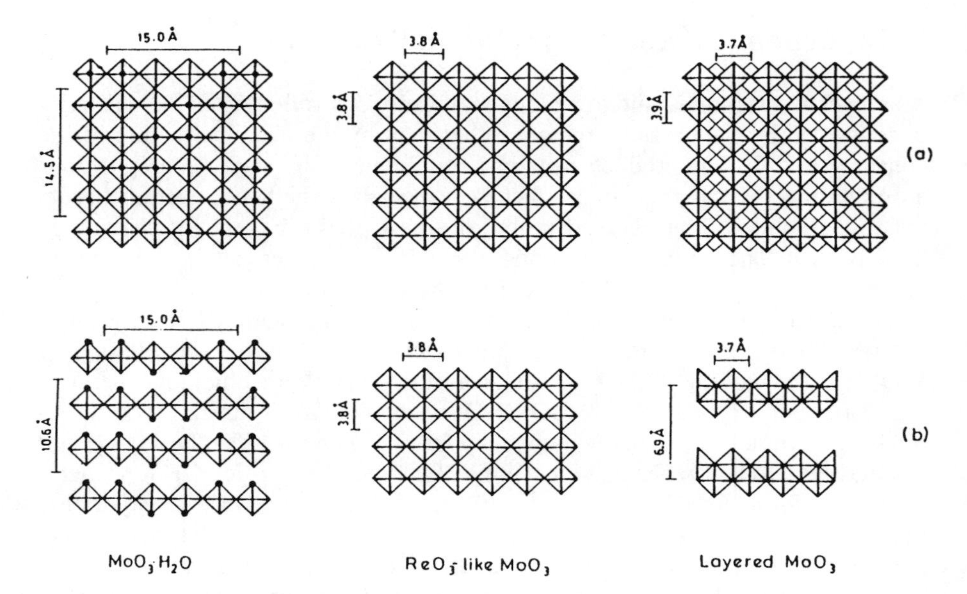

Figure 5. Schematic representation of $MoO_3 \cdot H_2O$, WO_3-like MoO_3 and the layered structure of MoO_3. (Courtesy of C.N.R. Rao.) Two different views are given in (a) and (b).

only by topotactic dehydration under mild conditions. Cubic MoO_3 has been prepared by the pyrolysis of H_2MoO_4. Figure 6 shows different WO_3 phases obtained by the dehydration of $WO_3 \cdot \frac{1}{3}H_2O$, reported by Figlarz and co-workers.[27] By careful deintercalation of amine intercalates by thermal or acid treatment, Rao et al.[29] have recently obtained cubic (of the ReO_3 type) WO_3, as well as tetragonal WO_3.

Reduction of ABO_3 perovskites to give $A_2B_2O_5$ and other defect oxides is often topochemical (e.g., $CaMnO_3$). We shall examine the reduction of $LaNiO_3$ and $LaCoO_3$, which crystallize in the rhombohedral perovskite structure.[30] The

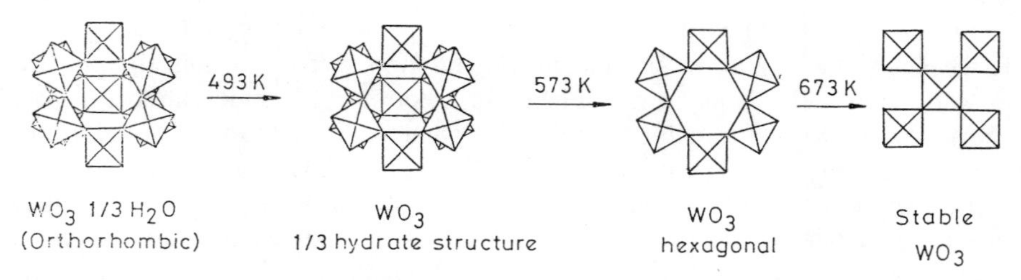

Figure 6. Different phases of WO_3 obtained by the dehydration of $WO \cdot \frac{1}{3}H_2O$. (Courtesy of M. Figlarz.)

possible occurrence of the $La_nNi_nO_{3n-1}$ homologous series had been suggested earlier on the basis of a thermogravimetric study of the decomposition of $LaNiO_3$. It was however not known whether a similar series exists in the case of $LaCoO_3$. Controlled reduction of $LaNiO_3$ and $LaCoO_3$ in hydrogen shows the formation of $La_2Ni_2O_5$ and $La_2Co_2O_5$ representing the $n = 2$ members of the homologous series $La_nB_nO_{3n-1}$ (B = Co or Ni). $La_2Ni_2O_5$ can be prepared by the reduction of $LaNiO_3$ at 600 K in pure or dilute hydrogen, but $La_2Co_2O_5$ can be prepared only by the reduction of $LaCoO_3$ in dilute hydrogen at 670 K. Both the oxides can be oxidized back to the parent perovskites at low temperatures. Neither $La_2Ni_2O_5$ nor $La_2Co_2O_5$ can be prepared by the solid state reaction of La_2O_3 and the transition metal oxide. The reduction of the high temperature superconductor $YBa_2Cu_3O_7$ to $YBa_2Cu_3O_6$ (Fig. 7) or the reverse oxidation is a topochemical process. Intercalation reactions are indeed generally topochemical.

Intercalation reactions of solids[9,15,31] involve the insertion of a guest species (ion or molecule) into a solid host lattice without any major rearrangement of the solid structure:

$$x(\text{guest}) + \square_x(\text{host}) \rightleftharpoons (\text{guest})_x + [\text{host}]$$

where $\square$ stands for a vacant lattice site. Redox intercalation reactions (e.g., Li_xTiS_2, where the lithium metal reduces the TiS_2 layers) can be written as follows:

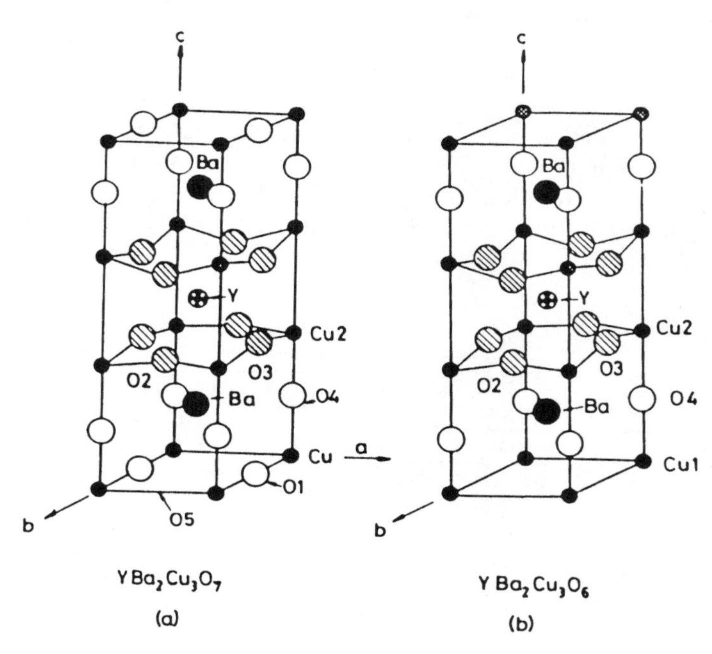

Figure 7. Structures of $YBa_2Cu_3O_7$ (a) and $YBa_2Cu_3O_6$ (b).

$$x(\text{guest})_x^+ + xe^- + \square_x[\text{host}] \rightleftharpoons (\text{guest})_x + [\text{host}]$$

A variety of layered structures can act as hosts. In general, however, their interlayer interactions are weak while intralayer bonding is strong. Intercalation compounds show interesting phase relations, staging (Fig. 8) being an important feature in some cases. Higher stages correspond to lower guest concentrations. Intercalation chemistry has been reviewed adequately, and we shall examine the essential features of these compounds.

Alkali metal intercalation involving a redox reaction is readily carried out electrochemically by using the host (MCh_2 dichalcogenide) as the cathode, the alkali metal as the anode and the nonaqeous solution of the alkali metal salt as the electrolyte:

Na/NaI—propylene carbonate/MCh_2 (Ch = S, Se, Te)
Li/$LiClO_4$—dioxolane/MCh_2

The reaction is spontaneous if a reverse potential is applied or if the cell is short-circuited. Low alkali metal concentrations are obtained by using solutions of salts such as Na or K naphthalide in tetrahydrofuran or *n*-butyllithium in hexane. Alkali metal intercalation in dichalcogenides is also achieved by direct reaction of the elements around 1070 K (e.g., A_xMCh_2, where M = V, Nb, or Ta) in sealed tubes. Metal oxyhalides (e.g., FeOCl) show intercalation reactions similar to dichalcogenides. In addition, they undergo irreversible substitution reactions wherein the halogens of the top layer are substituted by other groups such as $NHCH_3$ and CH_3. Layered transition metal oxides such as MoO_3, V_2O_5, $MOPO_4$, and $MoAsO_4$ (M = V, Nb, Ta) show reduction reactions similar to the dichalcogenides. Layered oxides of the types AMO_2 and $HTiNbO_5$ and $H_2Ti_4O_9$ undergo ion exchange and oxidative deintercalation reactions.

Ready deintercalation of Li from $LiMO_2$ enables these materials to be used as cathodes in lithium cells. Delithiation occurs not only by electrochemical methods, but also by reaction with I_2 or Br_2 in solution phase.[9,15] Lithium insertion in close-packed oxides such as TiO_2, ReO_3, Fe_2O_3, Fe_3O_4, and Mn_3O_4 results

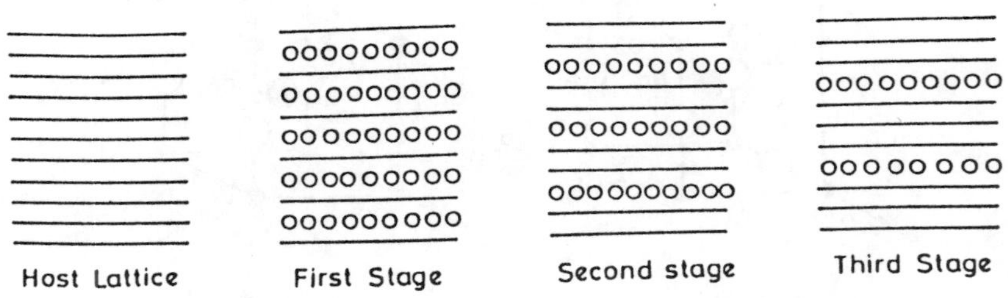

Host Lattice First Stage Second stage Third Stage

Figure 8. Staging in intercalation compounds (schematic): guest molecules (circles) lie between the layers (lines).

Table 2 Examples of Hosts and Guests in Intercalation Compounds

Neutral Layer Hosts	***Guests***
Graphite	$FeCl_3$, K, Br_2
MCh_2 (M = Ti, Zr, Nb, Ta, etc.; Ch = S, Se, Te)	Li, Na, NH_3, organic amines, $CoCp_2$
$MPCh_3$ (M = Mg, V, Fe, Zn, etc.)	Li, $CoCp_2$
WO_3, MoO_3, V_2O_5	H, alkali metal
$MOPO_4$ and $MOAsO_4$ (M = V, Nb, Ta)	H_2O, pyridine, Li
MOCl and MOBr (M = V, Fe, etc.)	Li, $FeCp_2$
Negatively Charged Layers	
(A) MX_2 (M = Ti, V, Cr, Fe; X = O, S)	A = group 1A (Li, Na, etc.)
Layered silicates and clays	Organics
$M(HPO_4)_2$ (M = Ti, Zr, etc.)	
$K_2Ti_4O_9$	

in interesting structural changes. Thus, each of the 12-coordinated cavities in the ReO_3 framework becomes two octahedral cavities occupied by lithium. Lithium insertion in Fe_2O_3 changes the anion array from hexagonal to cubic close packing. Jahn–Teller distortion of Mn_3O_4 is suppressed by Li intercalation. Lithium-intercalated anatase, $Li_{0.5}TiO_2$, transforms to superconducting $LiTi_2O_4$ at 770 K. Delithiation gives rise to oxides in unusual metastable structures (e.g., VO_2 obtained from delithiation of $LiVO_2$). Table 2 lists the important hosts and guests in intercalation compounds. Table 3 gives some oxides intercalated with

Table 3 Intercalation of Lithium in Oxide Hosts

Host	Description
MO_2	Li_xMO_2 ($x \geq 1$) (M = Mo, Ru, Os, or Ir); MO_2 of rutile structure
TiO_2 (anatase)	Li_xTiO_2 ($0 < x \leq 0.7$); $Li_{0.5}TiO_2$ irreversibly to $LiTi_2O_4$ spinel at 770 K
CoO_2	Li_xCoO_2 ($0 < x < 1$); phases obtained by electrochemical delithiation of $LiCoO_2$
VO_2	Li_xVO_2 ($0 < x < 1$); phases obtained by delithiation of $LiVO_2$ using Br_2/$CHCl_3$
	Li_xVO_2 ($0 < x < 0.67$); lithiation using *n*-butyllithium
Fe_2O_3	$Li_xFe_2O_3$ ($0 < x < 2$); anion array transforms from hcp to ccp upon lithiation
Fe_3O_4	$Li_xFe_3O_4$ ($0 < x < 2$); Fe_2O_4 subarray of the spinel structure remains intact
Mn_3O_4	$Li_xMn_3O_4$ ($0 < x < 1.2$); lithium insertion suppresses tetragonal distortion of Mn_3O_4
MoO_3	Li_xMoO_3 ($0 < x < 1.55$)
V_2O_5	$Li_xV_2O_5$ ($0 < x < 1.1$); intercalation of lithium by using LiI
ReO_3	Li_xReO_3 ($0 < x < 2$); three phases $0 < x \leq 0.35$; $x = 1$ and $1.8 \leq x \leq 2$

lithium to show the variety in this system. $V_2(PO_4)_3$ in nasicon structure has been prepared by oxidative deintercalation of $Na_3V_2(PO_4)_3$ by Gopalakrishnan et al.[32] Deintercalation does indeed provide a means of preparing metastable oxides with unusual structures. It is noteworthy that the lithium ion rocking chair batteries make use of $LiMn_2O_4$ or $LiCoO_2$ as the cathode and aqueous $LiNO_3$ as the electrolyte.[33]

Intercalation of sodium and potassium differs from that of lithium. In layered A_xMX_n, lithium is always octahedrally coordinated, while sodium and potassium occupy octahedral or trigonal prismatic sites; octahedral coordination is favored by large values of x and low formal oxidation states of M. For smaller x and higher oxidation states of M, the coordination of sodium and potassium is trigonal prismatic. Intercalated cesium in MX_n is always trigonal prismatic. Intercalation of sodium and potassium in layered MX_2 oxides results in structural transformations involving a change in the sequence of anion layer stacking.

Tungsten and molybdenum bronzes, A_xWO_3 and A_xMoO_3 (A = K, Rb, Cs), are generally prepared by the reaction of the alkali metals with the host oxide. Electrochemical methods are also employed for these preparations. A novel reaction yielding bronzes that are otherwise difficult to obtain involves the reaction of the oxide host with anhydrous alkali iodides, AI[34]:

$$Mo_{1-x}W_xO_3 + y(AI) \rightarrow A_yMo_{1-x}W_xO_3 + \frac{y}{2} I_2$$

Titration of the iodine directly gives the amount of alkali metal intercalated. Atomic hydrogen has been inserted into many binary and ternary oxides. Iodine has been intercalated into the superconducting cuprate $Bi_2CaSr_2Cu_2O_8$, causing an expansion of the c parameter of the unit cell, but without destroying the superconductivity.

Many layered compounds are intercalated by certain species so as to prop the layers and convert the two-dimensional layered structures into microporous ones. This process is called *pillaring*. MoO_3, Zr(IV) phosphates, and metal double hydroxides have been pillared by catonic and anionic species such as $Mo_7O_{24}^{6-}$, $V_{10}O_{28}^{6-}$, and $[Al_{13}O_4(OH)_{24}(H_2O)_{12}]^{7+}$.

6 Ion Exchange Method

Ion exchange in fast ion conductors such as β-alumina is well known. The exchange can be carried out in aqueous or in molten salt conditions. Thus, sodium β-alumina has been exchanged with H_3O^+, NH_4^+, and other monovalent and divalent cations, giving rise to different β-aluminas. Ion exchange in inorganic solids is a general phenomenon, not restricted to fast ion conductors. Thus, $Ag_2Si_2O_5$ with a sheet silicate structure is prepared by reaction of $Na_2Si_2O_5$ in molten $AgNO_3$. Goodenough and co-workers[35] have shown that the phenomenon of ion exchange occurs even when the diffusion coefficients are as small as $\sim 10^{-12}$ cm^2/s, at temperatures far below the sintering temperatures of solids. Ion exchange occurs at a considerable rate in stoichiometric solids as well. Mobile ion vacancies introduced by nonstoichiometry or doping seem to be unnecessary for exchange to occur. Since the exchange occurs topochemically, it enables the preparation of metastable phases that are inaccessible by high temperature reactions.

Goodenough and others have ion-exchanged several metal oxides possessing layered, tunnel, or close-packed structures in aqueous solutions or molten salt media to produce new phases. Typical examples are:

$$\alpha\text{-NaCrO}_2 \xrightarrow[570\ \text{K};\ 24\ \text{h}]{\text{LiNO}_3} \alpha\text{-LiCrO}_2$$

$$\text{KAl0}_2 \xrightarrow{\text{AgNO}_3(1)} \beta\text{-AgAl0}_2$$

$$\alpha\text{-LiFeO}_2 \xrightarrow{\text{CuCl(l)}} \text{CuFeO}_2$$

The structure of the framework is largely retained during ion exchange except for minor changes to accommodate the structural preferences of the incoming ion. Thus, when α-$LiFeO_2$ is converted to $CuFeO_2$ by exchange with molten CuCl, the structure changes from that of α-$NaCrO_2$ to that of delafossite to provide a linear anion coordination for Cu^+. Delafossites such as $AgNiO_2$ have been prepared by ion exchange. When $KAlO_2$ is converted to β-$AgAlO_2$ by ion exchange, there is a change in structure from the cristobalite type to the ordered wurtzite type. The change probably occurs to provide a tetrahedral coordination for Ag^+.

An interesting ion exchange reaction is the conversion of $LiNbO_3$ and $LiTaO_3$ to $HNbO_3$ and $HTaO_3$ respectively, by treatment with hot aqueous acid.[36] The exchange of Li^+ by protons is accompanied by a topotactic transformation of the rhombohedral $LiNbO_3$ structure to the cubic perovskite structure of $HNbO_3$. The mechanism suggested for the transformation is the reverse of the transformation of cubic ReO_3 to rhombohedral $LiReO_3$ and Li_2ReO_3,[37] involving the conversion of the 12-coordinated perovskite tunnel sites to two 6-coordinated sites in the rhombohedral structure by means of a twisting of the octahedra along the $\langle 111 \rangle$ cubic direction. Cations in oxides can often be exchanged with protons to give new phases exhibiting high protonic conduction. Typical examples are $HTaWO_6{\cdot}H_2O$, $HMO_3{\cdot}xH_2O$ (M = Sb, Nb, Ta), pyrochlores, and $HTiNbO_5$. An interesting structural change accompanying ion exchange is found in $Na_{0.7}CoO_2$, where the anion layer sequence is ABBAA[38]; cobalt ions occur in alternate interlayer octahedral sites and sodium ions in trigonal prismatic coordination in between the CoO_2 units. When this material is ion-exchanged with LiCl, a metastable form of $LiCoO_2$ with the layer sequence ABCBA is obtained. The phase transforms irreversibly to the stable $LiCoO_2$ (ABCABC) phase around 520 K.

7 Alkali Flux and Electrochemical Methods

Use of strong alkaline media in the form of either solid fluxes or molten (or aqueous) solutions has enabled the synthesis of novel oxides.[15] The alkali flux stabilizes higher oxidation states of the metal by providing an oxidizing atmosphere. Alkali carbonate fluxes have been traditionally used to prepare transition metal oxides such as $LaNiO_3$. A good example of an oxide synthesized in a strongly alkaline medium is provided by the bifunctional electrocatalyst $Pb_2(Ru_{2-x}Pb_x)O_{7-y}$, which has the pyrochlore structure with Pb in the +4 state. The procedure for preparation involves bubbling oxygen through a solution of Pb and Ru salts in strong KOH at 320 K.[39] The so-called alkaline hypochlorite method is used in many instances. For example, $La_4Ni_3O_{10}$ was prepared by bubbling Cl_2 gas through an NaOH solution of lanthanum and nickel nitrates. Alkaline hypobromite oxidation yields superconducting $La_2CuO_{4+\delta}$.

Stacy and co-workers have prepared superconducting $La_2CuO_{4+\delta}$ by reacting a mixture of La_2O_3 and CuO in molten KOH/NaOH around 520 K.[40] $LaCu_2O_4$ and related cuprates have been prepared from molten NaOH at 673 K.[40] $YBa_2Cu_4O_8$ can be prepared by using an Na_2CO_3/K_2CO_3 flux in a flowing oxygen atmosphere. A KOH melt has been used to prepare superconducting $Ba_{1-x}K_xBiO_3$.

Electrochemical methods have been employed to advantage for the synthesis of many solid materials, including oxides.[41–43] Vanadate spinels of the formula MV_2O_4, as well as tungsten bronzes (A_xWO_3), have been prepared by the electrochemical route.[44] Tungsten bronzes are obtained at the cathode when current is passed through two inert electrodes immersed in a molten solution of the alkali metal tungstate A_2WO_4 and WO_3; oxygen is liberated at the anode. Blue Mo bronzes have been prepared by fused salt electrolysis. Superconducting $Ba_{1-x}K_xBiO_3$ and $La_2CuO_{4+\delta}$ have been prepared electrochemically. Oxides containing metals in high oxidation states are conveniently prepared[45] by electrochemical methods (e.g., $La_{1-x}Sr_xFeO_3$). Mahesh et al.[46] have recently prepared a cubic ferromagnetic $LaMnO_3$ perovskite (with ~40% or more of Mn in the 4+ state) electrochemically. A typical electrode system appears in Figure 9. Demourges et al.[45] have recently prepared $La_2NiO_{4.25}$ ($La_8Ni_4O_{17}$) electrochemically; this oxide has an unusual peroxy or trioxy unit. Metallic $NdNiO_3$ has been prepared electrochemically[46]; nonstoichiometric $NdNiO_{3-\delta}$, on the other hand, shows a metal–insulator transition.

Intercalation of alkali metals in host solids is readily accomplished electrochemically. It is easy to see how both intercalation (reduction of the host) and deintercalation (oxidation of the host) are processes suited for this method. Thus, lithium intercalation is carried out by using a lithium anode and a lithium salt in a nonaqueous solvent.

Although the electrochemical method is old, the processes involved in the synthesis of solids are not entirely understood. Generally one uses solvents having high decomposition potentials (e.g., alkali metal phosphates, borates, fluorides). There is considerable scope to investigate the chemistry and applications of the electrochemical methods of synthesis.

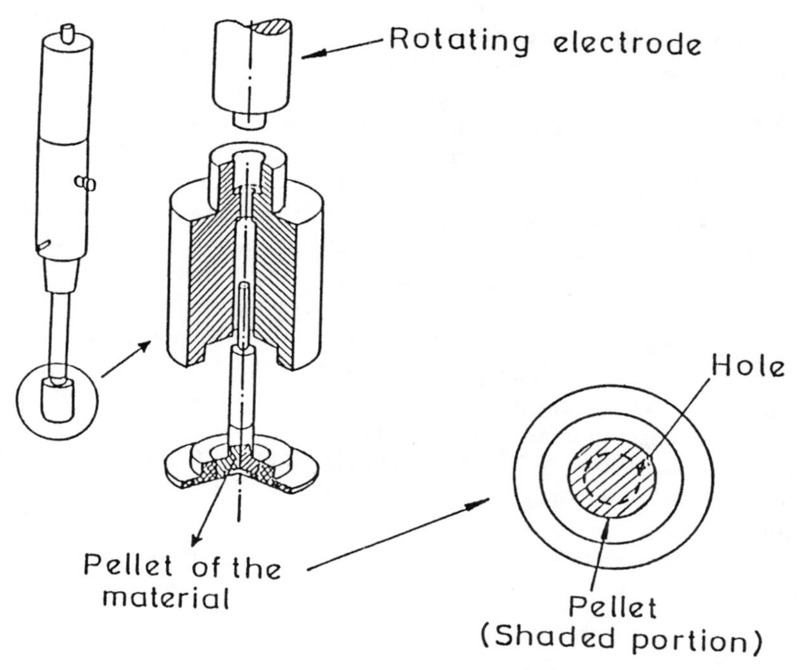

Figure 9. Electrode assembly employed for electrochemical oxidation. (Courtesy of C.N.R. Rao.)

8 Sol–Gel Method

The sol–gel method provides a highly useful means of preparing inorganic oxides.[47–50] It is a wet chemical method and a multistep process involving both chemical and physical processes such as hydrolysis, polymerization, drying, and densification. The name ''sol–gel'' is given to the process because of the distinctive viscosity increase that occurs at a particular point in the sequence of steps. A sudden increase in viscosity is the common feature in sol–gel processing, indicating the onset of gel formation. In the sol–gel process, synthesis of inorganic oxides is achieved from inorganic or organometallic precursors (generally metal alkoxides). Most of the sol–gel literature deals with synthesis from alkoxides. The important features of the sol–gel techniques are better homogeneity compared to the traditional ceramic method, high purity, lower processing temperature, more uniform phase distribution in multicomponent systems, better size and morphological control, the possibility of preparing new crystalline and noncrystalline materials, and easy preparation of thin films and coatings. The important processes in sol–gel synthesis are hydrolysis, polymerization, gelation, drying, dehydration, and densification. The sol–gel method is widely used in ceramic technology, and the subject has been extensively reviewed.

The various steps in the sol–gel technique may or may not be strictly followed in practice. Thus, many complex metal oxides are prepared by a modified sol–gel route without actually preparing metal alkoxides. For example, a transition metal salt solution is converted into a gel by the addition of an appropriate organic reagent. In the case of cuprate superconductors, an equimolar proportion of citric acid is added to a solution of metal nitrates, followed by ethylenediamine, until the solution attains a pH of 6–6.5. The blue sol is concentrated to obtain the gel. The xerogel is obtained by heating at ~420 K. The xerogel is decomposed at an appropriate temperature to get the cuprate.

The sol–gel technique has been used to prepare submicrometer metal oxide powders having a narrow particle size distribution and unique particle shapes (e.g., Al_2O_3, TiO_2, ZrO_2, Fe_2O_3). Uniform SiO_2 spheres have been grown from aqueous solutions of colloidal SiO_2. Variants of the basic sol–gel technique have been used to prepare a number of multicomponent oxide systems. Typical systems include SiO_2/B_2O_3, SiO_2/TiO_2, SiO/ZrO_2, SiO_2/Al_2O_3, and ThO_2/UO_2. A variety of ternary and more complex oxides, such as $PbTiO_3$, $PbTi_{1-x}Zr_xO_3$, and nasicon, have been prepared by this technique, as well as the different types of cuprate superconductors (e.g., $YBa_2Cu_3O_7$, $YBa_2Cu_4O_8$, $Bi_2CaSr_2Cu_2O_8$, $PbSr_2Ca_{1-x}Y_xCu_3O_8$).

9 Reactions at High Pressures

Use of high pressures for the synthesis of solids is quite common in recent years. Commercial equipment permitting use of both high pressure and high temperature conditions is available today. Various experimental aspects of high pressure techniques have been reviewed in the literature.[51–53] For the 1–10 kbar pressure range, the *hydrothermal method* is often employed. In this method, the reaction is carried out either in an open or a closed system. In the open system, the solid is in direct contact with the reacting gas (F_2, O_2, or N_2), which also serves as a pressure intensifier. A gold container is generally used in this type of synthesis. The method has been used for the synthesis of transition metal oxides such as RhO_2 and PtO_2 when the transition metal is in a high oxidation state. Hydrothermal high pressure synthesis under closed-system conditions is also employed for the preparation of higher valence metal oxides. To the reactants is added an internal oxidant (e.g., $KClO_3$) which upon decomposition under reaction conditions provides the necessary oxygen pressure. For example, pyrochlores of palladium(IV) and platinum(IV), $Ln_2M_2O_7$ (Ln = rare earth), have been prepared by this method (970 K, 3 kbar). $(H_3O)Zr_2(PO_4)_3$ and a family of zero thermal expansion ceramics (e.g., $Ca_{0.5}Ti_2P_3O_{12}$) have been prepared hydrothermally. Another good example is the synthesis of borates of aluminum, yttrium, and similar metals, wherein the sesquioxides are reacted with boric acid. Oxyfluorides have been prepared in HF medium. Haushalter and Mundi[53] have prepared shape-selective microporous solids involving molybdenum phosphates hydrothermally in aqueous H_3PO_4 in the presence of cationic templates. Typical examples are $(Et_4N)_6$ $Na[Na_{12}(H_3PO_4)\{Mo_6O_{15}(HPO_4)_3\}]\cdot xH_2O$, $Na_3[Mo_2O_4(HPO_4)(PO_4)]\cdot 2H_2O$, $Pr_4N(NH_4)[Mo_4O_8(PO_4)_2]$, and $Cs_3NH_3[Mo_2O_2(PO_4)_2(H_2PO_4)]$. A novel open-framework cobalt phosphate containing tetrahedrally coordinated Co(II), $CoPO_40.5C_2H_{10}N_2$ has been prepared recently by Chen et al.[54]

Pressures in the range 10–150 kbar are commonly used for solid state synthesis. In a piston–cylinder apparatus consisting of a tungsten carbide chamber and a piston assembly, the sample is contained in a suitable metal capsule surrounded by a pressure transducer (pyrophyllite). Pressure is generated by moving the piston through the blind hole in the cylinder, and the design incorporates a microfurnace made of graphite or molybdenum. Pressures up to 50 kbar and temperatures up to 1800 K are readily reached in a volume of 0.1 cm^3.

In the anvil apparatus, the sample is subjected to pressure by simply squeezing it between two opposed anvils. Although pressures of ~200 kbar and temperatures up to 1300 K are achievable, this technique is not popular for solid state synthesis because only milligram quantities can be handled. An extension of the opposed-anvil principle is the tetrahedral anvil design, in which four massively supported anvils disposed tetrahedrally ram toward the center, where the sample is located in a pyrophyllite medium together with a heating arrangement. The multianvil design has been extended to cubic geometry—specifically, six anvils act on the faces of a pyrophyllite cube located at the center. The belt apparatus provides the best high pressure–high temperature combination for solid state synthesis. This apparatus, which is a combination of the piston–cylinder and the opposed-anvil designs consists of two conical pistons made of tungsten carbide. The pistons ram through a specially shaped chamber from opposite directions. The chamber and pistons are laterally supported by several steel rings, making it possible routinely to reach fairly high pressures (~150 kbar) and high temperatures (~2300 K). The sample is contained in a noble metal capsule (a BN or MgO container is used for chalcogenides) and surrounded by pyrophyllite and a graphite sleeve, the latter serving as an internal heater.

High pressure methods are generally used for the synthesis of materials that cannot be made otherwise. As a rule, the formation of a compound from its components requires that the new composition have a lower free energy than the sum of the free energies of the components. Goodenough and co-workers[51] have discussed how pressure promotes the lowering of free energy in at least six different ways.

1. Pressure delocalizes outer *d* electrons in transition metal compounds by increasing the magnitude of coupling between the *d* electrons on neighboring cations, thereby lowering the free energy. A typical example is the synthesis of $ACrO_3$ (A = Ca, Sr, Pb) perovskites and CrO_2.
2. Pressure stabilizes higher valence states of transition metals, thus promoting the formation of a new phase. For example, under high oxygen pressures, iron is oxidized to Fe^{4+}, and $CaFeO_3$ with the perovskite structure can thus be prepared.

3. Pressure can suppress the ferroelectric displacement of cations, thereby aiding the synthesis of new phases. Synthesis of A_xMoO_3 bronzes, for example, requires populating the empty *d* orbitals centered on molybdenum; at ambient pressures, MoO_3 is stabilized by a ferroelectric distortion of MoO_6 octahedra up to the melting point.
4. Pressure alters site preference energies of cations. For example, it is not possible to synthesize $A^{2+}Mn^{4+}O_3$ (A = Mg, Co, Zn) ilmenites because of the strong tetrahedral site preference of the divalent cations. Instead of monophasic $AMnO_3$, one obtains a mixture of $A[AMn]O_4$(spinel) + MnO_2(rutile) under atmospheric pressure. This product is formed at high pressures with a corundum-type structure in which both the A and Mn ions are in octahedral coordination.
5. Pressure can suppress the $6s^2$ core polarization in oxides containing isoelectronic Tl^+, Pb^{2+}, and Bi^{3+} cations. Thus, $PbSnO_3$ cannot be made at atmospheric pressure because a mixture of PbO and SnO_2 is more stable than the perovskite.
6. Pressure often induces crystal structure transformations, the high pressure phase being generally more close-packed (having higher coordination). Pressure-induced phases can be quenched in some cases to retain the structures under ambient conditions. In the case of perovskite oxides, pressure decreases the tolerance factor.

Stabilization of unusual oxidation states and spin states of transition metals (e.g., $La_2Pd_2O_7$) is of considerable interest. Such stabilization can be rationalized by making use of correlations of structural factors with the electronic configuration. Hagenmuller, Damazeau, and co-workers[55] have used high oxygen pressures to stabilize hexacoordinated high spin iron(IV) in $La_{1.5}Sr_{0.5}Li_{0.5}Fe_{0.5}O_4$, possessing the K_2NiF_4 structure. The elongated FeO_6 octahedra and the presence of ionic Li—O bonds resulting from the K_2NiF_4 structure favor the high spin Fe(IV) state. The Li and Fe ions in this oxide are ordered in the *ab* plane, as evidenced by the supercell spots in the electron diffraction pattern. La_2LiFeO_6 prepared under high oxygen pressure has the perovskite structure, with the iron in the pentavalent state. $CaFeO_3$ and $SrFeO_3$ prepared under oxygen pressure also contain octahedral Fe(IV); while Fe(IV) in $SrFeO_3$ is in the high spin state with the e_g electron in the narrow σ^* band down to 4 K, Fe(IV) in $CaFeO_3$ disproportionates to Fe(III) and Fe(V) below 290 K.

Nickel in the +3 state is present in the perovskite $LaNiO_3$, which can be prepared at atmospheric pressure; other rare earth nickelates have been prepared at high oxygen pressures. Recently, however, $NdNiO_3$ has been prepared by sol–gel and other chemical routes. The oxides $MNiO_{3-x}$ (M = Ba or Sr) prepared under high pressure contain Ni(IV). In $La_2Li_{0.5}Co_{0.5}O_4$, prepared by Damazeau et al.,[55] there is evidence of the transformation of the low spin Co(III) to the intermediate as well as high spin states. The Li and Co ions are ordered in the *ab* plane of this oxide of K_2NiF_4 structure; the highly elongated CoO_6 octahedra seem to stabilize the intermediate spin state. Oxides in perovskite and K_2NiF_4 structures with trivalent Cu have been prepared under high oxygen pressure.

Pressure has a marked effect on the kinetics of reactions, reducing the reaction times considerably, and at the same time giving more homogeneous and crystalline products. For instance, $LnFeO_3$, $LnRhO_3$, and $LnNiO_3$ (Ln = rare earth) are prepared in a matter of hours under conditions of high pressure and temperature, whereas at ambient pressure the reactions require several days ($LnFeO_3$ and $LnRhO_3$) or do not occur at all ($LnNiO_3$). Thus $LnFeO_3$ is formed in 30 minutes at 50 kbar. $Sr_3Ru_2O_7$ and $Sr_4Ru_3O_{10}$ are formed in 15 minutes at 20 kbar and 1300 K.

High pressure methods have been employed in the synthesis of novel superconducting cuprates. A rudimentary example is the preparation of oxygen-excess La_2CuO_4 under high oxygen pressure. A more interesting example is the synthesis of the next homologue with two CuO_2 layers, $La_2Ca_{1-x}Sr_xCu_2O_6$ (earlier found to be an insulator), which was rendered superconducting by heating it under oxygen pressure. Superconducting cuprates with infinite CuO_2 layers of the type $Ca_{1-x}Sr_xCuO_2$ or $Sr_{1-x}Nd_xCuO_2$ have been prepared by Takano, Goodenough, and others under high hydrostatic pressures.[56] It should be noted that $Ca_{1-x}Sr_xCuO_2$ prepared at ambient pressure is an insulator.

10 Superconducting Cuprates

Bednorz and Müller discovered high T_c superconductivity (~30 K) in $La_{2-x}Ba_xCuO_4$. The discovery of a superconducting cuprate with a T_c above 77 K created sensation in early 1987. Wu et al.,[57] who announced this discovery, first made measurements on a mixture of oxides containing Y, Ba, and Cu obtained in their efforts to obtain the yttrium analogue of $La_{2-x}Ba_xCuO_4$. Independently, however, Rao et al.[58] worked on the Y/Ba/Cu/O system on the basis of their knowledge of solid state chemistry. These investigators knew that Y_2CuO_4 could not be made and that substituting Y by Ba in this cuprate was not the way to proceed. Therefore, they tried to make $Y_3Ba_3Cu_6O_{14}$ by analogy with the known $La_3Ba_3Cu_6O_{14}$ described earlier by Raveau and co-workers. Rao et al. varied the Y/Ba ratio as in $Y_{3-x}Ba_{3+x}Cu_6O_{14}$. By making $x = 1$, they obtained $YBa_2Cu_3O_7$ ($T_c \sim 90$ K). Because of the route adopted for the synthesis, they knew the structure had to be related to that of a perovskite.

The cuprates are ordinarily made by the ceramic method (mix, grind, and heat), which involves thoroughly mixing the various oxides and/or carbonates (or any other salt) in the desired proportion and heating the mixture (preferably in pellet form) at a high temperature. The mixture is ground again after some time and reheated until the desired product has formed, as indicated by X-ray diffraction. This method may not always yield the product with the desired structure, purity or oxygen stoichiometry. Variants of this method have been employed.[59] For example, decomposing a mixture of nitrates appears to yield a better product in the case of the 123 compounds. Some workers prefer to use BaO_2 in place of $BaCO_3$ for the synthesis.

One of the problems with the bismuth cuprates (e.g., $Bi_2CaSr_2Cu_2O_8$) is the difficulty in obtaining phasic purity—that is, minimizing the intergrowth of the different layered phases. The glass or the melt route has been employed to obtain better samples. The method involves preparing a glass by quenching the melt; the glass is then crystallized by heating it above the crystallization temperature. Thallium cuprates are best prepared in sealed tubes (gold or silver). Mercury cuprates are similarly best prepared in sealed tubes, since HgO decomposes at high temperatures, giving mercury. In the case of thallium cuprates, some workers prefer to heat Tl_2O_3 in a sealed tube with a matrix of the other oxides (already heated to 1100–1200 K). Because Tl_2O_3 (which readily sublimes) is highly toxic, it is important that thallium cuprates not be prepared in open furnaces. To obtain superconducting compositions corresponding to a particular copper content (number of CuO_2 sheets) by the ceramic method, one may have to start with arbitrary compositions in the case of the Tl cuprates. The real composition of a bismuth or a thallium cuprate superconductor is not exactly the starting composition. The actual composition must be determined by analytical electron microscopy and other methods.

Heating oxidic materials under high oxygen pressures or in flowing oxygen often becomes necessary to attain the desired oxygen stoichiometry. Thus, La_2CuO_4 and $La_2Ca_{1-x}Sr_xCu_2O_6$ heated under high oxygen pressures become superconducting with T_c of 40 and 60 K, respectively. The 123 compounds, however, lose oxygen easily. Note that superconducting $LnBa_2Cu_3O_7$ (Ln = Y, rare earth) is orthorhombic, while insulating $LnBa_2Cu_3O_6$ is tetragonal. It therefore becomes necessary to heat the material in an oxygen atmosphere below the orthorhombic–tetragonal transition temperature. Oxygen stoichiometry is not a serious problem with the bismuth cuprates. As-prepared samples of many of the thallium cuprates of the 1122 type tend to be oxygen-excess substances and show lower T_c or, they may not exhibit superconductivity at all. When such 1122-type compounds are annealed in vacuum or in an H_2 atmosphere, high T_c values can be attained. The real problem is to optimize the hole concentration by control of oxygen stoichiometry. Although the 124 superconductors (e.g., $YBa_2Cu_4O_8$) were first prepared under high oxygen pressures, it was later found that heating the oxide or the nitrate mixture in the presence of Na_2O_2 in flowing oxygen was sufficient to obtain 124 compounds. Superconducting Pb cuprates, on the other hand, can be prepared only in the presence of very little oxygen (N_2 with a small percentage of O_2); otherwise the Cu^+ ions are oxidized, destroying the superconductivity. In the case of the electron superconductor $Nd_{2-x}Ce_xCuO_4$, it is necessary to heat the material in an oxygen-deficient atmosphere; otherwise, the electron given by Ce will merely give rise to an oxygen-excess material. It may be best to prepare $Nd_{2-x}Ce_xCuO_4$ by a suitable method (say decomposition of mixed oxalates or nitrates) and then reduce it with hydrogen.

The sol–gel method has been conveniently employed for the synthesis of 123 compounds such as $YBa_2Cu_3O_7$ and the bismuth cuprates. Materials prepared by such low temperature methods must be annealed or heated under suitable conditions to obtain the desired oxygen stoichiometry as well as the characteristic high T_c. Thus 124 cuprates, lead cuprates, and even thallium cuprates have been made by the sol–gel method; the first two are particularly difficult to make by the ceramic method. Coprecipitation of all the cations in the form of a sparingly soluble salt such as carbonate or oxalate in a proper medium (e.g., using tetraethylammonium oxalate), followed by thermal decomposition of the dried precipitate, has been employed by many workers to prepare cuprates.

Several other strategies have been employed for the synthesis of superconducting cuprates,[59] and some of them were mentioned earlier during our discussion of the various methods. Rao et al[59] have reviewed cuprate synthesis. Both the combustion and the alkali flux methods have been employed for cuprate synthesis. Electrochemical oxidation of La_2CuO_4 to render it superconducting is noteworthy. Superconducting infinite-layered cuprates seem to be possible only when prepared under high pressures because of bonding (structural) considerations. Strategies that involve structural and bonding considerations in the synthesis are generally more interesting. One such example is the synthesis of modulation-free superconducting bismuth cuprates. Superconducting bismuth cuprates such as $Bi_2CaSr_2Cu_2O_8$ exhibit superlattice modulation. Since such modulation had something to do with the oxygen content in the Bi—O layers and lattice mismatch, one Bi^{3+} was substituted by Pb^{2+} to eliminate the modulation[60] without losing the superconductivity. It is interesting that oxyanions such as CO_3^{2-} and NO_3^- can be incorporated in superconducting cuprates. While CO_3^{2-} alone destroys superconductivity in $YBa_2Cu_3O_7$, it is possible to design superconducting oxyanion derivatives of cuprates of various families (see Parts I and II of the book for details).

11 Arc and Skull Techniques

The electric arc is effectively used for the preparation of oxides as well as for the growth of crystals.[12,61,62] For synthetic purposes, an arc is produced by passing high direct current from a tungsten cathode to a crucible anode, which acts as the container for the material to be synthesized (Fig. 10). To sustain high current density, the cathode tip is ground to a point. Typical operating conditions involve currents of the order of 70 A at 15 V. The arc is maintained in an inert (He, Ar, N_2) or reducing (H_2) atmosphere. Because even traces of oxygen attack the tungsten electrode, the gases are freed from oxygen (by gettering with heated titanium sponge) before being passed into the arc chamber. In an oxygen atmosphere, the arc can be maintained using graphite electrodes. The crucible (anode) is made of a cylindrical copper block and is water-cooled during operation. To perform synthesis, the starting materials are placed in the copper crucible and an arc is struck by allowing the cathode to touch the anode. The current is increased slowly while the cathode is simultaneously withdrawn, to maintain the

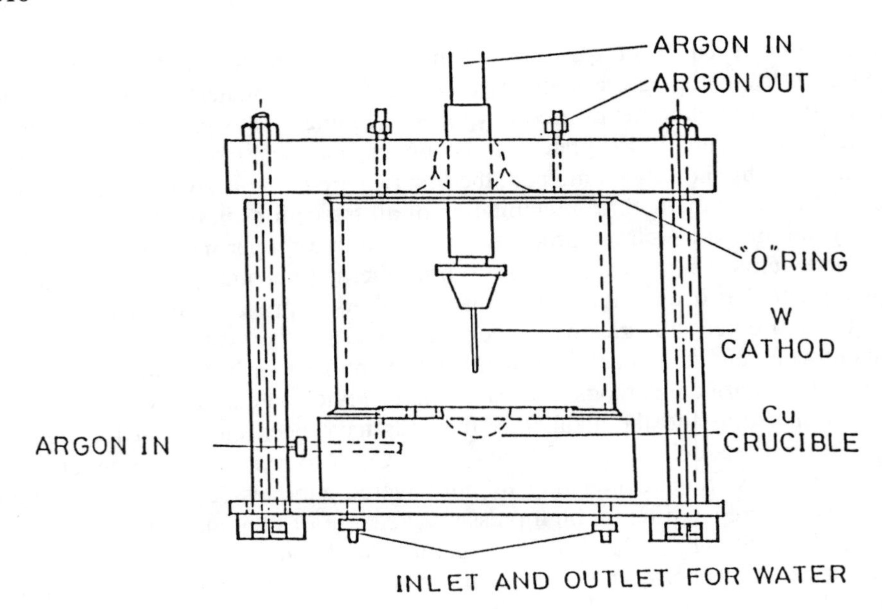

Figure 10. Dc arc furnace.

arc. The arc is then positioned so that it bathes the sample in the crucible, and the current is increased until the reactants melt. When the arc is turned off, the product solidifies in the form of a button. Because of the enormous temperature gradient between the melt and the water-cooled crucible, a thin solid layer of the sample usually separates the melt from the copper hearth. In this sense, the sample forms its own crucible; hence there is no contamination with copper. Contamination of the sample by tungsten vaporizing from the cathode can be avoided by using water-cooled cathodes. The arc method has been used for the synthesis of oxides of titanium, vanadium, and niobium. Lower valence rare earth oxides, $LnO_{1.5-x}$, can be prepared by arc fusion of Ln_2O_3 with Ln metal.

Another version of the arc method is the *tri-arc technique* of Fan and Reed[62] for growing single crystals of high-melting oxides. The main differences between the single and tri-arc furnace lie in the latter's use of three symmetrically arranged thoriated tungsten cathodes and a graphite crucible as anode, together with a water-cooled pulling rod. For crystal growth using the tri-arc, the polycrystalline sample is loaded in the crucible, the three arcs struck simultaneously, and the sample brought to just-melting condition. The pulling rod is lowered until the tip just touches the melt. Then the rod is slowly pulled up, with the result that a neck is formed between the frozen solid and the melt in the crucible. This is equivalent to using a single seed crystal in the conventional pulling technique for crystal growth. By carefully withdrawing the pulling rod, while maintaining the tri-arc, one obtains a single crystal boule. The method has been used to grow single crystals of Ti_3O_5, Ti_2O_3, NbO, and other oxides.

A further modification of the arc method is the *arc transport (transfer) technique* in which the material to be grown as a single crystal is transported by an electric arc. A vertical arc between the cathode (filled with the melt) and the anode carrying the seed crystal is kept at constant length by progressively raising the anode. The fused material is transported across the arc from cathode to anode, where it grows on the seed crystal. NiO crystals have been prepared by this technique by using Ni electrodes.

The skull melting technique[63,64] is useful not only for preparing metal oxides but also for growing crystals of these oxides, as demonstrated in recent years by Honig and co-workers. The technique involves coupling of the material to a radio frequency electromagnetic field (200 kHz–4 MHz, 20–50 kW). The material is placed in a container consisting of a set of water-cooled cold fingers mounted in a water-cooled base (all made of copper); the space between the fingers is large enough to permit penetration of the electromagnetic field into the interior, but small enough to avoid leakage of the melt. The process does not require a crucible, and a thin solid skull separates the melt from the water-cooled container. Large single crystal boules of oxides can be grown by this method, and the mass of the starting materials can be up to 1 kg. Temperatures up to 3600 K are reached in this technique, permitting growth of crystals of materials like ThO_2 and stabilized ZrO_2. The stoichiometry of the oxide is readily controlled by the use of an appropriate ambient gas (CO/CO_2 mixtures, air, or oxygen). Large crystals of CoO, MnO, and Fe_3O_4 have been grown by the skull method. In CoO and MnO, trivalent metal ions have been eliminated by heating the crystals in an appropriate CO/CO_2 mixture. Stoichiometric Fe_3O_4 crystals have been obtained similarly. Crystals of La_2NiO_4 and Nd_2NiO_4 have been grown by the skull method.

12 Crystal Growth

Single crystals of a variety of metal oxides have been grown. Besides crystals of pure oxides, crystals incorporating selective impurities (dopants) have been grown, with a view to achieving the required electronic and related properties. A number of methods are available for growing crystals (Table 4); see refs. 12 and 65–67 for details. The most common methods of growing crystals involve solidification from the melt (in the case of one-component systems) or crystallization from solution. Some of the methods for growing crystals from melt are described in Figure 11. In the *Czochralski method,* the material is melted by induction or resistance heating in a suitable nonreactive crucible (Fig. 11a). The melt temperature is adjusted to slightly above the melting point, and a seed crystal is dipped into the melt. After thermal equilibration has been attained, the seed is slowly lifted from the melt. As the seed is pulled, continuous growth occurs at the interface. The diameter of the growing crystal is controlled by adjusting the rate of pulling, the rate of melt level drop, and the heat fluxes into and out of the system. One advantage of the technique is that since the growth

Table 4 Crystal Growth Methods

Growth in one-component systems
1. Solid–solid
2. Liquid–solid
   (a) Directional solidification (Bridgman–Stockbarger)
   (b) Cooled seed (Kyropoulos)
   (c) Pulling (Czochralski and also tri-arc)
   (d) Horizontal and vertical zoning
   (e) Flame fusion (Verneuil)
   (f) Slow cooling in skull melter
3. Gas–solid: sublimation or sputtering

Growth involving more than one component
1. Solid–solid: precipitation from solid solution
2. Liquid–solid
   (a) Growth from solution
      (i) Aqueous and organic solvents, (ii) molten solvents, (iii) hydrothermal synthesis
   (b) Growth by reaction
   (c) Growth from melt
3. Gas–solid
   (a) Reversible reaction (chemical vapor transport)
   (b) Irreversible reaction (epitaxial process)

interface does not come into contact with the walls of the crucible, formation of unwanted nuclei is avoided. The method has been used for the growth of such ceramic oxides as Al_2O_3, and certain rare earth perovskites (e.g., $LnAlO_3$, $LnFeO_3$, garnets, and scheelites). In the *Bridgman–Stockbarger method* (Fig. 11b), a sharp temperature gradient is provided across the melt, which results in nucleation in the colder region. The conical geometry of the crucible at the bottom limits the number of nuclei formed.

Oxide materials can be grown in air, while others require closed systems and atmosphere control. The *Kyropoulos method* (Fig. 11c) is similar to the Czochralski method, but instead of pulling the seed crystal, the crystal–liquid interface moves into the melt as crystallization proceeds.

Crystals are grown from melt by the *floating zone technique* in which a section of the starting material, held vertically in the form of a rod, is melted by suitable heating. As the molten zone is moved along the rod, progressive melting of the sample at one end of the zone and crystal growth at the other end occur. If a seed crystal is provided at one end, the whole rod can be converted into a single crystal. The method has the advantage of avoiding contamination from the crucible. The method is similar to purification by *zone refining*.

In the *flame fusion* or *Verneuil method* (Fig. 11d), the powder sample is fed directly into an oxyhydrogen flame and the melt allowed to drip on a seed crystal. Because crystallization occurs on the top, the growing seed is lowered slowly, facilitating the growth of large crystals. The method has been used for the growth of high-melting oxides such as ruby and sapphire. In a variant of the flame fusion

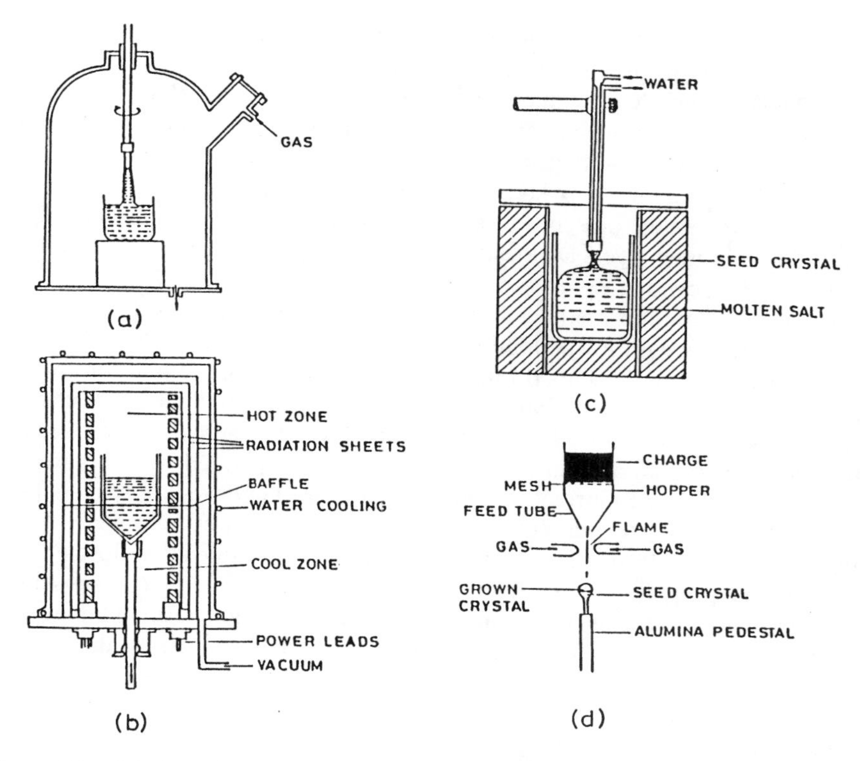

Figure 11. Methods for growing crystals from melt: (a) Czochralski, (b) Bridgeman–Stockbarger, (c) Kyropoulos, and (d) Verneuil.

technique, the *plasma torch method,* the powder is dropped through a hot plasma that is generated and maintained by high frequency current.

Growth of crystals from melt may involve more than one component, such as impurities or intentionally added dopants, in addition to the major component. It is therefore essential to know the distribution of the second component between the growing crystal and the melt. The distribution occurs according to the phase diagram relating the equilibrium solubilities of the second component (impurity) in the liquid and the solid phases.

Several crystal growth methods making use of the solubility of a solute in a suitable solvent are commonly employed. Crystallization requires supersaturation, which can be provided by a temperature difference between the dissolution and growth zones, by solvent evaporation, or by a chemical reaction. In the solution methods, it is rather difficult to avoid contamination by other components in the solution or in the flux. The flux technique has been used to grow crystals of oxides such as GeO_2, SiO_2, $BaTiO_3$, $KTaO_3$, α-Al_2O_3, $GdAlO_3$, $YBa_2Cu_3O_7$, and $Bi_2CaSr_2Cu_2O_8$. The role of the solvent is to depress the melting point of the solute to be grown as a crystal. Although supersaturation is required for crystal growth, the growth rate from solution is much faster than expected, considerable growth occurring even at low degrees of supersaturation (1%). This result has been explained by Frank and Cabrera in terms of the effect of dislocations on crystal growth. Whereas the growth of a perfect crystal requires nucleation of a new layer on a perfect surface after completion of the preceding layer, in the presence of a screw dislocation, growth does not require nucleation of a new layer. Dislocations provide stepped surfaces where growth occurs even at low degrees of supersaturation.

Among the solution methods, crystallization from aqueous solution is well known. Materials with low solubility in water may be brought into solution by the use of complexing agents (mineralizers). An important method for growing of crystals with low water solubility is the hydrothermal method. Hydrothermal synthesis is generally carried out with an autoclave. The autoclave consists of a lower nutrient region and an upper growth region separated by a baffle. The solute is placed in the nutrient region and a few seed crystals are suspended in the growth region. The vessel is filled with water to a predetermined volume (mineralizers are added if necessary), and the temperature of the closed vessel raised while providing a gradient. Since the solubility of most substances increases with increasing temperature, the nutrient region is maintained at a higher temperature than the growth region. The solvent saturates at the dissolving zone, moves by convection to the cooler growth zone (where it is supersaturated), and deposits the solute on the seed crystal. The solution is undersaturated when it reaches the nutrient region, thus dissolving more solute, and the cycle goes on. Large quartz crystals are grown by this method. If the solute (e.g., $AlPO_4$) has a retrograde solubility, the temperature gradient is reversed between the growth and the nutrient regions. In hydrothermal synthesis, the degree of filling the autoclave with solvent is an important factor: to provide a high density of the solvent, it must be more than 32% full. Conventionally, hydrothermal experiments are carried out in alkaline medium, the OH^- ion acting as the complexing agent. Hydrothermal synthesis has been used to stabilize unusual oxidation states in transition metal compounds and to synthesize low temperature and metastable phases.

Insoluble ionic solids such as $CaCO_3$ and $BaSO_4$ cannot be grown by the conventional solution method, or even by controlled chemical reaction, because the product precipitates instantaneously when the reactants are mixed. The *gel method* is useful for the growth of such solids; this simple method depends on the controlled diffusion of the reagents through the gel. A U-tube is filled with silica gel produced by acidifying a solution of sodium metasilicate. Upon being added to the two arms of the U-tube, the reagents diffuse slowly toward each other. In local regions where the concentration exceeds the solubility product, nuclei are formed and grow further into large crystals. Henisch and co-workers have made use of this method to grow crystals of calcium tartrate and calcium tungstate and have incorporated selective impurity ions (Mn^{2+}, Cu^{2+}, Cr^{3+}, etc.) during growth.

Solids such as KCl, KF, PbO, PbF_2, and B_2O_3 are powerful solvents (flux) in the molten state for many inorganic substances, hence can be used as media for the growth of crystals. The usual technique is to dissolve the solute in a suitable combination of flux contained in a platinum crucible while maintaining the temperature slightly above the saturation point. After the crucible has been cooled at a programmed rate, the flux is poured off or leached away by mineral acid, leaving behind the crystals. Yttrium iron garnet is a typical example of a solid grown by the *flux method* using PbO/PbF_2 fluxes. Some 123 cuprates and bismuth cuprates also have been grown by the flux method using KCl, PbO and other fluxes.

Electrolysis in molten salt solutions has been employed for crystal growth as well. The *electrolytic method* involves reduction, usually, of a cation, and deposition of a product containing the reduced cation at the cathode. Typical examples are vanadium spinels, alkali tungsten bronzes, and $Ba_{1-x}K_xBiO_3$.

The experimental setup for electrolytic growth can be exceedingly simple or complex, depending on the system studied. For example, electrolytic growth of alkali metal tungsten bronzes requires a Gooch crucible placed inside a bigger Gooch crucible, the inner crucible serving as the anode chamber and the outer one as the cathode chamber; electrodes of platinum or gold are used. No inert atmosphere is necessary, since atmospheric oxygen has no influence on the current–potential relationship.

Crystal growth from vapor may be divided into two categories depending on whether the vapor–crystal change is physical or chemical. When the composition is the same for the vapor and the crystal, the process is physical; examples are sublimation–condensation and sputtering. The process is termed chemical when a chemical reaction occurs during the growth; in such a case, the composition of the solid is different from the vapor. *Chemical vapor transport* (CVT) is useful for preparation of new solids as well as for growing them into crystals. In CVT, a condensed phase reacts with a gas to form volatile products. An equilibrium exists between the reactants and products:

$$iA(s) + kB(g) + \cdots \rightleftharpoons jC(g) + \cdots$$

CVT makes use of the temperature dependence of the foregoing heterogeneous equilibrium to transport solid A through the vapor phase by means of gaseous intermediate(s) C. Transport of A is not observed without the transporting agent B.

CVT is generally carried out by maintaining a temperature difference between the charge end and the growth end in a closed system. The forward reaction occurs in the charge zone (forming gaseous products), and the reverse reaction occurs in the growth zone (depositing crystals). The temperatures T_1 and T_2 chosen in the growth and the charge zones depend on whether the reaction is endothermic or exothermic. For endothermic reactions, transport requires that T_2 exceed T_1. The reverse is true for exothermic reactions. The factors that determine growth of crystals by this method are the choice of the chemical reaction chosen for transport (the CVT equilibrium should not lie at extremes,

Table 5 Some Crystals Grown by Chemical Vapor Transport Method

Starting Material	Final Product (crystals)	Transport Agent	Temperature (K)
SiO_2	SiO_2	HF	$470 \rightarrow 770$
Fe_3O_4	Fe_3O_4	HCl	$1270 \rightarrow 1070$
Cr_2O_3	Cr_2O_3	$Cl_2 + O_2$	$1070 \rightarrow 870$
$MO + Fe_2O_3$ (M = Mg, Co, Ni)	MFe_2O_4	HCl	
$Nb + Nb_2O_5$	NbO	Cl_2	

lest the reversal become difficult), the magnitude of T_1 and T_2 chosen, and the concentration of transport agent used. The examples listed in Table 5 show how the method can be used to prepare or grow crystals of almost any type of oxide material, provided a suitable transporting agent that can give volatile products can be found.

CVT has been employed to grow crystals of oxide phases such as V_8O_{15} free from contamination by the neighboring Magnéli phases V_7O_{13} and V_9O_{17}. The transporting agents commonly used are I_2, $TeCl_4$, and Cl_2. When a metal oxide can be volatilized readily (e.g., $SnO_2 \xrightarrow{\text{heat}} SnO + \frac{1}{2}O_2$), the vapor species formed can be recombined to form single crystals at the cooler end. Crystals of rare earth garnets have been grown by this method.

The finesse and sophistication of modern preparative solid state science is exemplified by the method of *molecular beam epitaxy*. This important technique for the preparation of semiconductors employs reactions of multiple molecular beams with a single crystal substrate. Several oxides, including superconducting cuprates, have been prepared by this technique. By depositing materials layer by layer, it is possible to synthesize unusual oxides.

13 Concluding Remarks

Chemical synthesis has emerged to become an important aspect of materials development, with newer types of materials being prepared quite frequently by employing novel methods. Some of the important methods and materials were reviewed in this Part, but there are many more. For example, we have not discussed methods such as the mechanochemical method, which uses mechanical grinding with high speed ball mills to prepare spinels and other oxide materials (often in nanometric dimensions), as demonstrated by Boldyrev and co-workers. We have touched only briefly on the chemistry involved with chemical transport. The ingenuity with which properties of oxides are modified drastically by appropriate substitutions or by the modification of the structure also forms part of synthetic strategies. Preparation of oxide materials in different forms (such as fine powders) involves chemical manipulations.

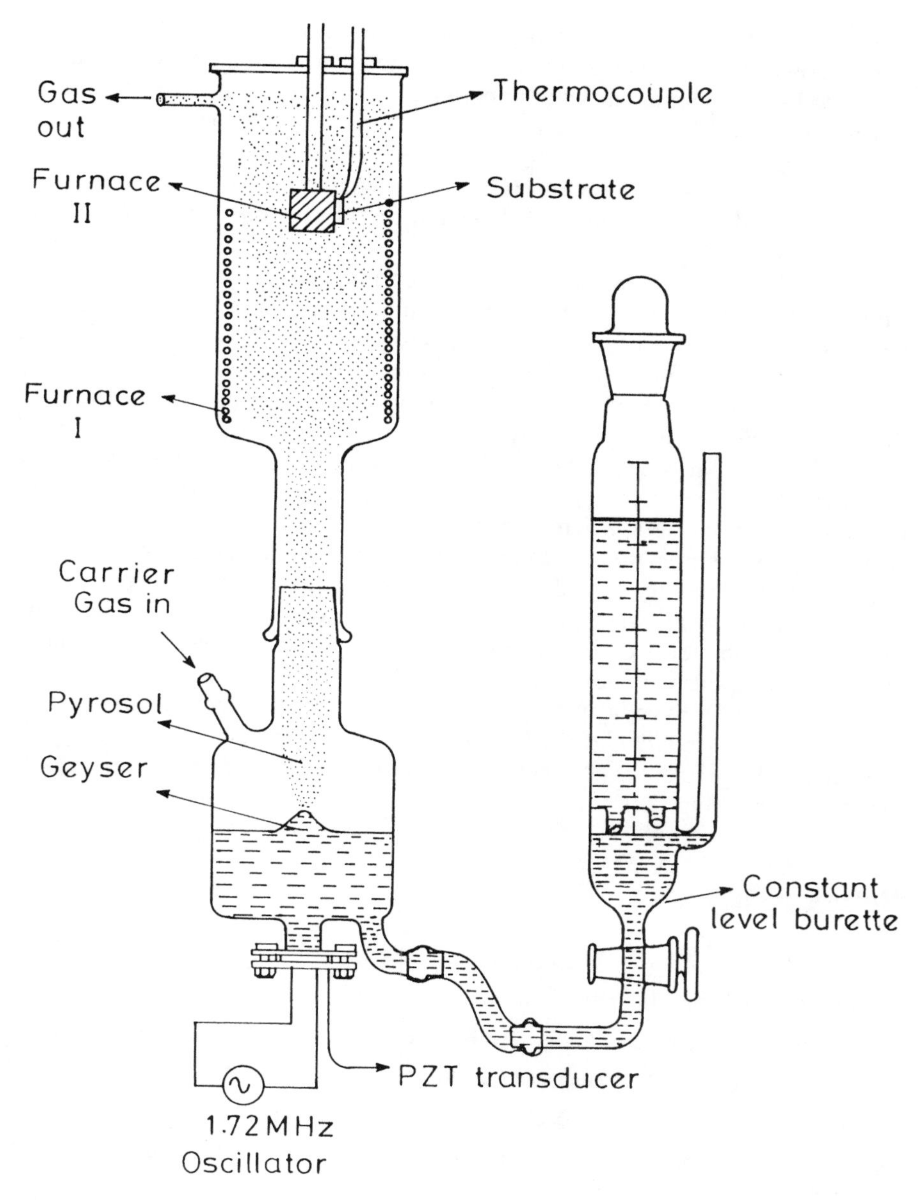

Figure 12. Apparatus for nebulized spray pyrolysis. (Raju et al.[69])

Some methods are in their infancy. Thus, electrochemical methods to prepare oxides, especially those containing transition metal ions in high oxidation states, remain to be fully exploited. More important, the so-called soft chemistry routes will provide not only alternative methods, but also new metastable oxides.

Intercalation in certain oxide hosts yields novel "composite" materials. For example, aniline has been polymerized in MoO_3 and α-$Zr(HPO_4)_2$ giving rise to conducting polyaniline inside the oxide host.[68] Exfoliation of layered materials yields single layer dispersions, which are then restacked in the presence of organic and inorganic guest species.[69,70] Such reactions have been carried out with MoO_3 and other layered compounds. Supramolecular organization involving nanoscale assembly of inorganic materials such as Fe_3O_4 in organic matrices has been carried out to prepare inorganic/organic composites.[71]

An area of considerable interest is the synthesis of nanoscale oxide particles.[15] These are prepared by vapor phase or solution methods (e.g., sol–gel). Thus, nanoparticles of $Mn_{0.8}Zn_{0.2}FeO_4$ and $PbTiO_3$ have been prepared by coprecipitation and the sol–gel method, respectively. Gel precipitation has been employed to prepare ZrO_2 clusters. Microemulsions and reversed micelles are employed effectively for the preparation of nanoparticles of oxide materials, besides techniques such as spray-drying and pyrolysis (especially nebulized spray pyrolysis).

Nebulized spray pyrolysis provides a route to the preparation of several oxide materials in film form or in the form of nanometric particles.[15,72,73] Besides simple binary oxides, oxides such as $PbTiO_3$, $LaNiO_3$, $La_4Ni_3O_{10}$, $YBa_2Cu_3O_7$, and (Ni, Zn)Fe_2O_4 have been prepared by this inexpensive but effective substitute for *metal-organic chemical vapor deposition.* Here also one employs organometallic precursors (β-diketonates, alkoxides) dissolved in an appropriate organic solvent (Fig. 12).

References

1. R. Roy, *Solid State Ionics, 32–33,* 3 (1989).
2. C. Torardi and R. E. McCarley, *J. Am. Chem. Soc., 101,* 3963 (1979).
3. J. P. Giroult, M. Goreaud, P. Labbé, and B. Raveau, *Acta Crystallogr. B36,* 2570 (1980).
4. J. B. Goodenough, H.Y.P. Hong, and J. A. Kafalas, *Mater. Res. Bull., 11,* 203 (1976).
5. R. C. Haushalter, K. G. Storhmaeu, and F. W. Lai, *Science, 246,* 1289 (1989).
6. R. Marchand, L. Brohan, and M. Tournoux, *Mater. Res. Bull., 15,* 1129 (1980); also see T. P. Fiest and P. K. Davis, *J. Solid State Chem., 101,* 275 (1992).
7. H. Rebbah, G. Desgardin, and B. Raveau, *Mater. Res. Bull. 14,* 1131 (1979).
8. C. Delmas and Y. Borthomieu, *J. Solid State Chem., 104,* 345 (1993).
9. C.N.R. Rao and J. Gopalakrishnan, *New Directions in Solid State Chemistry,* Cambridge University Press, Cambridge, 1986.
10. *Soft Chemistry Routes to New Materials: Proceedings of the International Symposium, Nantes,* 1993, Trans Tech Publications, Switzerland.

11. P. Hagenmuller (ed.), *Preparative Solid State Chemistry,* Academic Press, New York, 1972.

12. J. M. Honig and C.N.R. Rao (eds.), *Preparation and Characterization of Materials,* Academic Press, New York, 1981.

13. J. D. Corbett, in *Solid State Chemistry—Techniques* (A. K. Cheetham and P. Day, eds.), Clarendon Press, Oxford, 1987.

14. F. J. Di Salvo, *Science, 247,* 647 (1990).

15. C.N.R. Rao, *Chemical Approaches to the Synthesis of Inorganic Materials,* Wiley, New York, 1994.

16. J. B. Goodenough, *Prog. Solid State Chem., 5,* 149 (1971).

17. C.N.R. Rao, D. Bahadur, O. Parkash, P. Ganguly, and S. Nagabhushana, *J. Solid State Chem., 27,* 353 (1977).

18. K. Kasturirangan, B. R. Prasad, C. K. Subramanian, and J. Gopalakrishnan, *Inorg. Chem., 32,* 4291 (1993).

19. A. G. Merzhanov, Combustion method, in *Chemistry of Advanced Materials* (*IUPAC 21st Century Monograph series,* C.N.R. Rao, ed.), Blackwell, Oxford, 1992.

20. M.M.A. Sekar and K. C. Patil, *Mater. Res. Bull., 28,* 485 (1993).

21. R. Mahesh, V. A. Pavate, O. Parkash, and C.N.R. Rao, *Supercond. Sci. Technol, 5,* 174 (1992).

22. *Eur. J. Solid State Inorg. Chem., 29* (1992), Special Supplement on Precursors for CVD and MOCVD; also see P. Popper (ed.), *Special Ceramics* 1964, Academic Press, London, 1965.

23. C.N.R. Rao and J. Gopalakrishnan, *Acc. Chem. Res., 20,* 228 (1987).

24. C. D. Chandler, C. Roger, and M.I.H. Smith, *Chem. Rev., 93,* 1205 (1993).

25. C.N.R. Rao, *Mater. Sci. Eng., B18,* 1 (1993).

26. C.N.R. Rao (ed.) *Solid State Chemistry,* Dekker, New York, 1974.

27. M. Figlarz, B. Gerard, A. D. Vidal, B. Dumont, F. Harb, A. Cocou, and F. Fievet, *Solid State Ionics, 43* 143 (1990).

28. J. M. Thomas, *Phil Trans. R. Soc. London, A227,* 251 (1974).

29. S. Ayyappan, G. N. Subbanna, and C.N.R. Rao, *Chemistry-A European Journal* (1995), in print.

30. K. Vidyasagar, A. Reller, J. Gopalakrishnan, and C.N.R. Rao, *J. Chem. Soc., Chem. Commun.,* 336 (1986).

31. A. J. Jacobson, in *Solid State Chemistry* (A. K. Cheetham and P. Day, eds.), Clarendon Press, Oxford, 1992; M. S. Whittingham and A. J. Jacobson (eds.), *Intercalation Chemistry,* Academic Press, New York, 1982.

32. J. Gopalakrishnan and M. Kasturirangan, *Chem. Mater. 4,* 745 (1992).

33. D. Guyomard and J. M. Tarascon, *Adv. Mater., 6,* 408 (1994): W. Li, J. R. Dahn and D. S. Wainwright, *Science, 264,* 1115 (1994).

34. A. K. Ganguli, J. Gopalakrishnan and C.N.R. Rao, *J. Solid State Chem., 74,* 228 (1988).

35. W. A. England, J. B. Goodenough, and P. J. Wiseman, *J. Solid State Chem., 41,* 308 (1982).

36. C. E. Rice and J. L. Jackel, *J. Solid State Chem., 41,* 308 (1982).

37. R. J. Cava, A. Santoro, D. W. Murphy, S. Zaharute, and R. S. Roth, *J. Solid State Chem., 42,* 251 (1982).

38. C. Delmas, J. J. Braconnier, and P. Hagenmuller, *Mater. Res. Bull., 17,* 117 (1982).

39. H. S. Horowitz, J. M. Longo, and J. T. Lewandowski, *Mater. Res. Bull., 16,* 489 (1981).

40. W. K. Ham, G. F. Holland, and A. M. Stacy, *J. Am. Chem. Soc., 110,* 5214 (1988); Also see S. W. Keller et al., *J. Am. Chem. Soc., 116,* 8070 (1994).

41. E. Banks and A. Wold, Fused salt electrolysis, in *Solid State Chemistry* (C.N.R. Rao, ed.), Dekker, New York, 1974.

42. R. S. Feigelson, *Adv. Chem. Sci., 186,* 243 (1980).

43. A. Wold and D. Bellavance, in *Preparative Methods in Solid State Chemistry* (P. Hagenmuller, ed.), Academic Press, New York, 1972.

44. M. S. Whittingham and R. A. Huggins, in *Solid State Chemistry* (R. S. Roth and S. J. Schneider Jr., eds.), U.S. National Bureau of Standards, Washington, DC, 1972.

45. A. Wattiaux, J. C. Grenier, M. Pouchard, and P. Hagenmuller, *Rev. Chim. Miner. 22* (1985); see also J. C. Grenier et al., *Physica C, 173,* 139 (1991); A. Demourges et al., *J. Solid State Chem., 106,* 317, 330 (1993).

46. R. Mahesh, K. R. Kannan, and C.N.R. Rao, *J. Solid State Chem., 114,* 294 (1995).

47. L. L. Hench and D. R. Ulrich (eds.), *Science of Ceramic Chemical Processing,* Wiley, New York, 1986.

48. E. Matijevic, in *Ultrastructure Processing of Ceramics, Glasses and Composites* (L. L. Hench and D. Ulrich, eds.), Wiley, New York, 1984.

49. D. R. Uhlmann, B. J. Zelinski, and G. E. Wrek, in *Better Ceramics Through Chemistry,* North Holland, New York, 1984.

50. J. Livage, M. Henry and C. Sanchez, *Prog. Solid State Chem.,77,* 153 (1992).

51. J. B. Goodenough, J. A. Kafalas, and J. M. Longo, in *Preparative Methods in Solid State Chemistry* (P. Hagenmuller, ed.) Academic Press, New York, 1972.

52. J. C. Joubert and J. Chenavas, in *Treatise in Solid State Chemistry* Vol. 5, (N. B. Hannay, ed.), Plenum Press, New York, 1975; C.W.F.T. Pistorius, *Prog. Solid State Chem., 11,* 1 (1976).

53. R. C. Haushalter and L. A. Mundi, *Chem. Mater., 4,* 31 (1992).

54. J. Chen, R. H. Jones, S. Natarajan, M. B. Hursthouse, and J. M. Thomas, *Angew. Chem. Intl. Ed. Engl., 33,* 639 (1994).

55. G. Damazeau, B. Buffat, M. Pouchard, and P. Hagenmuller, *J. Solid State Chem., 45,* 881 (1982).

56. M. Takano, in *Chemistry of High Temperature Superconductors* (C.N.R. Rao, ed.), World Scientific, Singapore, 1991.

57. M. K. Wu et al., *Phys. Rev. Lett., 58,* 908 (1987).

58. C.N.R. Rao et al., *Nature, 326,* 856 (1987).

59. C.N.R. Rao, R. Nagarajan, and R. Vijayaraghavan, *Supercond. Sci. Technol. 6,* 1 (1993).

60. V. Manivannan, J. Gopalakrishnan, and C.N.R. Rao, *Phys. Rev., B43,* 8686 (1991).

61. R. E. Loehman, C.N.R. Rao, J. M. Honig, and C. E. Smith, *J. Sci. Ind. Res., 28,* 13 (1969); H. K. Muller-Buschbaum, *Angew. Chem. Int. Ed. Engl., 20,* 22 (1981).

62. J.C.C. Fan and T. B. Reed, *Mater. Res. Bull., 13,* 763 (1978).

63. J. M. Honig, *Proc. Indian Acad. Sci. (Chem. Sci.), 96,* 391 (1986); also see *J. Solid State Chem., 45,* 1 (1982).

64. H. R. Harrison, R. Aragon, and C. J. Sandberg, *Mater. Res. Bull., 15,* 571 (1989).

65. J. C. Brice, *The Growth of Crystals from Melt,* North-Holland, Amsterdam, 1973.

66. R. A. Laudise, *The Growth of Single Crystals,* Prentice-Hall, Englewood Cliffs, NJ, 1970.

67. S. Mroczkowski, *J. Chem. Educ., 57,* 537 (1980).

68. R. Bissessur, M. G. Kanatzidis, J. L. Schindler, and C. R. Kannewurf, *J. Chem. Soc. Chem. Commun.,* 1582 (1993).

69. W.M.R. Divigalpitiya, R. F. Frindt, and S. R. Morrison, *Science, 246,* 369 (1989).

70. A. J. Jacobson, *Mater. Sci. Forum, 152–153,* 1 (1994).

71. S. Mann, *Science, 261,* 1286 (1993).

72. M. Langlett and J. C. Joubert, in *Chemistry of Advanced Materials* (C.N.R. Rao, ed.), Blackwell, Oxford, 1992.

73. A. R. Raju, H. N. Aiyer, and C.N.R. Rao, *Chem. Mater., 7,* 225 (1995).

Index